Alexander M. Formalskii

Stabilisation and Motion Control of Unstable Objects

De Gruyter Studies in Mathematical Physics

Volume 33

Alexander M. Formalskii

Stabilisation and Motion Control of Unstable Objects

DE GRUYTER

Physics and Astronomy Classification Scheme 2010
02.30.Yy, 45.80.+r, 46.40.Ff, 45.40.Ln, 45.40.Cc, 46.40.-f

Author
Dr. Alexander M. Formalskii
Lomonosov Moscow State University
Institute of Mechanics
Mitchurinskii prospect, Dom 1
Moscow 119192
Russia

ISBN 978-3-11-037582-4
e-ISBN (PDF) 978-3-11-037589-3
e-ISBN (EPUB) 978-3-11-039282-1
Set-ISBN 978-3-11-037590-9
ISSN 2194-3532

Library of Congress Cataloging-in-Publication Data
A CIP catalog record for this book has been applied for at the Library of Congress.

Bibliographic information published by the Deutsche Nationalbibliothek
The Deutsche Nationalbibliothek lists this publication in the Deutsche Nationalbibliografie; detailed bibliographic data are available on the Internet at http://dnb.dnb.de.

Typesetting: le-tex publishing services GmbH, Leipzig
Printing and binding: CPI books GmbH, Leck
♾ Printed on acid-free paper
Printed in Germany

www.degruyter.com

Contents

Preface to English Edition

Current study is dedicated to an interesting and important, from the author's point of view, topic. The book discusses systems that require a certain control impact in order to behave in a desired way. To be precise, these are systems that are unstable if no controlling force is affecting their motion. Such systems can be found in nature, as well as among man-made machines. Bipedal animals (including humans) keep their bodies in vertical, unstable position. Speaking of sports, like gymnastics, athletics, the sportsman often has to maintain postures that are supported by controlling muscle strain. People often make devices that are unstable by their nature, or contain unstable parts. Such devices include, for example, modern double- and single-wheeled transportation vehicles, bipedal walking machines. Flying vehicles that have center of pressure located in front of the center of mass are also unstable. To make these devices function properly, certain control laws must be designed and implemented. At that, the resources that can be assigned to develop these control forces are often limited in one way or another. Therefore, not all deviations of an unstable object from a desired functioning mode can be countered and corrected.

Humans, animals can only relocate a part of their bodies with respect to other parts. But this relocation is done in such a way that external forces, like forces that arise from interaction with the environment, or the gravity forces, make the body as a whole move in a desired way, and in a desired direction. However, it often happens that the number of control inputs in the system is less than its number of degrees of freedom, or, in other words, the system is underactuated. This case is also discussed in this study.

There are many Russian sources, as well as English articles and books that consider problems of controlling various systems that include unstable elements. Current study references to a number of them, but the author understands that such reference list just cannot be complete. Current book is entirely dedicated to mentioned problems. It investigates some particular unstable mechanical systems; each one is used as an example in providing a general approach to designing control algorithms that would be applicable to similar systems.

The monograph was first written and published in Russian language. During translation into English, it has gone through some corrections in an attempt to improve the contents, or to make it easier to understand for an English-language reader. Besides, some new references were included. A new (fifth) part has been added. It considers undesired vibrations in systems that contain compliant elements, and proposes means to avoid these vibrations.

The Russian text, of course, contains many references to articles and books issued in Russian. However, a number of these articles are printed in journals that have English editions. Many books have also been translated. Therefore, in current study the author tries to refer to English sources wherever possible. Yet there are some publications, including books, that only appear in Russian. The author still includes them in the list of literature, having their titles translated into English.

This edition would not be possible without skills, patience and extensive theoretical knowledge of Vadim Belotelov, who translated the book from Russian; and the author expresses his immense gratitude to him for this work.

Alexander Formalskii — Moscow, May 2015

Preface

This monograph is dedicated to the designing motion control for objects that may have desired working regimes unstable without additional guidance. For example, an object of such kind is the recently invented vehicle "Segway" that, having a passenger on board, is an unstable inverted pendulum mounted on a platform with a wheel pair. There exists also a single-wheel "Segway" (one-wheel scooter) called "Solowheel". This single-wheel scooter is even less stable than the two-wheel "Segway". There are many more unstable devices. Most links of a two-legged anthropomorphic walker (like the "links" of its prototype, the human being) are unstable inverted pendulums. Flying vehicles are unstable if the center of pressure is located in front of the center of mass. If the center of pressure is "far" behind the mass center, the flying vehicle has a large static stability margin and can not be agile. In order to increase its agility, it is necessary to reduce the stability margin and, for some special purposes (like military aircraft), even "enter" the instability domain. The problems of stabilizing a desired operating mode also arise in relation with design of an exoskeleton and of different unstable systems that have magnetic or electrostatic support.

If the desired operating regime is stable without control, i.e., the object of control is naturally stable [105], then the purpose of the control system is to improve the quality of transient processes. These transients may occur when the object deviates from the desired mode due to external disturbances. But if the desired operating regime is unstable without control, i.e., the object is naturally unstable, the primary goal of the control system becomes stabilization of this regime. In this case, it is virtually impossible to implement the desired operating regime without any control, and the matters of transient quality fade into insignificance.

The problems of designing a control law for an unstable object and stabilizing a desired operating regime are concerned with certain difficulties. Actually, control resources are in some way or another limited in any real system. Hence an unstable object cannot be translated to a desired operating regime from any state. In other words, the set of states from which the object can be driven to the desired operation mode for a set of control resource limitations occupies an area of the phase space. This set is referred to as "controllability domain". The domain of attraction of the desired operating regime, appearing when some specific control law is designed, for example, as a feedback, belongs to the controllability domain; mostly, it occupies only a part of this domain. The domain of attraction usually means a set of initial states from which the system can be moved asymptotically to the desired regime by means of the control law (this definition is used further in this book; some authors refer to it as "basin of attraction"). If the domain of attraction is small with respect to practically possible disturbances of the object motion, then the desired operating regime is practically unrealizable. This means that the control resources are insufficient, or the control law is not efficient. Thus, if limitations are imposed on control resources, the problem of constructing a control law that would maximize the domain of attraction becomes crucial. This problem is discussed thoroughly in this book.

If the number of control inputs is less than the number of degrees of freedom, then designing a control law is typically associated with some difficulties. Such objects are referred to as "underactuated". Examples of such objects are pendulum systems; walking mechanisms with drives installed in inter-link joints. Besides, animals and humans can move "links" of their body only relatively to each other. But they do it in such a way that external forces arising during this relative motion (forces of interaction with the environment, gravity forces) realize the desired motion of the body as a whole. For instance, walking, running of animals, crawling of reptiles all is possible because of the friction against the supporting surface; swimming of animals is thanks to the pressure of the fluid, flying of the birds is supported by aerodynamic forces. Animals organize proper effect of these external forces during the *relative* motion of parts of their bodies. Human controls oscillations of a swing about the suspension point by moving the body properly, while there is no external control torque in the swing suspension point. A gymnast swings on a bar controlling primarily the angle in the hip joint, the torque in the wrist joint being very small. In both of the latter cases a person makes proper use of the gravity force. Remember also Pablo Picasso's picture with a girl balancing on a ball without any external stabilizing actions.

In current book that consist of five *parts*, primarily the problems of controlling unstable underactuated systems are considered. A group of particular problems is studied. However, the approaches used in control law design that are used for solving these problems, in author's opinion, can be also used in other cases. Some conclusions based on investigation of relatively simple tasks are applicable to more complicated ones.

Approaches to control law design for unstable systems developed in this monograph are based on methods of optimal control theory. Concepts of controllability domains and domains of attraction are used. Methods of control law design for systems with degree of instability equal to one or two are developed. In a majority of considered problems the control law used to stabilize a desired working regime of a discussed object is constructed in such a way that all resources of the system are used for suppressing unstable modes of motion. Such method of control law design provides maximal domain of attraction of the desired working regime. Problems of both local and global stabilization are investigated in the book. When solving global stabilization problems, it is necessary to construct essentially nonlinear control functions. In some cases, intuitive consideration is used to create a control law.

Below, a brief description of the book contents is given chapter by chapter.

Part 1 is dedicated to problems of motion control of objects that contain a single-link pendulum. It begins (chapter 1) with investigation of a separate inverted single-link pendulum with fixed suspension point. A control torque is applied at that point; this torque is limited in absolute value. Stabilization of the inverted pendulum is discussed. Domains of controllability and control law are constructed; the control law provides maximal domain of attraction of the unstable equilibrium. Effect of time delay in the feedback loop upon the stability of the stabilization process is studied. Evaluation of this effect is of interest not only for the system discussed in the first chapter, but also for those discussed in the later chapters. After discussing the control problem stated for this simplest pendulum with fixed suspension point, chapter 2 concerns the question of stabilizing an inverted pendulum with its suspension point located in the center of a wheel. This question resembles a well-known problem of balancing an inverted rod on a palm. The wheel can roll without slipping over a supporting surface. The pendulum is controlled by means of a motor; its rotor is rigidly attached to the pendulum and the stator is attached to the wheel, or vice versa. This is a system with two degrees of freedom and one "internal" control torque that is assumed to be limited in absolute value (like in the previous problem). It turns out that the controllability domain of a pendulum on a wheel is larger than for the similar pendulum with fixed suspension point. Relations are obtained that allow estimating the influence of different parameters (both of the pendulum and of the wheel) on the size of the attraction domain of the unstable equilibrium. The problem of controlling the wheel-based pendulum is not of solely theoretical value. Apparently, it is helpful for designing a control law for a Segway-type vehicle (both two-wheeled and one-wheeled).

The third chapter again discusses a pendulum with fixed suspension point. However, a flywheel is mounted at the end of the pendulum, so that its rotor is rigidly fixed to the pendulum and its stator moves with the flywheel. In some studies, the pendulum with flywheel is referred to as "inertia wheel pendulum". As opposed to chapter 2, where the motion of a wheel-based pendulum is considered, in chapter 3 motion of a wheel (flywheel) on a pendulum is studied. Both systems have two degrees of freedom, and their mathematical models are alike. The control parameter in system "pendulum + flywheel" is the DC voltage applied to the electric drive. The magnitude of this voltage is limited. A flywheel control algorithm is designed that makes the pendulum move from the bottom, stable equilibrium to the top, unstable one, and stabilizes it there. This solves the problem of global stability of the inverted pendulum, because one can easily translate the pendulum to the bottom equilibrium from any initial state. The chapter provides results of both mathematical simulation and experiments. The latter confirm efficiency of the constructed control law. Flywheel systems can be used for suppressing oscillations of payload attached to the arm of a crane with a rope. Such stabilization systems can be used also for increasing the stability domain of tower cranes. Gyrodynes, that are essentially flywheel systems, are used for controlling orientation of Earth-orbiting satellites.

The possibility of making a wheel roll along the supporting surface by deflecting the pendulum suspended in the center of the wheel is studied in chapter 4. It is supposed that the pendulum is controlled by a motor. Its rotor is rigidly attached to the pendulum and the stator is attached to the wheel. Deflecting the pendulum from the vertical, it is possible to make the wheel roll not only along a horizontal plane, but also upwards along a slope. The maximum slope angle is determined for which the wheel can still move upwards. Naturally, this angle depends on the parameters of the pendulum and the wheel. Device similar to the one studied in chapter 4 was invented as early as in XIX century. The photos of such devices, as a number of them were built, can be found on the Internet.

Chapter 5 is dedicated to the problem of optimal behavior of a human on a swing. Optimal means that this motion provides maximal deviation of the swing from the vertical at the end of each semiperiod of oscillations. At the same time, the optimal motion of the human is determined that minimizes this swing deviation at the end of each semiperiod. In other words, the problem of the best damping of swing oscillations is also solved.

In chapter 6, problems of minimizing mechanical energy consumption while controlling a physical pendulum are discussed. Tasks of driving the pendulum both to the bottom, stable equilibrium and to the top, unstable one are considered. Pendulum motion is investigated on the phase cylinder. Optimal control proves to be impulse-based, and it is described by Dirac's delta-functions.

In Part 2, problems of controlling a two-link pendulum are considered. The control torque can be applied in the inter-link joint or in the suspension point. The control torque is again assumed to have limited absolute value. This part first (chapter 7) gives solution to the problem of local stabilization of the pendulum equilibrium where both links are directed upwards. Controllability domains for this unstable equilibrium are determined. The control law that stabilizes this equilibrium is obtained in form of feedback. It is possible to design the feedback so that the corresponding domain of attraction is largest possible (in linear approximation). The problem of maximizing the domain of attraction for the double pendulum turns out to be more complicated than for the single-link pendulum, as the double pendulum has two unstable motion modes, and the single pendulum only has one. At the same time, this system is underactuated, because the two-link pendulum has two degrees of freedom and only one control input (applied either in the inter-link joint or in the pivot).

In chapter 8, a different problem of controlling the two-link pendulum is considered. It is assumed that the control parameter is the angle in the inter-link joint. In this case, it is possible to reduce the original fourth-order system of nonlinear differential equations to a single equation. After that, it becomes possible to design a control law that optimally swings or damps the pendulum.

The next chapter 9 concerns the problem of controlling a two-link pendulum by means of a torque applied in the interlink joint; in some articles such pendulum is called the "acrobot". The control algorithm in form of feedback is created that provides global stability of the device. The algorithm is designed based on results obtained in chapters 7 and 8.

In chapter 10, the problem of controlling two-link pendulum using a torque applied at the suspension point is discussed; in some papers such system is referred to as "pendubot". A control algorithm is proposed that provides global stability of the inverted pendulum. Results obtained in chapter 7 (those concerning the algorithm of local stabilization that maximizes the domain of attraction) are used for construction of this algorithm.

In chapter 11, equations of plane motion of a multi-link (with arbitrary number of links) pendulum mounted on a moving base (a wheel or a cart) are proposed. The control torque applied between the base and the first pendulum link does not depend on the position and the velocity of the base. In this case, it is possible to decompose the mathematical model of the system and to extract equations that describe exclusively pendulum motion. The segregated equations are different from the known motion equations of a pendulum with fixed suspension point. It is particularly remarkable that they include parameters that characterize the moving base. The phase picture representing motion of a free (uncontrolled) single-link pendulum on a wheel or a cart is developed. A pendulum control law that uses limited resources is constructed. This law provides global stabilization of the top, unstable equilibrium. The time-optimal control is obtained in form of feedback.

Part 3 studies a problem of stabilizing a ball on a beam. It consists of two chapters, 12 and 13. In chapter 12, the beam is straight, and the ball can roll along the beam without slipping (in some articles such kind of systems is referred to as "beam-and-ball system"). The beam suspension point is located below the beam itself; therefore, the horizontal position of the beam is unstable. It is supposed that an electric drive is installed at the beam suspension point. Voltage limitation for DC electric motor is again taken into account. The studied system thus has two degrees of freedom, but only one control input is applied. The goal is to control the voltage applied to the motor so that to stabilize the unstable (without control) equilibrium of the system. Additionally, it is required not only to stabilize the unstable equilibrium, but also to maximize its domain of attraction.

Chapter 13, as well as chapter 12, considers stabilization of a ball on a beam. However, in that chapter, as opposed to chapter 12, it is assumed that the beam is curvilinear (an arc). The author does not know any studies that mention systems of such kind. The linearized model of the system with a straight beam has only one unstable mode of motion, while the linearized model of the system with curvilinear beam (when its curvature is "large") contains two unstable modes. Therefore, the problem of stabilizing a ball on a curvilinear beam is more complicated than that with a straight one.

Part 4 investigates the problem of gyroscopic stabilization of the vertical (unstable) position of a two-wheel bicycle robot. This part contains two chapters: 14 and 15. In chapter 14, design schemes of two models of bicycle robot created in the Institute of Mechanics of Lomonosov Moscow State University are provided. In one bicycle, the front wheel is driving and steering at the same time (like in the three-wheel bicycle used by the smallest children), and the rear wheel is passive. In the other bicycle, each wheel is both driving and steering. The design of gyrostabilizer is discussed. The mathematical model of device oscillations is developed. It concerns two variables – the bicycle roll angle and gyroscope inclination angle. In chapter 15, the control system for the gyrostabilizer is discussed. The means of measuring the roll angle of the bicycle by two accelerometers is provided. This angle is required to design the system of vehicle stabilization. The law of controlling the torque applied to the precession axis of gyroscopes is designed. The purpose of this law is to stabilize the vertical position of the bicycle. Results of numerical and practical experiments are provided along with the theory.

Part 5, the closing one, investigates mechanical systems that contain compliant elements. A compliant element can be, for example, a platform where the operated object (plant) is installed; or/and an elastic gear that connects a motor with this moving plant. Only the first (lowest) resonance frequency of the system is taken into account; and the considered mathematical model of the system has two degrees of freedom. Control laws that drive the plant from a given initial state to a given final state in a finite time are designed. These laws prevent vibrations of the system. This part contains two chapters: 16 and 17. In chapter 16, the mathematical model, the statement of the problem and the fluent control law are considered. A fluent control of trapezoid shape is studied in chapter 17. This control is compared with a bang-bang control. Then, a control law in form of a feedback with a feedforward term is suggested. Numerical examples are provided.

For a variety of problems, the results of numerical simulations are provided in this book along with the theoretical formulations. These results have been obtained by means of programs developed in MathWorks Matlab® and Wolfram Research Mathematica® environments. The developed programs contain motion animations for the considered problems; however, it is impossible to make these animations available for readers of the paper edition. For the pendulum with flywheel, for two-wheel bicycles with gyroscopic stabilization systems, experiments performed with them were recorded on video. These movies, along with articles and clips related to other mechanical systems that are not discussed in frame of this book, are available at www.msurobot.com.

Chapters in this book are numbered consecutively. Formulas and figures are numbered by double numbers, separated by dots: the first part is the number of the chapter, the second one is the sequential number of the formula or figure in this chapter.

The author hopes that this work would be useful for scientific researchers and engineers who work in spheres related to control theory, robotics, mechatronics. It is also helpful for students and post-graduates specializing in the said subjects.

The author will always remember Anatolii Lenskii, whom he was blessed to work with on various problems. These include the control problems discussed in this book – controlling a pendulum with a flywheel, and bicycles with gyroscopic stabilization systems. Some results were obtained and published in collaboration with Yuri Martynenko, whom the author will treasure memory about.

Part I: **Devices containing a single-link pendulum**

Many unstable mechanical systems have links that can be described as inverted pendulums. For example, an individual transportation device like "Segway" [165], platforms on paired coaxial wheels, bipedal walking machines [19, 20, 55, 65]. In a complex problem of controlling unstable pendulum systems, the elementary problems of controlling a *single*-link pendulum and stabilizing its unstable top equilibrium can be called the core ones. They are considered classical in theoretical mechanics and control theory. In most cases, such problems are solved by controlled movement of the pivot. In studies [93, 149] an inverted pendulum is stabilized by vertical movements of its pivot. When stabilized in such a way, it is often called Kapitza's pendulum. Study [137] considers stabilization of a multi-link pendulum, and a single-link pendulum as a most simple case, by horizontal movements of its pivot. Such a device can be mounted on a cart moving on a horizontal plane (see [29]). It is well known that one can hold a vertical rod standing on an open palm of his hand and prevent it from falling by moving his hand. Study [74] discusses a robot that can maintain its balance on a cylinder. See also article [118] on the control of unstable mechanical systems in particular on pendulum, and article [153] on the stabilization of linear systems with bounded control parameters.

This part introduces several problems related with controlling a single-link, plane physical pendulum [22–24, 54, 55, 57, 58, 64, 111, 112, 122, and 123]. First, a regular physical pendulum is considered, with control torque applied to it in its pivot point. The absolute value of this control torque is assumed limited. Then, a pendulum with its pivot located in the center of a wheel is discussed (see [57, 58, 64]). Here the wheel rolls over a straight horizontal line. The control torque is again applied at the pivot point of the pendulum. Therefore, it turns the wheel at the same time as it turns the pendulum. The task is to stabilize the pendulum in its topmost point. Yet another device considered in this part is a pendulum with a flywheel attached to its end [7, 22–24, 64], or, in other words, an "inertia wheel pendulum" [1, 25, 49, 147]. A control algorithm is suggested for relocating the pendulum from its bottom position, which is a stable equilibrium, into the top position. The suggested algorithm can stabilize the top equilibrium that is naturally unstable. Further, this part discusses the problem of controlling the motion of a wheel by means of a pendulum attached to its center. The conditions for the wheel to be able to move uphill are formulated. The last chapter of this part, once again a simple single-link physical pendulum is considered, with control torque applied at its stationary pivot point [54, 55]. The problem investigated there concerns transferring the pendulum to its bottom – stable – equilibrium, and to its top – unstable – equilibrium. Among all possible control algorithms, the optimal ones are chosen, that provide minimal energy consumption for the translation process. These optimal algorithms are proved to be pulse-based.

One more interesting device that is investigated in stabilization studies is worth mentioning here. It is not fully related to pendulum systems, though. In article [31] and patent article [32] a long body is discussed; in particular, it can be a beam. This beam is being compressed by forces acting along its axial direction. The beam is connected to a support by a joint. To maintainin stability of the static position of the beam, a stabilizing torque is suggested that is applied in the joint. This stabilizing torque depends on the beam strain measured in some of its points. The suggested torque pattern helps significantly increase admissible loads applied to the beam.

1 A pendulum with stationary pivot

... dear Fagot, the audience is starting to get bored.
Show us something simple, for a start.
M.A. Bulgakov. The Master and Margarita.

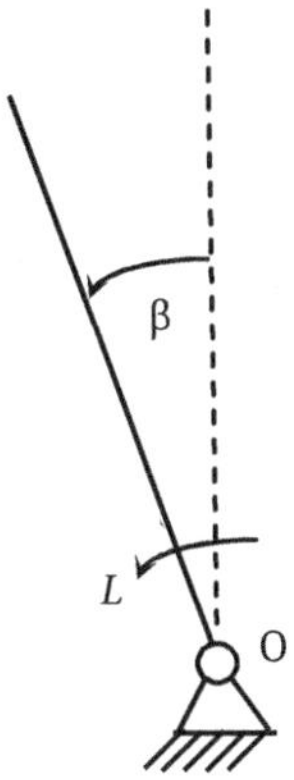

Fig. 1.1. Pendulum with a stationary pivot.

Figure 1.1 illustrates a single-link, physical pendulum with a stationary pivot point O. Torque L is applied in this point (the pivot) O. The positive direction of the torque is the counter-clockwise one. The absolute value of the torque L is limited by a constant L_0,

$$|L| \leq L_0 \tag{1.1}$$

1.1 Equations of motion

The equation of motion of a pendulum is well known. It can be written in the following form:

$$mr^2\ddot{\beta} - mgb\sin\beta = L \tag{1.2}$$

Here m denotes the mass of the pendulum, b – the distance from the pivot point O to the center of mass C of the pendulum, r – the radius of gyration of the pendulum about point O. The angle between segment OC and the line containing the same segment, when the pendulum is in its top equilibrium, is denotes as β. The value of this angle increases in counter-clockwise direction. Gravity acceleration is designated as g. Dot stands for time derivative. For this particular problem, the friction force in pivot is neglected.

After introducing nondimensional time τ and nondimensional torque μ according to expressions

$$\tau = t\sqrt{gb/r}, \quad \mu = L/mgb \tag{1.3}$$

equation of motion (1.2) can be transposed to a simple form

$$\beta'' - \sin\beta = \mu \tag{1.4}$$

Prime mark $'$ means differentiating with respect to nondimensional time τ. The nondimensional torque μ is limited in its absolute value by a constant μ_0,

$$|\mu| \le \mu_0, \quad \mu_0 = L_0/mgb\,. \tag{1.5}$$

The set of piecewise-continuous functions $\mu(\tau)$, each of them complying with inequality (1.5), will further be denoted as W, so $W = \{\mu(\tau) \colon |\mu(\tau)| \le \mu_0\}$. For $\mu_0 > 1$, the pendulum can be relocated from any initial state

$$\beta(0)\,, \quad \beta'(0) = 0 \tag{1.6}$$

to unstable equilibrium

$$\beta = 0,\, 2\pi\,; \quad \beta' = 0 \tag{1.7}$$

and in this process of relocating, angle β will monotonically change, remaining within interval $0 \le \beta \le \pi$ or $\pi \le \beta \le 2\pi$. For $\mu_0 < 1$, i.e. when there is a tight limitation for the torque, there are some special initial states (1.6). From such states, the pendulum can be translated into unstable equilibrium (1.7) only after a certain number of oscillations. These oscillations have increasing magnitude, and they occur about the bottom, stable equilibrium

$$\beta = \pi\,, \quad \beta' = 0\,. \tag{1.8}$$

In the latter case, the state of equilibrium (1.8) is one of such special initial states.

In further discussion, it will be assumed that $\mu_0 < 1$.

Linearization of equation (1.4) about its equilibrium $\beta = 0$, $\beta' = 0$ (1.7) yields the following equation:

$$\beta'' - \beta = \mu\,. \tag{1.9}$$

Here it is assumed that the deviation of the pendulum from its vertical position is "small".

It is easy to prove (see [69, 98]) that system (1.9) is completely controllable in Kalman's sense [89–92].

Equation (1.9) can be transposed to its Jordan form:

$$y' = y + \mu\,, \quad y = \beta + \beta'\,, \tag{1.10}$$

$$z' = -z - \mu\,, \quad z = \beta - \beta'\,. \tag{1.11}$$

Equations (1.10) and (1.11) describe respectively unstable and stable motion modes of the pendulum.

1.2 Domain of controllability

Consider a set Q of initial states, where for each state there exists a control function $\mu(\tau) \in W$, that moves the system (1.10), (1.11) into its equilibrium $y = 0$, $z = 0$ ($\beta = 0, \beta' = 0$). Such set Q is called *domain of controllability* [53]. This set is bounded only with respect to the unstable coordinate y [53]. It is described by the following inequality:

$$|y| < \mu_0 \quad \text{or} \quad |\beta + \beta'| < \mu_0 \,. \tag{1.12}$$

Indeed, let $|y(0)| < \mu_0$. Control function $\mu = -\mu_0 \operatorname{sign}[y(0)]$ will translate variable y to zero in a finite time. After that, assume $\mu = 0$. With $\mu = 0$, variable y will remain equal to zero, and the solution $z(\tau)$ to equation (1.11) for *any* initial state $z(0)$ converges to zero with $\tau \to \infty$. Thus, with $\mu = 0$ system (1.10), (1.11) on a finite interval of time will reach any pre-defined, arbitrarily small neighborhood of the origin $y = z = 0$. Then, from this sufficiently small neighbourhood the system can be translated to the origin $y = z = 0$ itself, in finite time, involving admissible control function $|\mu(\tau)| \le \mu_0$. This is possible since system (1.10), (1.11) is completely controllable in Kalman's sense [89–92].

Interval (1.12) of axis y is illustrated in Figure 1.2.

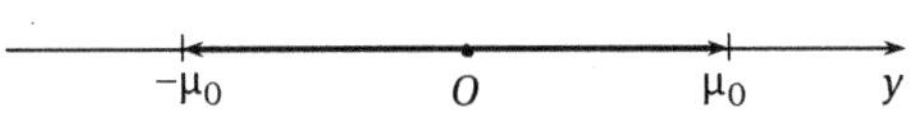

Fig. 1.2. Domain of controllability with respect to variable y.

On the coordinate plane of variables y, z domain of controllability Q is represented by a *stripe* that is not bounded with respect to variable z. The boundaries of this stripe are parallel to axis y (that is, vertical). This stripe is shown in Figure 1.3.

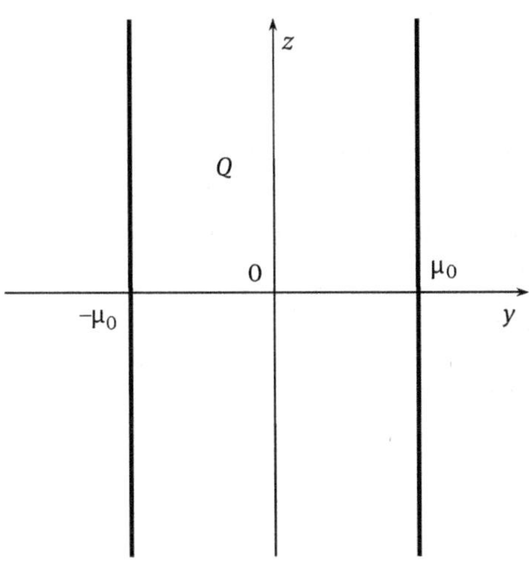

Fig. 1.3. Domain of controllability Q in plane y, z.

Figure 1.4 shows domain Q (1.12) represented in coordinate plane β, β'. Domain Q is an open region. That means, Q does not contain its boundary. Consider a control function applied to the system, that is maximal in its absolute value, $\mu(\tau) = \mp\mu_0$. In this case, there exist states of equilibrium of equation (1.9), namely, $\beta = \pm\mu_0$, $\beta' = 0$. Such states of equilibrium lie on the boundary of domain Q (1.12). At the same time, for nonlinear equation (1.4) with $\mu(\tau) = \mp\mu_0$, there exist similar equilibriums, $\beta = \pm \arcsin\mu_0$, $\beta' = 0$. Note that $\arcsin\mu_0 > \mu_0$, so, if to consider the original nonlinear system, the angle between the pendulum in such equilibrium and its vertical position is greater than the same angle in the linearized system.

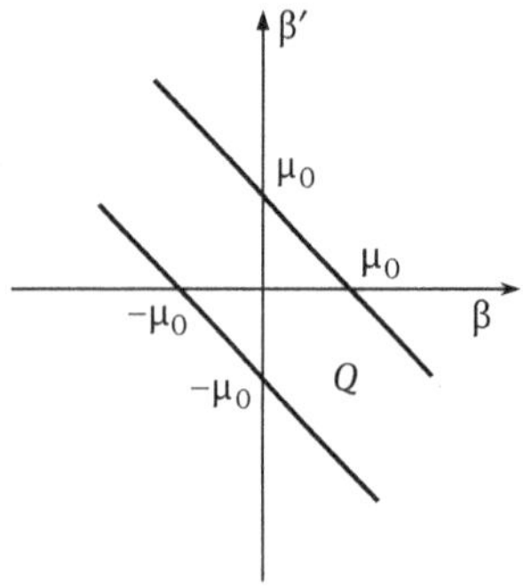

Fig. 1.4. Domain of controllability Q in plane β, β'.

Assuming $\beta'(0) = 0$ or $\beta(0) = 0$ in relation (1.12) yields inequalities that limit initial values of $\beta(0)$ and $\beta'(0)$. With $\beta'(0) = 0$, the limitation imposed on initial angle $\beta(0)$ from which the pendulum can be translated into the unstable equilibrium, $\beta = 0$, $\beta' = 0$ (1.7), looks as follows:

$$|\beta(0)| < \mu_0 \quad \left(|\beta(0)| < \frac{L_0}{mgb}\right). \tag{1.13}$$

There is a clear physical interpretation to this inequality. When the pendulum deviates an angle β from vertical line, the gravitational moment (its linear approximation) is less than maximum possible torque L_0, developed by the motor in the joint O (see Figure 1).

With $\beta(0) = 0$, there is a limitation for initial angular velocity $\beta'(0)$. This limitation is as follows:

$$|\beta'(0)| < \mu_0 \quad \left(|\dot{\beta}(0)| < \frac{L_0}{mr\sqrt{gb}}\right). \tag{1.14}$$

Having such admissible angular velocity at the initial moment, the pendulum can be translated into unstable equilibrium $\beta = 0$, $\beta' = 0$ (1.7).

1.3 Maximizing domain of attraction

The unstable motion mode of system (1.10), (1.11) can be "suppressed" by means of control function that is effectively a feedback:

$$\mu = \gamma y, \quad \gamma < -1 \tag{1.15}$$

To prove this statement, make substitution (1.15) in equation (1.10). The result looks as follows:

$$y' - (1 + \gamma)y = 0 \tag{1.16}$$

With $\gamma < -1$, the sum $1+\gamma < 0$, and for any initial value of $y(0)$ the solution to equation (1.16) $y(\tau) \to 0$ as $\tau \to \infty$. And if $y(\tau) \to 0$, then $\mu(\tau) \to 0$, since $\mu = \gamma y$. And then again, solution $z(\tau)$ to equation (1.11) for *any* initial value $z(0)$ converges to zero.

When control functions are limited as in (1.5), then, instead of linear feedback (1.15), a linear *saturated* feedback should be considered, i.e., a *nonlinear* feedback like

$$\mu = \mu(y) = \begin{cases} -\mu_0 & \text{when} \quad \gamma y \le -\mu_0\,, \\ \gamma y & \text{when} \quad |\gamma y| \le \mu_0\,, \qquad \gamma < -1 \\ \mu_0 & \text{when} \quad \gamma y \ge \mu_0\,. \end{cases} \tag{1.17}$$

Function (1.17) is plotted in heavy line in Figure 1.5.

With control function like in (1.17), equation (1.10) is transposed to

$$y' = y + \mu(y) \tag{1.18}$$

Function $y + \mu(y)$, that is the right-hand side of equation (1.18), is plotted in heavy line on Figure 1.6. On the same figure, thin lines represent function (1.17), and also the bisector of the first and the third quadrants.

On interval $-\mu_0 < y < 0$, the value of $y+\mu(y)$ is positive, and on interval $0 < y < \mu_0$ it is negative ($y + \mu(y) < 0$). Thus $y' > 0$ for $-\mu_0 < y < 0$ and $y' < 0$ for $0 < y < \mu_0$. Hence under all initial conditions $-\mu_0 < y(0) < \mu_0$ the solution to equation (1.18) $y(\tau) \to 0$ as $\tau \to \infty$. And when $y(\tau) \to 0$, then, in accordance with relation (1.17), $\mu(\tau) \to 0$. And then solution $z(\tau)$ to equation (1.11) for *any* initial value $z(0)$ converges to zero [45], because the solution to homogeneous equation (1.11) converges to zero exponentially for any initial value $z(0)$. Thus, the trivial solution $y(\tau) = 0$, $z(\tau) = 0$ to system (1.10), (1.11) with applied control function (1.17) is asymptotically stable under all initial conditions $-\mu_0 < y(0) < \mu_0$, regardless of initial condition $z(0)$. Hence the domain of attraction B of the coordinates origin $y = 0$, $z = 0$ for control function (1.17) coincides with domain of controllability Q: $B = Q$. As usual, domain of attraction is understood as a set of initial coordinate pairs $y(0)$, $z(0)$, from which a controllable system converges asymptotically to a desired state. Looking closer at Figure 1.6, one can notice among other things that $y' > 0$ when $y > \mu_0$ and $y' < 0$ when $y < -\mu_0$. That means, as $\tau \to \infty$, solution $y(\tau) \to +\infty$, if $y(0) > \mu_0$, and $y(\tau) \to -\infty$, if $y(0) < -\mu_0$.

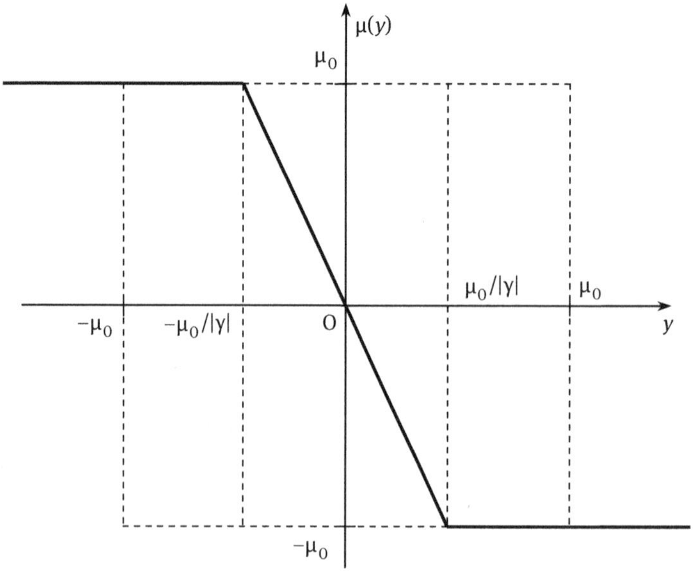

Fig. 1.5. Function $\mu(y)$.

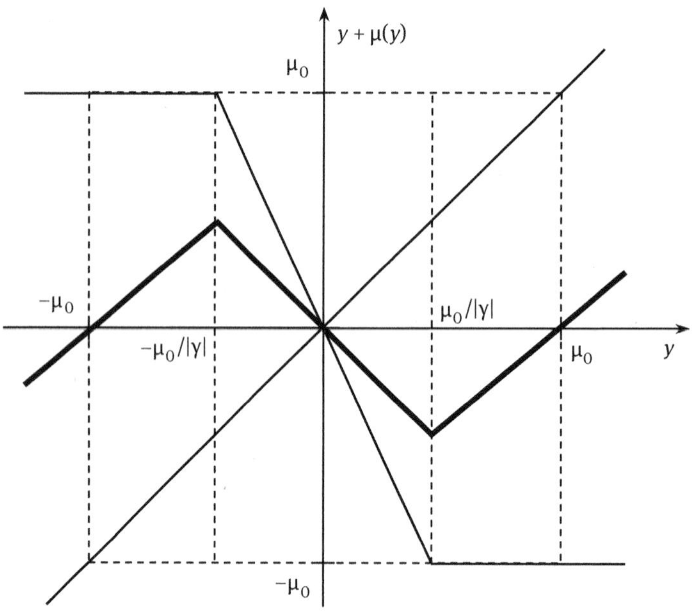

Fig. 1.6. Function $y + \mu(y)$.

It is worth mentioning that the larger the absolute value of feedback gain γ, the faster will variable y converge to zero. This happens for two reasons. First, with large absolute value of γ control function μ will longer stay saturated, because relation $\mu_0/|\gamma|$ is smaller (see. Figure 1.5). Second, the linear segment of control function (1.17) is steeper. The next section of this chapter, however, illustrates how "large" feedback gain γ can render the system unstable should any delay occur in the feedback chain.

Consider a linear saturated feedback that involves not only unstable variable y but stable variable z as well:

$$\mu = \mu(y) = \begin{cases} -\mu_0 & \text{when} \quad \gamma y + \delta z \leq -\mu_0 \,, \\ \gamma y + \delta z & \text{when} \quad |\gamma y + \delta z| \leq \mu_0 \,, \\ \mu_0 & \text{when} \quad \gamma y + \delta z \geq \mu_0 \,. \end{cases} \tag{1.19}$$

Here δ = const. Criteria of asymptotic stability for system (1.10), (1.11) with feedback control (1.19) are formulated as inequalities: $\gamma + \delta < -1$, $\delta > \gamma$. It will be shown below that within domain of controllability there exist such initial states $y(0)$, $z(0)$, from which system (1.10), (1.11) with control function (1.19) does not translate to the origin $y = z = 0$. Moreover, it even goes outside the boundaries of domain Q.

Figure 1.7 shows domain of controllability Q (1.12) in terms of variables y, z, along with its boundaries $y = \pm\mu_0$ and two straight lines $\gamma y + \delta z = -\mu_0$, $\gamma y + \delta z = \mu_0$. In between these straight lines, in domain

$$|\gamma y + \delta z| \leq \mu_0 \tag{1.20}$$

control function (1.19) is linear. The boundaries of domain (1.20) are shown for $\gamma < 0$ and $\delta < 0$.

To the right (to the top) of the right (top) border of stripe (1.20) $\mu = -\mu_0$, and to the left (to the bottom) of the left (bottom) border of this stripe $\mu = \mu_0$. If $\delta \neq 0$, then stripe (1.20) intersects stripe (1.12). The right (top) border of domain (1.20) $\gamma y + \delta z = -\mu_0$ intersects straight line $y = -\mu_0$ at point $y = -\mu_0$, $z = \mu_0(\gamma - 1)/\delta$. Let the initial state $y(0)$, $z(0)$ of system (1.10), (1.11), (1.19) be chosen so that $y(0) = 0$, i.e. inside domain Q (1.12) (note the point marked by an asterisk at Figure 1.7). Let $z(0)$ be large enough so that state $y(0)$, $z(0)$ lies outside domain (1.20), particularly, to the top of it. Then the value of control function (1.19) at initial moment $\tau = 0$ is equal to $-\mu_0$. If control function $\mu(\tau) \equiv -\mu_0$ in equation (1.10), then, provided $y(0) = 0$, its solution $y(\tau)$ will become less than the value $-\mu_0$ in some finite time period $\bar{\tau}$. At the same time, initial value $z(0)$ can be chosen large enough for $z(\tau)$, that is a solution to equation (1.11) with $\mu(\tau) \equiv -\mu_0$, to remain larger than $z = \mu_0(\gamma - 1)/\delta$ at all times within interval $0 \leq \tau \leq \bar{\tau}$. That means, the trajectory of system (1.10), (1.11) starts from its initial state 0, $z(0)$, and the value of $z(0)$ is large enough. Then, actuated by control function (1.19), it leaves domain of controllability Q (passing through its left border $y = -\mu_0$). To conclude, trajectories of system (1.10), (1.11), (1.19) that begin at some states $y(0), z(0) \in Q$ will go outside domain of controllability Q. It follows from the theorem about continuous

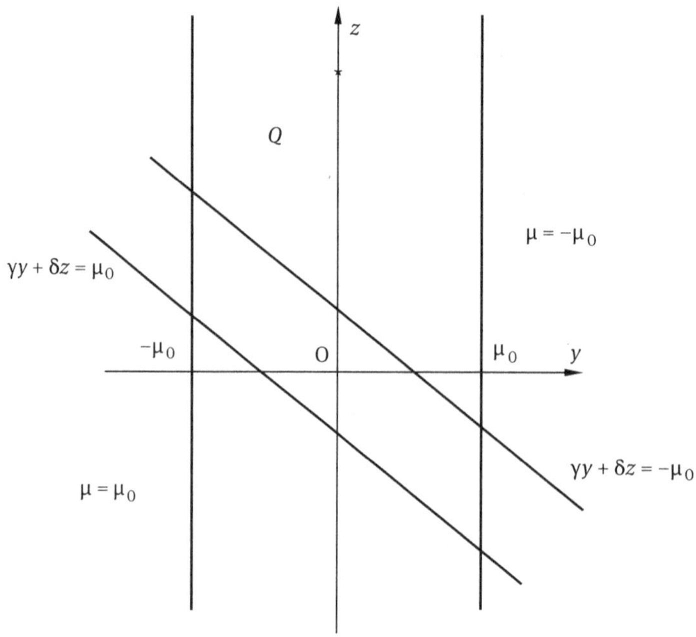

Fig. 1.7. Domain of controllability Q (1.12) and stripe (1.20).

dependence of solution to a differential equation on its initial condition, that the set of such initial states has its measure function not equal to zero. Hence if $\delta \neq 0$, then for control function (1.19) domain of attraction B can fill up only a part of domain of controllability Q.

Consequently, the following theorem is proved to be valid.

Theorem. *For control function (1.19), domain of attraction B coincides with domain of controllability Q, that means, it is the largest possible, if and only if gain* $\delta = 0$, *and gain* $\gamma < -1$.

To summarize, for maximizing domain of attraction for system (1.10), (1.11), *all resources* of control must be spent on suppressing the *unstable* mode.

Control function (1.17), in terms of variables β and β', looks as follows:

$$\mu = \begin{cases} -\mu_0 & \text{when} \quad \gamma(\beta + \beta') \leq -\mu_0\,, \\ \gamma(\beta + \beta') & \text{when} \quad |\gamma(\beta + \beta')| \leq \mu_0\,, \qquad \gamma < -1 \\ \mu_0 & \text{when} \quad \gamma(\beta + \beta') \geq \mu_0\,. \end{cases} \tag{1.21}$$

Control function (1.21) maximizes domain of attraction for system (1.9), and in this sense it is optimal. Note that for controlling a real pendulum, it is required to measure angle β and angular velocity $\dot{\beta}$. The latter is related to derivative β' by equation $\dot{\beta} = \beta'\sqrt{gb}/r$ (see the first of equations (1.3)).

Control law (1.21), stated in original variables, looks as follows:

$$L = \begin{cases} -L_0 & \text{when} \quad \gamma mgb\left(\beta + \dot{\beta} r/\sqrt{gb}\right) \le -L_0 \,, \\ \gamma mgb\left(\beta + \dot{\beta} r/\sqrt{gb}\right) & \text{when} \quad \left|\gamma mgb\left(\beta + \dot{\beta} r/\sqrt{gb}\right)\right| \le L_0 \,, \qquad \gamma < -1 \\ L_0 & \text{when} \quad \gamma mgb\left(\beta + \dot{\beta} r/\sqrt{gb}\right) \ge L_0 \,. \end{cases} \tag{1.22}$$

To put this function (1.22) into practical algorithm, one needs to know the pendulum mass m, the distance b between the pivot point O and the center of mass of the pendulum, and its radius of gyration about the pivot point.

1.4 Delay in feedback loop

It was stated above that the larger the absolute value of feedback gain γ, the faster will variable y converge to zero. Now it will be illustrated that whenever a delay is present in the feedback loop, the feedback gain should not be too large. Otherwise, the system may become unstable. Delays may arise due to control circuit lags or due to lags in data acquisition from the sensors. The problem of estimating acceptable delay in feedback loop is among important questions for any task of stabilizing an unstable object. Even with a "short" delay stabilization may become impossible.

To determine stability of system (1.10), (1.11) with control function (1.17) and delay in the feedback loop (consider it a pure delay), it is sufficient to investigate equations (1.10), (1.11) with linear feedback (1.15). With pure delay in feedback loop, equations (1.10), (1.11), (1.15) are transposed to

$$y'(\tau) = y(\tau) + \mu(\tau) \,, \tag{1.23}$$

$$z'(\tau) = -z(\tau) - \mu(\tau) \,, \tag{1.24}$$

$$\mu(\tau) = \gamma y(\tau - \zeta) \,. \tag{1.25}$$

Here $\zeta = \text{const} > 0$ is the constant value of pure delay.

The task here is to estimate parameters of system (1.23)–(1.25) that provide asymptotic stability of equilibrium $y = 0$, $z = 0$. Because of the specific form of control function (1.25), a feedback that involves only the unstable variable y, the stability problem of system (1.23)–(1.25), as it is shown below, is reduced to investigation of stability of solution $y = 0$ for differential equation

$$y'(\tau) = y(\tau) + \gamma y(\tau - \zeta) \,, \tag{1.26}$$

derived from equations (1.23) and (1.25).

Characteristic quasipolynomial that corresponds to equation (1.26) is quite simple (λ is a spectral parameter)

$$\lambda - 1 = \gamma e^{-\lambda\zeta} \tag{1.27}$$

Having in mind the method of D-decomposition [120], consider in equation (1.27) substitutions $\lambda = 0$ and $\lambda = i\rho$, where i is unit imaginary number, and ρ is a parameter that runs from 0 to ∞. In resulting expressions, separate real and imaginary parts. This yields equations that can be used to describe the boundary of stability domain

$$\gamma = -1, \quad -1 = \gamma\cos\rho\zeta, \quad \rho = -\gamma\sin\rho\zeta \quad (0 < \rho < \infty) \tag{1.28}$$

Let $\rho\zeta = \varepsilon$, then equations (1.28) can be transposed to

$$\gamma = -1, \quad \gamma = -\frac{1}{\cos\varepsilon}, \quad \zeta = \varepsilon \operatorname{ctg}\varepsilon \quad (0 < \varepsilon < \infty) \tag{1.29}$$

Equations (1.29) are used to construct the domain of asymptotic stability in the plane of coordinates ζ and γ. To build up this domain, it is sufficient to vary parameter ε within the range from 0 to $\pi/2$ in two last equations of (1.29) [47, 72, 129]. The corresponding curve Γ starts with $\varepsilon = 0$ at point $\zeta = 1$, $\gamma = -1$. Upon calculating derivatives of functions in (1.29) with respect to ε, it is easy to confirm that, with ε increasing, variables ζ and γ decrease strictly monotonically. With $\varepsilon \to \pi/2$ the convergence is $\zeta \to 0$, $\gamma \to -\infty$, that means, with $\varepsilon \to \pi/2$ curve Γ approaches axis $\zeta = 0$. The domain of asymptotic stability is shown in grey color at Figure 1.8. It is bounded by straight lines $\zeta = 0$, $\gamma = -1$ and curve Γ

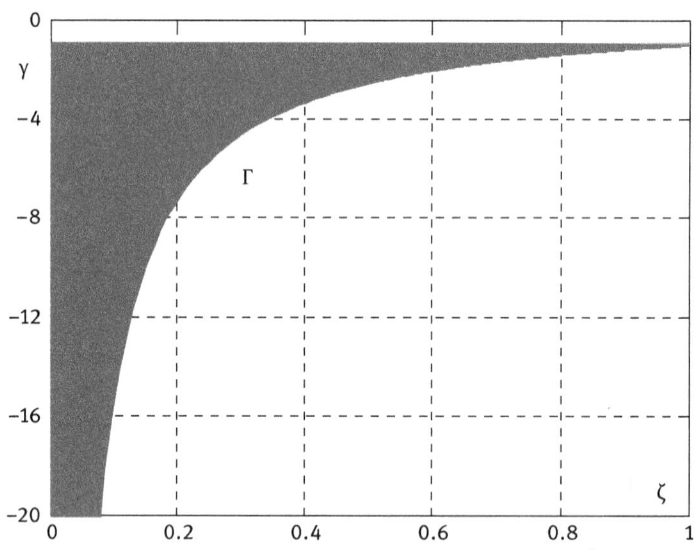

Fig. 1.8. Domain of asymptotic stability in terms of variables: ζ – delay period, γ – feedback gain.

The constructed domain of stability refers to equations (1.23), (1.25) (or, to equation (1.26)). However, if solution to equation (1.26) $y(\tau) \to 0$ as $\tau \to \infty$, then corresponding control function determined as in (1.25), $\mu(\tau) \to 0$. Now consider equation

(1.24) as a nonhomogeneous one. With $\mu = 0$, any solution to this equation exponentially approaches zero. Therefore (see [45]), if $\mu(\tau) \to 0$ as $\tau \to \infty$, then any solution to equation (1.24), $z(\tau) \to 0$, under any initial condition $z(0)$. Hence the constructed domain of asymptotic stability is related to the whole system (1.23)–(1.25). Then it is related to nonlinear system (1.10), (1.11), (1.17) with pure delay, and thus, it is as well related to system (1.9), (1.21) with delay.

Investigation of constructed domain of stability, leads to, among other things, the fact that with feedback gain γ increasing (in absolute value), admissible values of delay time ζ become shorter. In other words, admissible values of $\zeta \to 0$ as $|\gamma| \to \infty$, and it looks natural. Note that nondimensional value of delay time ζ is related to dimensional value, that will be denoted as θ, by expression $r\zeta = \theta\sqrt{gb}$ (see the first expression in (1.3)). Applying this relation, from known value of ζ one can evaluate admissible delay time θ and see how it depends on lengths b and r.

Characteristic equation similar to (1.27) is discussed, for example, in studies [47, 72, 129]. However, these studies do not construct domain of stability in coordinate plane of variables: delay time ζ and feedback gain γ.

It is worth mentioning that, in general, it is impossible to find analytic expressions for domain of stability for a system of arbitrary order with feedback control having pure delay. This study provides a domain of asymptotic stability for such system in analytic form only by virtue of the specific form of feedback law.

One should to keep in mind that inequalities (1.12)–(1.14) and control function (1.21) are sensible only for linearized equation (1.9). It is worth investigating only for the task of *local* stabilization of the pendulum about its top, unstable equilibrium (1.7).

The results obtained above are related to the full extent to a system that is described by a matrix equation of the order n:

$$x' = Ax + b\mu\,, \tag{1.30}$$

where A and b are constant matrices of orders $(n \times n)$ and $(n \times 1)$, respectively; the control parameter μ is compliant with inequality (1.5), and the spectrum of matrix A has *one real positive* eigenvalue and $n - 1$ eigenvalues with negative real components [79]. After separating from system (1.30) the variable that corresponds to the positive eigenvalue, it is possible to design a control law that can suppress this instability in the same manner as demonstrated above. Such control function will provide maximum possible domain of attraction. The conclusions concerning system stability when pure delay is present also relate fully to system (1.30).

Feedback of type (1.15) suppresses the unstable motion mode. When such feedback is provided to system (1.30), all eigenvalues that lie to the left of imaginary axis remain unchanged ("are not forced to move"). If some of these eigenvalues are "close" to imaginary axis, then the transient process in closed-loop system appear to be stretched in time. In this case it may be reasonable, for example, to design a linear feedback law that assigns eigenvalues of closed-loop system. Such feedback law is possible to create if system (1.30) is completely controllable [89–92]. However, if the feedback output is limited by inequality (1.5), then such way of feedback generation results in reduced domain of attraction.

To conclude, the discussed way to design feedback control is applicable to systems that have degree of instability equal to one.

1.5 Nonlinear control

In this section, a complete nonlinear model (1.2) or (1.4) will be considered. It will be shown that equilibrium (1.7), one or the other of them, can be reached (see [94]) from any initial state by applying a limited in absolute value control function $\mu(\tau)$, see inequalities (1.1) or (1.5).

The total energy E of the system (1.4) is represented by expression

$$E = \frac{1}{2}\beta'^2 + \cos\beta \,. \tag{1.31}$$

The full time derivative of E (1.31) (the time is nondimensional) with respect to system (1.4) looks as follows:

$$\frac{dE}{d\tau} = \frac{d}{d\tau}\left(\frac{1}{2}\beta'^2 + \cos\beta\right) = \beta'\beta'' - \beta'\sin\beta = \beta'\mu \,. \tag{1.32}$$

For derivative (1.32) to have maximum value at all times, the control function (that is limited in its absolute value) should look as follows:

$$\mu = \mu_0 \operatorname{sign}\beta' = \mu_0 \operatorname{sign}\dot{\beta} \,. \tag{1.33}$$

With control function as in (1.33), energy E increases monotonically, assuming $\beta' \neq 0$. The total energy of the system with the pendulum resting in the top equilibrium – state (1.7) – equals its potential energy in this state, that is, one: $E = 1$. Let the initial state of the system be such that the full energy in this state is less than one, $E < 1$. Control law (1.33) increases the magnitude of pendulum oscillations, along with the increase of the energy. To put it in other words, at the end of each semiperiod of oscillation, when angular velocity β' becomes zero, the pendulum gets farther from its bottom equilibrium than it was at the end of the previous semiperiod.

Let the control function become zero, $\mu = 0$, when the total energy of the system becomes equal to one at some moment of time during the oscillation process. After this moment, the pendulum will asymptotically approach one of its equilibriums (1.7) in its free, "passive" motion. To negate possible perturbances, it is appropriate to provide a stabilizing torque (1.21) acting on the pendulum. This can be done as soon as the pendulum is close enough to its equilibrium (1.7). Instead of torque (1.21), it is possible to apply torque

$$\mu = -\sin\beta - k_\beta\beta - k_{\beta'}\beta' \,. \tag{1.34}$$

With control function (1.34), equation (1.4) is transposed as follows:

$$\beta'' + k_{\beta'}\beta' + k_\beta \beta = 0 \,. \tag{1.35}$$

Expressions (1.34) and (1.21) for stabilizing control torque, as well as equation (1.35), should have value β changed to $\beta - 2\pi$, if in process of oscillations, and later – in its free motion, the pendulum approached the state $\beta = 2\pi$, $\beta' = 0$, and not the state $\beta = 0$, $\beta' = 0$. Gain values k_β, $k_{\beta'}$ must be chosen so that the eigenvalues corresponding to equation (1.35) would have their real parts negative.

With control function (1.34), criterion $|\mu| < \mu_0$ (1.5) will be satisfied if this control is “switched on” when the pendulum is close enough to state (1.7) (either of them). This is true by virtue of the fact that the closer initial state $\beta(0)$, $\beta'(0)$ to equilibrium (1.7), the closer to zero is solution $\beta(\tau)$ of equation (1.35), along with its derivative $\beta'(\tau)$. However, at the same time with functions $\beta(\tau)$, $\beta'(\tau)$, function (1.34) is close to zero. This leads to the fact that there exists such neighbourhood of the origin $\beta = 0$, $\beta' = 0$, that all trajectories of system (1.4), (1.34) comply with criterion $|\mu| < \mu_0$ (1.5).

The motion of the pendulum driven by control function (1.33) does not depend on the way the value of $\operatorname{sign}\beta'$ is defined at point $\beta' = 0$. Yet for initial state $\beta(0) = \pi$ and $\beta'(0) = 0$, if it is defined that $\operatorname{sign} 0 = 1$, angle β will increase at first (the pendulum will start its motion counter-clockwise), and if $\operatorname{sign} 0 = -1$, angle β at first will decrease (the pendulum will move clockwise). Therefore, with definition $\operatorname{sign} 0 = 1$ the pendulum will finish its motion at one of the states (1.7), and with definition $\operatorname{sign} 0 = -1$ it will end up at the other one. The plots of time functions $\beta_1(\tau)$ and $\beta_2(\tau)$ that correspond to these motions, are symmetric with respect to the straight line $\beta = \pi$: $[\beta_1(\tau) - \pi] = -[\beta_2(\tau) - \pi]$, i.e. $\beta_1(\tau) = -\beta_2(\tau) + 2\pi$.

Studies [132–134] consider the problem that involves complete nonlinear equation in terms of phase coordinates that comprise a cylinder. In those studies, a control function is defined that translates the pendulum from any initial state to the top equilibrium in the least possible time.

1.6 Controllability domain of the nonlinear model

Previously, in Section 2 of the current chapter, domain of controllability Q was constructed for linearized model (1.9). Now, the task will be to determine a set D of initial states located within a stripe, $-\pi \le \beta \le \pi$, so that from each of these states *nonlinear* model (1.4) can be translated to its equilibrium $\beta = 0$, $\omega = \beta' = 0$ (1.7). This motion should occur *without oscillations* about position $\beta = \pi$. In this discussion, it will be assumed that $\mu_0 < 1$.

The set D is apparently symmetric with respect to the origin $\beta = 0$, $\omega = \beta' = 0$. This means that it is enough to determine only one border of this area in the phase plane (β, ω). To construct the right border, consider equation (1.4) with $\mu = -\mu_0$. Phase plot of the solutions to this equation includes a *saddle point* on axis β, its abscissa is

$\beta = -\arcsin\mu_0$. Into this saddle point, two separatrices come with $t \to +\infty$, and two more with $t \to -\infty$. For equation (1.4) with μ = const a first integral can be found

$$\frac{1}{2}\omega^2 + \cos\beta - \mu\beta = \text{const}\,. \tag{1.36}$$

Assuming in integral (1.36) $\mu = -\mu_0$, one can find the four separatrices. Figure 1.9 illustrates these curves obtained numerically for $\mu_0 = 0.5$. The arrow marks indicate direction in which phase point moves as time increases. Heavy lines are the separatrices that come into the saddle point with $t \to +\infty$, thin lines show the separatrices that come into the saddle point with $t \to -\infty$. Other thin lines in Figure 1.9 also show trajectories of equation (1.4) for $\mu = -\mu_0$ that come "close" to the saddle point. These trajectories "resemble" hyperbolas. It is possible to prove [104] that the separatrices that come into the saddle point with $t \to +\infty$ constitute the *right* border of domain D. In Figure 1.9, also by heavy line, the *left* border of domain D is shown. It results from rotating the right one 180° about the origin $\beta = 0$, $\omega = 0$.

Consider trajectories of the nonlinear system with $\mu = -\mu_0$. From analyzing Figure 1.9 it can be noticed that the trajectories that begin to the left of the right border of domain D will remain to the left of this border. The trajectories that begin to the right of the right border will "go away" from it. It will be shown that domain D does not contain its boundary, i.e. D is an open domain. Consider an arbitrary phase point with $\omega > 0$, lying on the separatrix that comes into the saddle point with $t \to +\infty$. Phase velocity vector in this point looks like

$$(\omega, \mu + \sin\beta)\,. \tag{1.37}$$

With $\mu = -\mu_0$, vector (1.37) is directed right-down towards the saddle point, tangent to the separatrix. With $-\mu_0 < \mu \le \mu_0$, the ordinate of (1.37) is larger than with $\mu = -\mu_0$, and this vector becomes directed to the right of the right border of domain D. That means, of all the phase velocity vectors that start on the separatrix with $-\mu_0 \le \mu \le \mu_0$, none is directed inside domain D. Consider now the separatrix that comes into the saddle point with $t \to +\infty$ from semiplane $\omega < 0$. In the same way as above, it can be shown that none of the phase velocity vectors (1.37) on this curve with $-\mu_0 < \mu \le \mu_0$ is directed inside domain D. Thus it is impossible to get inside domain D from any point on its border. This concludes the fact that D is an open domain.

Figure 1.10 illustrates domain D, that is constructed for nonlinear equation (1.4) with $\mu_0 = 0.5$. Thin straight lines mark the boundaries of controllability domain Q, that is constructed with $\mu_0 = 0.5$ for linearized equation (1.9). It follows from examining Figure 1.10 that on interval $-\pi/3 < \beta < \pi/3$ domain Q constructed for linearized equation is "close enough" to domain D constructed for exact nonlinear equation.

Applying principle of maximum [27, 130], it is possible to construct inside domain D a picture of time-optimal control design, where control function is limited in its absolute value. In studies [133, 134] a control design picture is constructed, for precise nonlinear problem, for the whole phase space that is topologically a cylinder. The con-

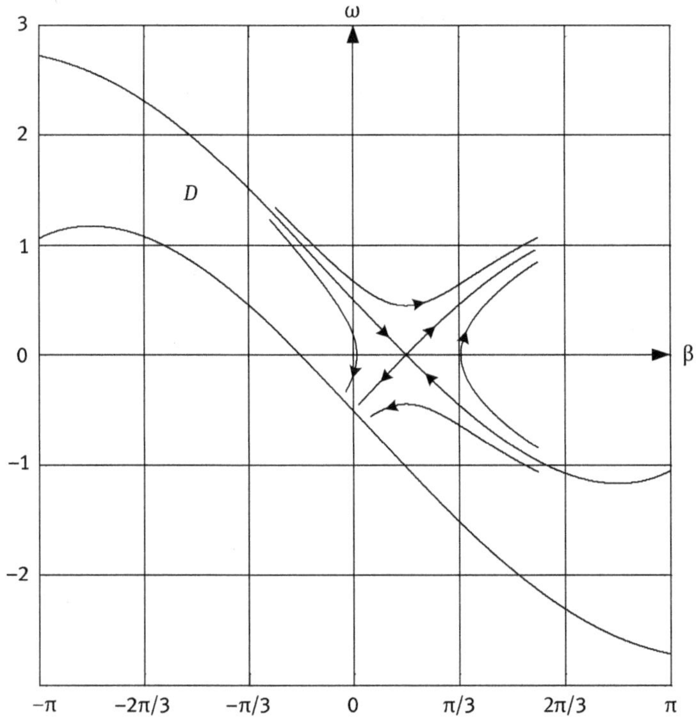

Fig. 1.9. Controllability domain D of the nonlinear model.

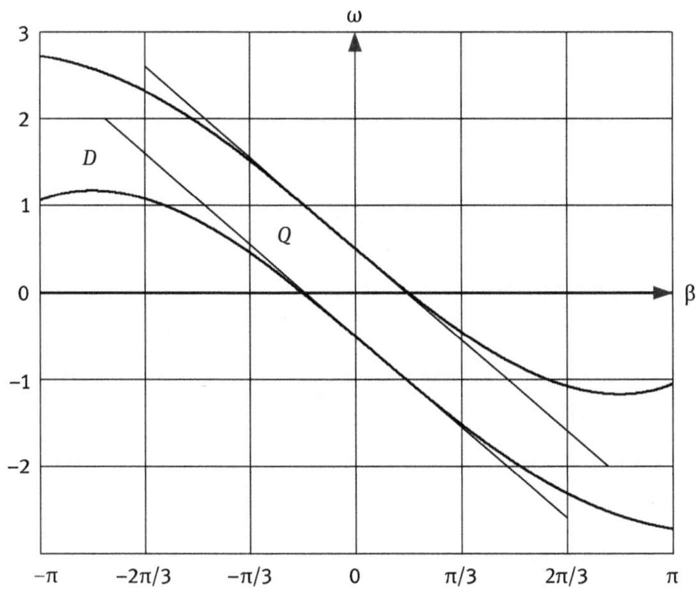

Fig. 1.10. Controllability domains D and Q for nonlinear and linearized models, respectively.

trol designed there translates the pendulum to its top equilibrium in minimum possible time. The oscillations of the pendulum about its downmost state $\beta = \pi$ are admitted there, thus enabling the pendulum translation to its top equilibrium from any initial state.

In current chapter, the pendulum with stationary pivot is discussed. Such pendulum has one degree of freedom and a single control parameter. The next chapter considers a pendulum with a mobile pivot – it is connected to the center of a wheel. Such a device has two degrees of freedom and one control parameter. That means it is underactuated.

2 A pendulum with wheel-based pivot

Consider a single-link pendulum with its pivot located in the center point O of a wheel (Figure 2.1). The wheel is symmetric about its axis O, and it can roll on a flat horizontal surface along a straight line. It is assumed that the wheel is vertical, and that no slipping takes place at all times. Similar system was discussed in study [94]. The wheel mass is denoted as M, its radius as R, radius of gyration about its center O as ρ. The value of an angle that the wheel turns counter-clockwise is denoted as φ. Wheel turning means that a particular, distinguished wheel radius line that was horizontal and aligned with axis X at initial moment, turns a particular angle φ about wheel center O. Horizontal translation of the wheel center of mass O is represented as x, and thus the following relation takes place: $\dot{x} = -\dot{\varphi}R$.

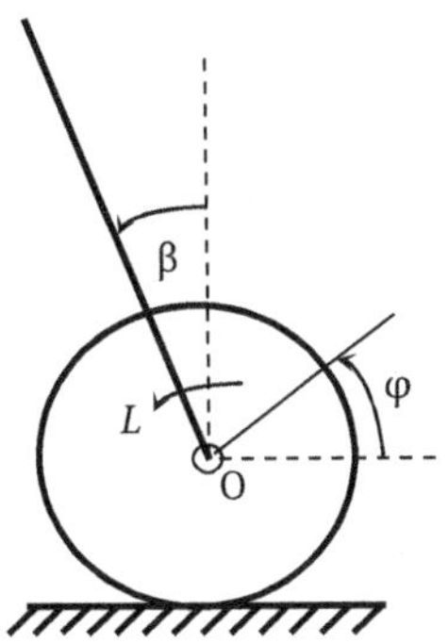

Fig. 2.1. The pendulum with wheel-based pivot.

Like in chapter 1, we consider β as the angle of deviation of the pendulum from its vertical position, m as its mass, b as the distance between the pivot point O and the center of mass of the pendulum, r as the radius of gyration of the pendulum about its pivot point O. We assume that an electric direct current motor is installed at the wheel axis. Its stator is attached firmly to the wheel, and its armature is attached to the pendulum body. Torque generated by the motor is designated as L. Torque L forces the pendulum to rotate counter-clockwise; the opposite torque $-L$ forces the wheel to rotate clockwise.

2.1 Equations of motion

For the specified system of two bodies, kinetic energy looks as follows:

$$T = \frac{1}{2}\left(a_{11}\dot{\varphi}^2 + 2a_{12}\cos\beta\dot{\varphi}\dot{\beta} + a_{22}\dot{\beta}^2\right) , \tag{2.1}$$

here

$$a_{11} = M(R^2 + \rho^2) + mR^2, \quad a_{12} = mRb, \quad a_{22} = mr^2 . \tag{2.2}$$

All coefficients represented as letters in relations (2.2) are considered positive. Expressions for potential energy Π and for virtual work δW look as follows:

$$\Pi = mgb\cos\beta, \quad \delta W = L(\delta\beta - \delta\varphi)\,. \tag{2.3}$$

Here $\delta\beta$ and $\delta\varphi$ stand for virtual displacements of angles β and φ.

Applying Lagrangian approach for the system in consideration (see studies [16, 41, 73]), one can derive the equations of motion of the second kind, taking into account expressions (2.1), (2.3):

$$\begin{aligned} a_{11}\dot\omega + a_{12}\cos\beta\ddot\beta - a_{12}\sin\beta\dot\beta^2 &= -L\,, \\ a_{12}\cos\beta\dot\omega + a_{22}\ddot\beta - mgb\sin\beta &= L\,. \end{aligned} \tag{2.4}$$

In equations (2.4), variable $\omega = \dot\varphi$ represents the angular velocity of the wheel. The mechanical system described above has two degrees of freedom. However, its wheel rotation angle φ is a *cyclic* coordinate, thus its motion can be represented by equation system (2.4) of the third order. Whenever ω is a known time function $\omega = \omega(t)$, one can integrate equation $\dot x = -\omega R$ and get the function for coordinate x of the wheel. Adding together the two equations in (2.4) yields an equation that describes the dynamics of system angular momentum about the point where its wheel contacts with the supporting surface.

The equations in (2.4) can be solved for the highest-order derivatives. This will yield the following expressions:

$$\left(a_{11}a_{22} - a_{12}^2\cos^2\beta\right)\ddot\beta + a_{12}^2\dot\beta^2\sin\beta\cos\beta - a_{11}mgb\sin\beta = (a_{11} + a_{12}\cos\beta)L\,, \tag{2.5}$$

$$\left(a_{11}a_{22} - a_{12}^2\cos^2\beta\right)\dot\omega - a_{12}a_{22}\dot\beta^2\sin\beta + a_{12}mgb\sin\beta\cos\beta = -(a_{22} + a_{12}\cos\beta)L\,. \tag{2.6}$$

Coefficient of the highest-order derivatives in equations (2.5), (2.6) is positive for any value of angle β, because it is effectively a determinant of positive definite matrix of kinetic energy of the system (see expression (2.1)).

Consider the force that resists the wheel rolling. If this force is a function of velocity $\dot x$, then equations (2.4) will involve angular velocity ω along with angular acceleration $\dot\omega$. The wheel angular velocity will also appear in equations (2.4), should counter-electromotive force (counter-EMF) be taken into account in the expression for motor torque L. Counter-EMF is proportional to algebraic difference $\dot\beta - \omega$ [42, 76, 108]. In these cases, when the equations of motion contain ω, a single equation like (2.5), one that describes exclusively the pendulum body oscillations, cannot be derived.

Introducing nondimensional time τ and torque μ as indicated in expressions (1.3), one can transpose equations (2.5), (2.6) to get their form

$$(1 - d^2\cos^2\beta)\beta'' + d^2\beta'^2\sin\beta\cos\beta - \sin\beta = (1 + e^2\cos\beta)\mu\,, \tag{2.7}$$

$$(1 - d^2\cos^2\beta)\sigma' - e^2\beta'^2\sin\beta + e^2\sin\beta\cos\beta = -e^2\left(\frac{e^2}{d^2} + \cos\beta\right)\mu\,. \tag{2.8}$$

Here $\sigma = \varphi'$ is a nondimensional variable for the wheel angular velocity, and

$$d^2 = \frac{a_{12}^2}{a_{11}a_{22}} = \frac{mR^2b^2}{r^2[M(R^2+\rho^2)+mR^2]} < 1, \quad e^2 = \frac{a_{12}}{a_{11}} = \frac{mRb}{M(R^2+\rho^2)+mR^2} .$$

Inequality $d^2 < 1$ can be verified directly. Otherwise, it follows from the fact that the determinant of positive-definite matrix of kinetic energy (2.1), being positive for all values of β, is also positive for $\beta = 0$. Note that system (2.7), (2.8) has only two nondimensional parameters, namely d and e.

Equation (2.7), describes oscillations of a pendulum with its pivot in the center of a wheel. It involves angle β and its first two derivatives, and it does not involve ω, the angular velocity of the wheel. Hence the equation (2.7) can be separated from the system (2.7), (2.8). At the same time, inertial and geometric attributes of the wheel do occur in equation (2.7), and they influence the pendulum behavior at any applied control torque $\mu(\tau)$. The reason is that when any torque is applied to the pendulum, at the same time the opposite value torque is applied to the wheel. The movement of the wheel, in turn, influences the movement of the pendulum. If the pendulum behavior is of concern, and the wheel motion is not, then equation (2.7) can be analyzed apart from equation (2.8). The structure of equation (2.7) is significantly more complicated than that of equation (1.4), the one that describes oscillations of a pendulum with fixed pivot.

Note that a mathematical model for a pendulum with its pivot fixed on a cart also includes a *cyclic* coordinate. This coordinate depicts the cart position on a horizontal axis. For this model, one equation, similar to equation (2.7), can be separated. However, unlike equation (2.7), the right-hand side of such equation does not depend on the angle of deviation of the pendulum from its vertical position. It only involves the torque applied at the pivot.

With $\mu = 0$, position $\beta = 0, \beta' = 0$ (1.7) is an unstable equilibrium not only for equation (1.4), but for equation (2.7) as well. Below, the problem of stabilization of this equilibrium of the pendulum is discussed. In particular, a specific domain of controllability should be determined for equation (2.7) (linearized). Next, expressions for control input should be determined in such a way that the domain of attraction would be coincident with the domain of controllability. Such domain of attraction is maximum possible. Finally, such domain should be compared to the domain of controllability determined in (1.12) above for a pendulum with fixed pivot.

Equation (2.7), being linearized at its equilibrium $\beta = 0, \beta' = 0$ (1.7), looks as follows:

$$a^2\beta'' - \beta = c\mu \tag{2.9}$$

where $a^2 = 1 - d^2 > 0, c = 1 + e^2 > 1$.

Differential equation (2.9) coincides with (1.9) when $a = 1, c = 1$. With no control applied, i.e. when $\mu = 0$, one of the corresponding eigenvalues is positive $(1/a)$, the other one is negative $(-1/a)$.

A second order equation (2.9) can be transposed into a system of two first order equations in Jordan form

$$y' = \frac{y}{a} + \frac{c}{a}\mu\,, \qquad y = \beta + a\beta'\,; \tag{2.10}$$

$$z' = -\frac{z}{a} - \frac{c}{a}\mu\,, \quad z = \beta - a\beta'\,. \tag{2.11}$$

Differential equation (2.10) describes the behavior of the "unstable" variable y, that corresponds to positive eigenvalue $1/a$. Equation (2.11) describes the behavior of the "stable" variable z, that corresponds to negative eigenvalue $(-1/a)$. When $a = 1$, $c = 1$ equations (2.10), (2.11) respectively lead to relations (1.10), (1.11).

2.2 Domain of controllability

Let $W = \{\mu(\tau) \colon |\mu(\tau)| \le \mu_0\}$, like it was defined earlier in chapter 1. Consider a set of initial states, denoted as P. For each initial state of this set P there exists a control function $\mu(\tau) \in W$, such that the solution to equation (2.9) (or, similarly, to system of (2.10) and (2.11)) with control function $\mu(\tau)$ comes to equilibrium $\beta = 0$, $\beta' = 0$. This domain of controllability P is described (see [53]) by inequality:

$$|y| < c\mu_0 \quad \text{or} \quad |\beta + a\beta'| < c\mu_0\,. \tag{2.12}$$

Domain of controllability P on the plane of variables β, β' is illustrated by Figure 2.2. To compare, Figure 2.2 shows domain of controllability Q of system (1.10), (1.11).

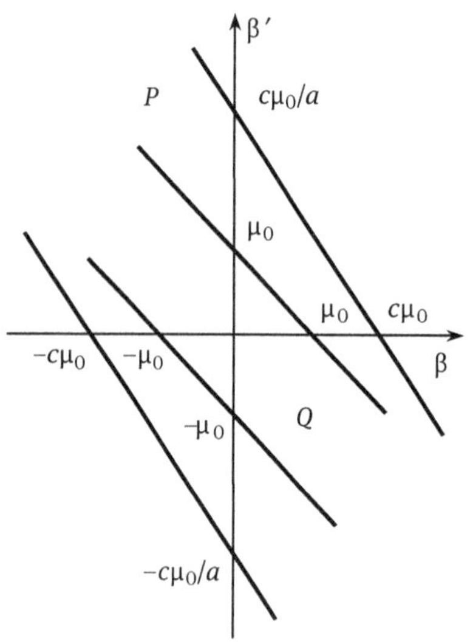

Fig. 2.2. Domains of controllability Q and P, for the pendulum with stationary pivot and for wheel-based pendulum, respectively.

In case when initial velocity $\beta'(0) = 0$ the pendulum can be translated to its equilibrium $\beta = 0$, $\beta' = 0$ (1.7) when the initial angle lies inside the following interval:

$$|\beta(0)| < c\mu_0 \,. \tag{2.13}$$

Again, in case when initial angle $\beta(0) = 0$, the interval for initial velocity that the pendulum can have to be able to come to equilibrium $\beta = 0$, $\beta' = 0$ (1.7) looks like

$$|\beta'(0)| < \frac{c}{a}\mu_0 \,. \tag{2.14}$$

Writing (2.12) in terms of original parameters of the mechanical system yields inequality

$$\left|\beta + \sqrt{1 - \frac{m(Rb)^2}{r^2[M(R^2+\rho^2)+mR^2]}}\beta'\right| < \left[1 + \frac{mRb}{M(R^2+\rho^2)+mR^2}\right]\frac{L_0}{mgb} \,. \tag{2.15}$$

Inequalities (2.13), (2.14) in terms of original parameters are transposed as

$$|\beta(0)| < \left[1 + \frac{mRb}{M(R^2+\rho^2)+mR^2}\right]\frac{L_0}{mgb}, \quad |\beta'(0)| < \frac{1 + \frac{mRb}{M(R^2+\rho^2)+mR^2}}{\sqrt{1 - \frac{mR^2b^2}{r^2[M(R^2+\rho^2)+mR^2]}}}\frac{L_0}{mgb} \,. \tag{2.16}$$

It is worth to compare domains of controllability Q and P, that correspond respectively to the pendulum with stationary pivot and to the pendulum with wheel-based pivot (consider Figure 2.2).

First, note that stripe P (2.12) is wider than stripe Q (1.12), because $c > 1$ and $0 < a < 1$ (see Figure 2.2). Therefore, it may be considered larger, though the areas of both stripes are infinite. However, domain Q does not lie inside domain P, because the straight lines that confine domain Q form a smaller angle with horizontal axis than lines that enclose P. Remember also, that domains Q and P were built for linear equations (1.9) and (2.9), respectively, and linearization is valid only in a range of values β and β' (the angle and its velocity) that is close enough to zero. And for close-to-zero values of β and (or) β' domain Q lies inside domain P. Besides, intervals (1.13) and (1.14) lie inside respective intervals (2.13) and (2.14), because $c > 1$ and $0 < a < 1$. That means, rhomb Π_Q, formed by end points of intervals (1.13) and (1.14), lies entirely inside rhomb Π_P, formed by end points of intervals (2.13) and (2.14) (see Figures 2.2, 2.3). Thus domain of controllability of the pendulum with stationary pivot is in a sense smaller than domain of controllability of the pendulum with wheel-based pivot. In other words, wheel-based pendulum is easier to stabilize. This is reasonable, because in the latter case the torque generated by the motor actuates not only the body of the pendulum, but the wheel as well, accelerating the pivot (the wheel center) and contributing even more to system stabilization.

Rhombs Π_Q and Π_P are drawn to illustrate (symbolically), that in the neighbourhood of the origin $\beta = \beta' = 0$ domain Q lies inside domain P.

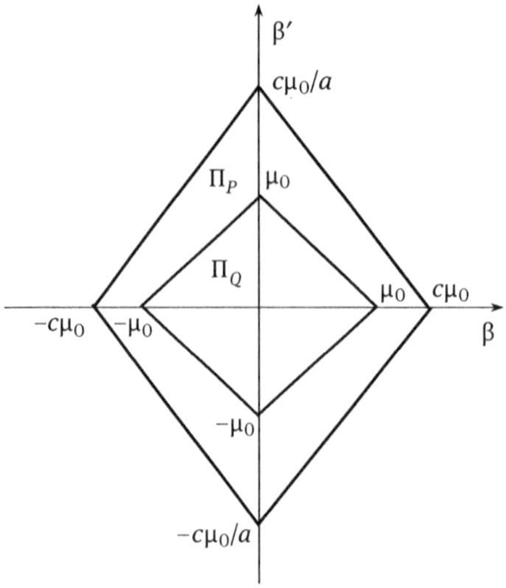

Fig. 2.3. Rhombs Π_Q and Π_P, inscribed into controllability domains Q and P.

As the wheel mass M increases, value

$$a = \sqrt{1 - \frac{m(Rb)^2}{r^2[M(R^2 + \rho^2) + mR^2]}} \tag{2.17}$$

also increases strictly monotonically, and as $M \to \infty$, it approximates one. At the same time, value

$$c = 1 + \frac{mRb}{M(R^2 + \rho^2) + mR^2} \tag{2.18}$$

decreases strictly monotonically as M grows, however, as $M \to \infty$, it also approximates one. The lengths of intervals (2.13) and (2.14), and therefore the size of rhomb Π_P, shrink as the mass increases. As $M \to \infty$, these intervals approach respectively intervals (1.13) and (1.14), and rhomb Π_P approaches rhomb Π_Q. That means, as mass M grows, domain (2.12) (or (2.15)) shrinks, and as $M \to \infty$, it converges to domain (1.12). This can be explained by the fact that a heavier wheel is the less mobile, and thus it makes pivot O of the pendulum more difficult to move.

The radius of gyration of the wheel $\rho \le R$, so if $R \to 0$, then $\rho \to 0$ and, as it follows from expression (2.18), $c \to \infty$. That means, the smaller the wheel, the wider is range (2.13) of deviation angles that are "affordable" from the point of view of stabilizability. This conclusion, based on linear model investigation, is verified further in a full nonlinear model.

Making use of expressions (2.15)–(2.18), it is possible to evaluate the influence of various system parameters on the controllability domain. These evaluations are helpful, for example, when constructing a transportation device like Segway [165], as well as other coaxial wheel based devices [20].

2.3 Maximizing domain of attraction

Likewise in chapter 1, it can be shown that linear feedback

$$\mu = \gamma y = \gamma(\beta + a\beta'), \quad \gamma < -1/c\,, \tag{2.19}$$

and thus linear feedback with saturation

$$\mu = \begin{cases} -\mu_0 & \text{when} \quad \gamma(\beta + a\beta') \le -\mu_0 \\ \gamma(\beta + a\beta') & \text{when} \quad |\gamma(\beta + a\beta')| \le \mu_0 \\ \mu_0 & \text{when} \quad \gamma(\beta + a\beta') \ge \mu_0 \end{cases}, \quad \gamma < -1/c \tag{2.20}$$

provides asymptotic stability of solution (1.7) to equation (2.9) (or to system (2.10), (2.11)). When feedback (2.20) is applied, the domain of attraction of desired equilibrium $\beta = 0$, $\beta' = 0$ (1.7) coincides with the entire domain of controllability P (2.12). That means, feedback (2.20) provides maximum possible domain of attraction (in linear approximation). In this sense, such feedback is optimal. By virtue of Lyapunov's theorem for stability and asymptotic stability of equilibrium on the basis of linear approximation, solution $\beta = 0$, $\beta' = 0$ (1.7) to nonlinear system (2.7), (2.20) is asymptotically stable.

Influence of delays in feedback loop (2.19) or (2.20) on the stability of equilibrium $\beta = 0$, $\beta' = 0$ for wheel-based pendulum can be investigated in the same manner like for the pendulum with a stationary pivot (see chapter 1).

With terminal control input $\mu(t) = -\mu_0$ the steady state of nonlinear equation (2.7) is in accordance with conditions

$$\sin\beta = (1 + e^2\cos\beta)\mu_0, \quad \beta' = 0\,. \tag{2.21}$$

Under such conditions, the wheel rolls leftwards (see Figure 2.1), having constant angular acceleration σ'. Its value can be found from equation (2.8), by substituting expression (2.21) into it, and it equals

$$\sigma' = \frac{e^4}{d^2}\mu_0\,. \tag{2.22}$$

Equation (2.21) is nonlinear with respect to variable β. If $\mu_0 < 1$, then this equation has a solution on interval $0 < \beta < \pi/2$. It can be shown that this solution is greater than the value $\beta = \arcsin\mu_0$. Indeed, left-hand side of equation (2.21) increases strictly monotonically as β grows from 0 to $\pi/2$, and right-hand side strictly monotonically decreases. When $\beta = \arcsin\mu_0$ ($\sin\beta = \mu_0$), left-hand side of equation (2.21) is less than its right-hand side, and when $\beta = \pi/2$, on the other hand, left-hand side is greater than right-hand side, because $\mu_0 < 1$.

To mention again, values $\beta = \pm \arcsin \mu_0$ express stationary motion modes of the pendulum with a stationary pivot. They are stationary values of the angle that the pendulum is away from its vertical position when terminal control torque $\mu(t) = \mp\mu_0$ is applied to its body in pivot.

In other words, the pendulum with a wheel-based pivot can be kept in balance when it deviates a greater angle from its vertical position, as compared to the pendulum with a stationary pivot. This can be explained from the physical point of view. The reason is, the torque applied to the pendulum in its pivot not only tends to rotate it around the pivot, pushing it back to its equilibrium. This torque at the same time accelerates the center of the wheel, contributing more to pendulum stabilization. In the steady-state mode, the value x'' of such acceleration can be found by multiplying value (2.22) by wheel radius R: $x'' = R\sigma'$.

If $R \to 0$, then $e \to \infty$, and the solution to equation (2.21) approaches $\pi/2$. Therefore, the smaller the wheel radius R, the larger angles can the pendulum deviate from its vertical position, up to $\pi/2$, as $R \to 0$.

The state

$$\beta = 0, \quad \beta' = 0, \quad \sigma = 0 \tag{2.23}$$

is a state of equilibrium for system of equations (2.7), (2.8) with $\mu = 0$. Equation (2.8), being linearized about its solution (2.23), is transposed to

$$a^2\sigma' + e^2\beta = -e^2\left(1 + \frac{e^2}{d^2}\right)\mu \tag{2.24}$$

Without control (with $\mu = 0$) third-order system (2.9), (2.24) has one corresponding positive eigenvalue $1/a$, one negative one $(-1/a)$, and one zero eigenvalue. It can be easily verified that this system is completely controllable in Kalman's sense. The domain of controllability of this third-order system in terms of three-dimensional phase space β, β', σ is a set of full measure. This set is described by the same inequality (2.12) as the domain of controllability for second-order equation (2.9). Inequality (2.12) does not include angular velocity σ; thus system (2.9), (2.24) can (or cannot) be translated into its equilibrium (2.23) irrespectively of the current value of angular velocity σ.

With control law (2.20) applied, equilibrium $\beta = 0, \beta' = 0$ (1.7) of linear second-order equation (2.9), as well as the same equilibrium of nonlinear equation (2.5), is asymptotically stable. However, control law (2.20) does not provide asymptotic stability of equilibrium (2.23) of third-order system of equations (2.9), (2.24). At the same time, it is possible to stabilize state (2.23), because system (2.9), (2.24) is controllable in Kalman's sense. Designing a feedback that stabilizes equilibrium (2.23) is an easy task. One can, for instance, do it by assigning the eigenvalues for a closed-loop system. Such feedback has to include information of current values of angle β, its velocity β', and angular velocity σ.

2.4 Nonlinear control

This section discusses full nonlinear model (2.7). It will be shown that equilibriums (1.7) (one of them or the other) may be reached [94] from any initial state of the system.

It will be assumed that $e^2 < 1$. This means, $1+e^2 \cos\beta > 0$ for any value of angle β. Taking equation (2.7) alone, consider the following control law:

$$\mu = \frac{1}{1+e^2\cos\beta}\Big\{(1-d^2\cos^2\beta)\left[\beta_d'' + k_{\beta'}\left(\beta_d' - \beta'\right) + k_\beta(\beta_d - \beta)\right] + \\ + d^2\beta'^2\sin\beta\cos\beta - \sin\beta\Big\}\,. \quad (2.25)$$

Here $\beta_d = \beta_d(\tau)$ is the desired time function, it describes how angle β should behave. Substituting expression (2.25) into motion equation (2.7) yields

$$(1-d^2\cos^2\beta)\beta'' = (1-d^2\cos^2\beta)\left[\beta_d'' + k_{\beta'}(\beta_d' - \beta') + k_\beta(\beta_d - \beta)\right]\,. \quad (2.26)$$

Factor $1 - d^2\cos^2\beta$ does not become zero for any value of angle β, because $d^2 < 1$. Therefore it can be cancelled in both sides of equation (2.26), to get

$$\beta'' - \beta_d'' + k_{\beta'}\left(\beta' - \beta_d'\right) + k_\beta(\beta - \beta_d) = 0\,. \quad (2.27)$$

If both coefficients $k_\beta > 0$ and $k_{\beta'} > 0$, then as $\tau \to \infty$, solution to equation (2.27) $\beta(\tau) \to \beta_d(\tau)$ under any initial conditions $\beta(0)$, $\beta'(0)$. If $\beta_d(\tau) \equiv 0$, then control law (2.25) has a simpler form:

$$\mu = \frac{1}{1+e^2\cos\beta}\left\{\left(d^2\cos^2\beta - 1\right)\left(k_{\beta'}\beta' + k_\beta\beta\right) + d^2\beta'^2\sin\beta\cos\beta - \sin\beta\right\}\,. \quad (2.28)$$

and equation (2.27) becomes simpler as well:

$$\beta'' + k_{\beta'}\beta' + k_\beta\beta = 0\,. \quad (2.29)$$

Solution to equation (2.29), $\beta(\tau) \to 0$ under any initial conditions, and the top equilibrium $\beta = 0$, $\beta' = 0$ (1.7) becomes globally asymptotically stable.

To realize control law (2.25) or (2.28), one has to know two nondimensional parameters d, e, and in the whole process of motion angle β and its velocity β' must be measured.

However, it cannot be guaranteed that control functions (2.25) or (2.28) will satisfy constraint $|\mu| < \mu_0$ (1.5). Nevertheless, this inequality is satisfied if $\beta_d(\tau) \equiv 0$ and initial state $\beta(0)$, $\beta'(0)$ is close enough to state (1.7) (one or the other of them). The closer the initial state is to equilibrium (1.7), the closer will solution to equation (2.29) – function $\beta(\tau)$ together with its derivative $\beta'(\tau)$ – be to zero. And at the same time with $\beta(\tau)$, $\beta'(\tau)$, function (2.28) is close to zero as well. This proves that there exists a neighbourhood of the origin $\beta = 0$, $\beta' = 0$, such that all trajectories of system (2.7), (2.28), that begin within that neighbourhood, satisfy inequality $|\mu| < \mu_0$ (1.5).

It is reasonable to try to stay within limits (1.5) by specifically choosing function $\beta_d = \beta_d(\tau)$: it should converge to zero together with its first and second derivatives. This function is appropriate to construct as a combination of polynomials and trigonometric functions. The uncertain parameters (coefficients) may be chosen so that to minimize the maximum value of control function $\mu(\tau)$, as it is evaluated by solving equations (2.7), (2.25).

3 A pendulum with a flywheel

This chapter enunciates the results of theoretical and experimental investigation of motion of a plane, single-link pendulum with a *stationary* pivot. The pendulum is influenced by the gravity field. In the pivot point, i.e. in the joint, no forces are applied except friction and reaction force of the support. The other end of the pendulum has an electric DC motor attached to it. The armature of the motor is firmly connected to a flywheel. A motor control algorithm is proposed that is capable of translating the pendulum from any initial state into the top, unstable equilibrium, and then this algorithm is used to stabilize the pendulum in that position. Other algorithms are proposed that realize other modes of pendulum motion, for example, one that rotates the pendulum to either of the directions, making it stop and stabilize in top or bottom equilibrium. Such device with a flywheel is often referred to as "inertia wheel pendulum" [1, 25, 124–126, 138, 147].

Rotating the flywheel to stabilize the pendulum in the top equilibrium is similar to swinging arms by a person trying to keep from falling backwards or forwards. In study [123] such way of stabilization is proposed to sustain vertical position of a two-wheel bicycle. The flywheel plays the same role like gyrodynes that sustain orientation of a satellite.

Such device was developed into practice in the Institute of Mechanics of Lomonosov Moscow State University by A. V. Lenskii, at that time head of Mechatronics Lab. The device includes a pendulum with a flywheel and a personal computer to provide feedback. Currently this device is used by students of Department of Mechanics and Mathematics who want to practice control theory.

3.1 The arrangement of the pendulum with a flywheel

Figure 3.1 illustrates the pendulum with a flywheel designed in the Institute of Mechanics. It consists of link 1 that is connected to flywheel 3. The pendulum can rotate in vertical plane. Its support that holds its rotation axle 2 is stationary. Flywheel axle 4 is firmly attached to the pendulum, and it is parallel to the pendulum axle 2. The flywheel is actuated by a DC motor 5. The motor and the flywheel are both attached to the pendulum. The stator is firmly attached to the pendulum, the armature is attached to the flywheel axle 4. The system of motor control includes a personal computer, a controller and a power amplifier used to provide power to the motor. The data used in the control system includes angles and angular velocities of the link rotation with respect to the support and of the flywheel rotation with respect to the link. To measure these data, angle encoders are used.

Figure 3.2 illustrates the structure of the pendulum. Link OC is connected to the stationary support at point O. The axis of the joint is perpendicular to the pendulum

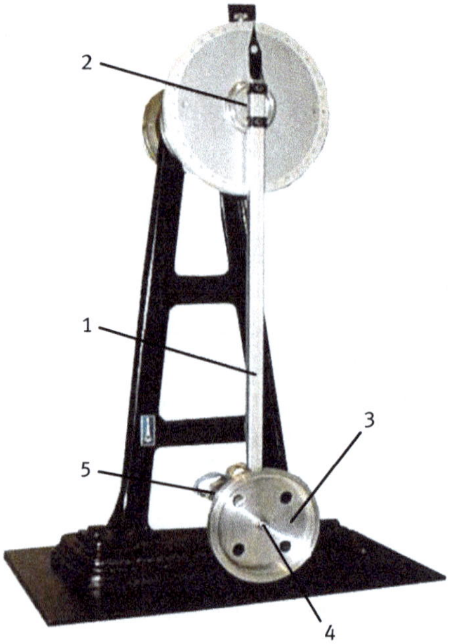

Fig. 3.1. A photo of the pendulum: 1 – pendulum link, 2 – pendulum axle, 3 – flywheel, 4 – flywheel axle, 5 – electric DC motor.

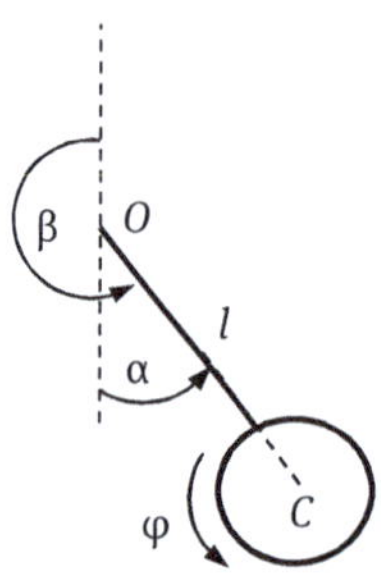

Fig. 3.2. Arrangement of the pendulum with flywheel: O – pendulum support joint, C – flywheel center, $l = OC$.

movement plane (that is the drawing plane). The following notation will be used: l – link length ($l = OC$), m – link mass, b – the distance between joint O and link center of mass, J_m – its moment of inertia with respect to joint O.

The flywheel is symmetric with respect to its rotation axis. It is installed at the end point of link OC, namely, at point C. Figure 3.2 shows the flywheel as a circle with its center point C. The flywheel can rotate clockwise or counter-clockwise about a horizontal axis that goes through point C and is perpendicular to the pendulum swing plane. This axis is parallel to the axis O of the support joint. The axis of the motor that rotates the flywheel is also parallel to joint axis O. Thus the flywheel together with the motor armature has one degree of freedom with respect to the pendulum link. The only control parameter that is present in the system is u – the voltage applied to the motor that rotates the flywheel.

The parameters of the system will be denoted as follows. Let M designate the total mass of the flywheel together with the motor, J_M and J_r – the moments of inertia of the flywheel and the motor armature with respect to their rotation axes. It is assumed that the center of mass of the flywheel together with the motor is located at point C of the pendulum OC. To maximize the moment of inertia of the flywheel (while keeping its mass the same), the material of the flywheel should be spread away from its center as far as possible. A dumbbell shape can be used instead of a wheel. Such shape can be easily built so that its moment of inertia will be much larger than of a wheel having the same mass [111, 112].

3.2 Equations of motion

Behavior of the mechanical system described above can be characterized by two generalized coordinates. Angle β is the angle that the pendulum swings away from its vertical position, and angle φ is the angle that the flywheel turns with respect to the pendulum. Both angles are measured in counter-clockwise direction (see Figure 3.2). However, angle φ of flywheel rotation is a *cyclic* variable, because the center of mass of the flywheel together with motor is located at the end of the pendulum link, at point C. Thus equations of motion provided below contain only the angular velocity of the flywheel $\omega = \dot{\varphi}$ with respect to the pendulum link – the segment OC. The motor armature rotates with respect to its stator at velocity Ω, and $\omega = \chi\Omega$, where χ denotes reduction ratio. Note that the value of angle φ is of no interest for the problem of controlling pendulum oscillations. So the equation that describes this variable, $\dot{\varphi} = \omega$, is not present in motion equations provided below.

Kinetic energy of the system is denoted as T, potential energy – as Π, and virtual work – as δW. These values can be evaluated by means of the following expressions (g is gravity acceleration):

$$
\begin{aligned}
&2T = \left(J_m + Ml^2\right)\dot{\beta}^2 + J_r\left(\dot{\beta} + \Omega\right)^2 + J_M\left(\dot{\beta} + \omega\right)^2 , \quad \Pi = (mb + Ml)g\cos\beta , \\
&\delta W = L\delta\varphi/\chi - L_f\delta\beta .
\end{aligned}
\tag{3.1}
$$

Here L is electromagnetic torque applied to the armature of the motor from its stator, L_f is the counter-torque generated by forces acting in joint O. If these forces are represented by a viscous friction, then this moment can be written as $L_f = \kappa\dot{\beta}$. Positive coefficient κ is proportional to viscosity factor. In case of dry (Coulomb) friction, this moment is written as $L_f = f \operatorname{sign}\dot{\beta}$, where positive constant coefficient f characterizes the threshold of dry friction force. Torque L acts (via the gearbox) on the flywheel. At the same time, torque $-L$ is applied to the pendulum. This is similar to the scheme discussed in the previous chapter, where the pendulum *installed on a wheel* is discussed, as opposed to the pendulum with a *flywheel installed on it*, as discussed in current chapter.

Applying Lagrangian approach of the second kind [16, 41, 73], considering expressions (3.1), one can derive the motion equations of the system

$$J\chi\ddot{\beta} + (J_r + \chi J_M)\dot{\omega} = (mb + Ml)g\chi \sin\beta - L_f \,, \tag{3.2}$$

$$(J_r + \chi J_M)\chi\ddot{\beta} + (J_r + \chi^2 J_M)\dot{\omega} = \chi L \,. \tag{3.3}$$

Here $J = J_m + Ml^2 + J_r + J_M$. Equations (3.2), (3.3) can also be derived by using the principle of angular momentum [16, 41, 73]. Equation (3.2) describes dynamics of the angular momentum of the system with respect to point O under the influence of gravitational moment and friction torque. Extracting angular acceleration $\dot{\omega}$ from equation (3.3) and substituting it in equation (3.2) will yield an equation that contains β and its first and second derivatives, and does *not* contain angular velocity ω. If counter-torque L does not depend on angular velocity ω, then the resulting equation that describes only the oscillations of the pendulum can be separated from the system (3.2), (3.3). It should be noted, however, that this equation *does* contain the parameters of the flywheel. This is similar to the problem of controlling a wheel-based pendulum, as described in chapter 2.

If torque L is developed by a DC motor, then it is proportional to the electric current flowing in the armature coil. Considering inductivity of this coil negligible (it is related to electromagnetic time constant), this torque can be represented (see [42, 76, 108]) as

$$L = c_u u - c_v \Omega \,. \tag{3.4}$$

Product $c_v\Omega$ describes torques arising due to counter-electromagnetic effects. Taking into account counter-EMF, further, instead of limitation (1.1) imposed on torque L, a limitation imposed on voltage u supplied to the motor will be considered. This limitation is

$$|u(t)| \le u_0, \quad u_0 = \text{const} \,. \tag{3.5}$$

Positive constant coefficients c_u and c_v can be derived from technical data provided for each particular DC motor – its stall torque, no-load torque and no-load rotation speed, and nominal voltage of the motor [76]. These values can also be identified as uncertain parameters of the motor, by solving the appropriate problem of identification. A set of piecewise-continuous functions $u(t)$, each complying with inequality (3.5), will be denoted as U.

When building the pendulum, the friction in its pivot axle was made as small as possible. The friction torque L_f in pivot thus can be neglected, and equations (3.2)–(3.4) can be transposed (keeping in mind that $\omega = \chi\Omega$) to

$$\begin{aligned} J\chi\ddot{\beta} + (J_r + \chi J_M)\dot{\omega} &= (mb + Ml)g\chi \sin\beta \,, \\ (J_r + \chi J_M)\chi\ddot{\beta} + (J_r + \chi^2 J_M)\dot{\omega} &= \chi c_u u - c_v \omega \,. \end{aligned} \tag{3.6}$$

Nondimensional time τ is introduced according to the formula

$$t = \vartheta\tau \quad \left(\vartheta^2 = \frac{(J_m + Ml^2)(J_r + \chi^2 J_M) + J_r J_M (1-\chi)^2}{(mb + Ml)g\,(J_r + \chi^2 J_M)}\right) , \tag{3.7}$$

and equations (3.6) can then be transposed into nondimensional form as follows:

$$\begin{aligned} \beta'' &= \sin\beta + j_M e\sigma/\chi - j_M v\,, \\ \sigma' &= -j_M\chi\sin\beta - j_m e\sigma + j_m\chi v\,. \end{aligned} \tag{3.8}$$

Like before, prime mark $'$ means differentiating with respect to nondimensional time τ, values

$$e = \frac{c_v}{\vartheta(mb+Ml)g}, \quad j_M = \frac{J_r + \chi J_M}{J_r + \chi^2 J_M}, \quad j_m = \frac{J}{J_r + \chi^2 J_M} \tag{3.9}$$

are nondimensional parameters of the system, and

$$\sigma = \vartheta\omega = \frac{d\varphi}{d\tau}, \quad v = \frac{c_u}{(mb+Ml)g}u = \frac{c_u\vartheta e\chi}{c_v}u \tag{3.10}$$

are nondimensional angular velocity of the flywheel and nondimensional voltage.

In new nondimensional notation, inequality (3.5) is transposed to

$$|v(\tau)| \le v_0, \quad v_0 = c_u\vartheta e\chi u_0/c_v = \text{const}\,. \tag{3.11}$$

System of equations (3.8), formulated in terms of nondimensional variables, is considerably simpler than original system (3.6). It includes only three parameters, that are nondimensional ones, see (3.9). The device with pendulum and flywheel built in the Institute of Mechanics has the armature of the motor connected directly to the flywheel, i.e. no gearbox is used. This means that $\chi = 1$, and relations (3.6)–(3.11) are even simpler. In particular,

$$\vartheta^2 = \frac{J_m + Ml^2}{(mb+Ml)g}, \quad j_M = 1\,.$$

However, the formulas developed in current chapter do include reduction ratio, providing a general view on the problem.

3.3 Local stabilization of the pendulum in the top equilibrium

Current section discusses the problem of stabilization of the pendulum in its top, unstable equilibrium $\beta = 0$, $\beta' = 0$, assuming that at the beginning of this process the pendulum is already within some neighbourhood of the desired position.

a) *Linearized equations.* Equation system of third order (3.8) has the only nonlinear component – $\sin\beta$. The round swings of the pendulum will not be considered, and also angle β will be assumed close to zero during the process of stabilization. Then substituting for function $\sin\beta$ its argument β transposes (3.8) into an approximate linear system of equations

$$\begin{aligned} \beta'' &= \beta + j_M e\sigma/\chi - j_M v\,, \\ \sigma' &= -j_M\chi\beta - j_m e\sigma + j_m\chi v\,. \end{aligned} \tag{3.12}$$

With $v = 0$ nonlinear system (3.8), as well as linear one (3.12), both have a trivial solution

$$\beta = 0, \quad \beta' = 0, \quad \sigma = 0 \,, \tag{3.13}$$

that corresponds to the top, unstable equilibrium of the pendulum and motionless flywheel. The problem of stabilization of equilibrium (3.13) is worth discussing in detail.

The matrix form of system (3.12) can be represented as follows:

$$x' = Ax + bv \,, \tag{3.14}$$

Here (asterisk meaning transposed matrix)

$$x = \|\beta, \beta', \sigma\|^* \,, \quad A = \|a_{ij}\| = \begin{Vmatrix} 0 & 1 & 0 \\ 1 & 0 & j_M e/\chi \\ -j_M\chi & 0 & -j_m e \end{Vmatrix} \,, \quad b = \|b_i\| = \begin{Vmatrix} 0 \\ -j_M \\ j_m\chi \end{Vmatrix} \quad (i = 1, 2, 3). \tag{3.15}$$

Trivial solution (3.13) is written in terms of new variables as

$$x = \|\beta, \beta', \sigma\|^* = 0 \,. \tag{3.16}$$

b) *Open-loop system eigenvalues.* The open-loop system can be derived from (3.14), (3.15) by assuming $v = 0$ ($u = 0$). The task is to determine the positions on the complex plane of the eigenvalues that correspond to this system, i.e. the eigenvalues of matrix A. It is a third-order matrix, and its characteristic polynomial looks like (λ – spectral parameter)

$$F(\lambda) = \lambda^3 + j_m e\lambda^2 - \lambda + (j_M^2 - j_m)e = (\lambda + j_m e)(\lambda^2 - 1) + j_M^2 e = 0 \,. \tag{3.17}$$

First, let $c_v = 0$, then it follows from (3.9) that $e = 0$. Equation (3.17) with $e = 0$ has two nonzero real roots, $\lambda_1 = 1$, $\lambda_2 = -1$, that only have opposite signs, and one zero root $\lambda_3 = 0$. In other words, the spectrum of the open-loop system with $e = 0$ is symmetric with respect to imaginary axis. This is natural, because with $c_v = 0$ the open-loop system is conservative. When counter-EMF is "added" into it, ($c_v > 0$, $e > 0$), the zero eigenvalue "drifts" leftwards, the other two eigenvalues also drift, but still remain positive and negative at all values of $e > 0$. This statement proves to be true since function $F(\lambda)$ changes its sign three times when its argument λ variates from $-\infty$ to $+\infty$. Its plot intersects negative semiaxis λ twice and positive semiaxis once. Indeed,

$$F(-\infty) = -\infty < 0, \quad F(-j_m e) = j_M^2 e > 0, \quad F(0) = (j_M^2 - j_m)e < 0, \quad F(+\infty) = +\infty > 0. \tag{3.18}$$

The second last inequality in (3.18) is true because $j_m - j_M^2 > 0$ (see notation (3.9)). This inequality can be proved directly, it is also related to the fact that the matrix of kinetic energy is positive-definite (see the first of relations (3.1)). The determinant of this matrix is formed of the coefficients at highest-order derivatives in equations (3.6).

Thus equation (3.17) has three real roots. One positive: $\lambda_1 > 0$, and two negative: $\lambda_2, \lambda_3 < 0$. That means, the open-loop system, i.e. system without control, is unstable.

It is worth to emphasize that only one of its eigenvalues is located in the right semiplane, and it is a real one. The other two, also the real ones, lie in the left semiplane.

If $e = 0$ ($c_v = 0$), then, as mentioned above,

$$\lambda_1 = 1\,. \tag{3.19}$$

If parameter e (or coefficient c_v) is small, the approximate values of roots $\lambda_1, \lambda_2, \lambda_3$. can be found from equation (3.17). The linear approximation with respect to parameter e (or, parameter c_v) for λ_1 looks like

$$\lambda_1 = 1 - \frac{1}{2} j_M^2 e\,. \tag{3.20}$$

The form of expression (3.20) shows that, when counter-EMF is "added" ($c_v > 0, e > 0$) eigenvalue λ_1, while remaining positive, "drifts" leftwards (in linear approximation). Linear approximated values for eigenvalues λ_2 and λ_3 with respect to parameter e are given by expressions $\lambda_2 = -1 - j_M^2 e/2$, $\lambda_3 = (j_M^2 - j_m)e < 0$, i.e. these values also "drift" leftwards as counter-EMF is added.

c) *Separating unstable coordinate and building domain of controllability.* A linear transformation of variables with constant nondegenerate matrix K,

$$y = Kx \tag{3.21}$$

transposes matrix equation (3.14), (3.15) into Jordan form, and the system breaks into three scalar equations that are cross-related only by control input v:

$$y_1' = \lambda_1 y_1 + d_1 v, \quad y_2' = \lambda_2 y_2 + d_2 v, \quad y_3' = \lambda_3 y_3 + d_3 v\,. \tag{3.22}$$

Here $y = \|y_i\|$, $d = \|d_i\| = Kb$, $KA = \Lambda K$, $\Lambda = \operatorname{diag}\|\lambda_i\|$ $(i = 1, 2, 3)$. The elements of matrix $K = \|k_{ij}\|$ $(i, j = 1, 2, 3)$ can be written as

$$\begin{aligned} k_{i1} &= (a_{22} - \lambda_i)(a_{33} - \lambda_i) - a_{32}a_{23} = \lambda_i(\lambda_i + j_m e)\,, \\ k_{i2} &= a_{32}a_{13} - a_{12}(a_{33} - \lambda_i) = \lambda_i + j_m e\,, \\ k_{i3} &= a_{12}a_{23} - (a_{22} - \lambda_i)a_{13} = j_M e/\chi\,. \end{aligned} \tag{3.23}$$

All elements k_{11}, k_{12}, k_{13} of the first row k_1 of matrix K are positive.

Variables d_i are of simple form (k_i is the i-th row of matrix K)

$$d_i = k_i b = -j_M \lambda_i \quad (i = 1, 2, 3)\,. \tag{3.24}$$

Since $j_M > 0$, $\lambda_1 > 0$, $\lambda_2 < 0$, $\lambda_3 < 0$, inequalities $d_1 < 0$, $d_2 > 0$, $d_3 > 0$ take place.

It follows from inequalities $d_i \neq 0$ $(i = 1, 2, 3)$, $\lambda_i \neq \lambda_j$ $(i \neq j)$ that system (3.22), and therefore the original system (3.14), (3.15), is completely controllable in Kalman's sense [89–92].

In terms of state space $Y(y_1, y_2, y_3)$ the set of initial states from which the system (3.22) can be translated into the origin while inequality (3.11) is maintained (that implies limitation on control input v), is a stripe (see [53] and also inequality (1.12))

$$|y_1| < |d_1| v_0/\lambda_1 \quad \text{or} \quad |y_1| < j_M v_0 . \tag{3.25}$$

This domain of controllability Q is limited only in its "unstable" variable y_1. In terms of nondimensional variables β, β', σ inequality (3.25) that describes domain of controllability Q, looks as follows:

$$\left|k_{11}\beta + k_{12}\beta' + k_{13}\sigma\right| < j_M v_0 . \tag{3.26}$$

Elements k_{11}, k_{12}, k_{13} of the first row of transformation (3.21) are calculated according to formulas (3.23) where $i = 1$, so inequality (3.26) can be transposed to

$$\left|\lambda_1(\lambda_1 + j_m e)\beta + (\lambda_1 + j_m e)\beta' + j_M e\sigma/\chi\right| < j_M v_0 . \tag{3.27}$$

Using inequality (3.27), it is possible to derive limitations that each of the three phase variables $\beta(0)$, $\beta'(0)$ and $\sigma(0)$ must comply with, assuming that the other two variables both are zero. That means, if two of the variables are zero, the third one must belong to one of the following intervals:

$$|\beta(0)| < \frac{j_M v_0}{\lambda_1(\lambda_1 + j_m e)}, \quad |\beta'(0)| < \frac{j_M v_0}{\lambda_1 + j_m e}, \quad |\sigma(0)| < \frac{\chi v_0}{e} . \tag{3.28}$$

Rewriting inequalities (3.28) in terms of original variables $\beta(0)$, $\dot\beta(0)$ and $\omega(0)$ yields

$$|\beta(0)| < \frac{c_u \vartheta e \chi j_M u_0}{c_v \lambda_1(\lambda_1 + j_m e)}, \quad |\dot\beta(0)| < \frac{c_u e \chi j_M u_0}{c_v(\lambda_1 + j_m e)}, \quad |\omega(0)| < \frac{c_u \chi^2 u_0}{c_v} . \tag{3.29}$$

Inequalities (3.29) mean, in particular, that the size of domain of initial states that allow stabilization of the top, unstable equilibrium of the pendulum is proportional to c_u and u_0. The value u_0 is the maximum voltage that can be supplied to the motor. This is natural from the physical point of view.

d) *Design of stabilization algorithm.* The task of designing control law that stabilizes the unstable equilibrium (3.13) (or (3.16)), is to find an algorithm of voltage control as a feedback function

$$v = v(\beta, \beta', \sigma) \quad (\text{or} \quad u = u(\beta, \dot\beta, \omega)) , \tag{3.30}$$

that makes the equilibrium asymptotically stable.

In engineering practice, often a linear feedback is used. So it is worth discussing a way to find a control law (3.30) as a linear feedback

$$v = g_1\beta + g_2\beta' + g_3\sigma , \tag{3.31}$$

where g_1, g_2, g_3 are the uncertain constant coefficients that must be determined. However, it is impossible to realize the "purely" linear feedback (3.31) due to limitation

(3.11) implied on control voltage v. Constraint (3.11) turns control law (3.31) into a linear feedback with saturation

$$v = \begin{cases} -v_0 & \text{with} \quad g_1\beta + g_2\beta' + g_3\sigma \le -v_0 \\ g_1\beta + g_2\beta' + g_3\sigma & \text{with} \quad |g_1\beta + g_2\beta' + g_3\sigma| \le v_0 \\ v_0 & \text{with} \quad g_1\beta + g_2\beta' + g_3\sigma \ge v_0 . \end{cases} \tag{3.32}$$

Coefficients g_1, g_2, g_3 can be determined in accordance with Routh–Hurwitz criterion. It specifies an area of allowable values in a three-dimensional space of these coefficients. To take a particular point from within this area, some extra reasoning must be applied. For example, one can try to achieve better transients shape, or (and) to maximize the domain of attraction. To remind, by domain of attraction, as usually, a set B of initial states is understood; from each of these states the system asymptotically (as $\tau \to \infty$) approaches the origin (3.13). It will now be described a way of "nearly unambiguous" selection of feedback coefficients that provide asymptotic stability of equilibrium (3.13), and at the same time realize maximum possible domain of attraction B.

The open-loop system (3.22) with $v = 0$ has only one "unstable" coordinate y_1. Its unstability can be "suppressed" by a linear feedback

$$v = \gamma y_1 = \gamma(k_{11}x_1 + k_{12}x_2 + k_{13}x_3) = \gamma(k_{11}\beta + k_{12}\beta' + k_{13}\sigma) \tag{3.33}$$

where constant coefficient γ must follow the criterion (see (3.22), (3.24))

$$\lambda_1 + d_1\gamma < 0 \quad \text{or} \quad \gamma > 1/j_M . \tag{3.34}$$

Coefficients k_{11}, k_{12}, k_{13} in feedback loop (3.33) are defined as in expressions (3.23). The positive eigenvalue λ_1 of the open-loop (with $v = 0$) system (3.14), (3.15) "turns" into negative $\lambda_1 + d_1\gamma = \lambda_1(1 - j_M\gamma)$, when the feedback loop is closed by (3.33), as $\gamma > 1/j_M$. The negative eigenvalues λ_2 and λ_3 remain unchanged, they only depend on engineering parameters of the system.

When limitation (3.11) to control voltage v takes place, linear feedback (3.33) turns into

$$v = \begin{cases} -v_0 & \text{when} \quad \gamma y_1 \le -v_0 \\ \gamma y_1 & \text{when} \quad |\gamma y_1| \le v_0 \\ v_0 & \text{when} \quad \gamma y_1 \ge v_0 . \end{cases} \tag{3.35}$$

It can be easily verified that if inequality (3.34) takes place, the interval on axis y_1, where feedback (3.35) is linear, lies inside interval (3.25). So function (3.35) has a linear segment on interval (3.25), and also two constant segments. As the absolute value of coefficient γ grows, the segment of axis y_1, where function (3.35) remains linear, becomes more narrow, and as $|\gamma| \to \infty$, it shrinks to a single point $y_1 = 0$.

Consider the first of the three scalar differential equations (3.22). The sign of its right-hand side with control law (3.35) is opposite to the sign of variable y_1 on interval (3.25). It can be shown easily (see chapter 1) that from any initial state $y_1(0)$

within interval (3.25) the solution to this first equation $y_1(\tau) \to 0$ as $\tau \to \infty$. And if $y_1(\tau) \to 0$, then, as defined by expression (3.35) control function $v(\tau) \to 0$ as $\tau \to \infty$. Eigenvalues $\lambda_2, \lambda_3 < 0$. So if $v(\tau) \to 0$ as $\tau \to \infty$, then the solutions of the second and the third equations of (3.22), $y_2(\tau) \to 0$ and $y_3(\tau) \to 0$ as $\tau \to \infty$, regardless of initial values $y_2(0)$ and $y_3(0)$ [45]. So from any initial state that belongs to domain of controllability Q system (3.22), closed by control law (3.35), asymptotically approaches equilibrium $y = 0$. That means, with control (3.35), domain of attraction B coincides with domain of controllability, i.e. $B = Q$. Thus the suggested control law realizes the largest possible domain of attraction for linear system (3.14), (3.15), and in this sense it is optimal. Such maximum possible domain of attraction is described by inequality (3.26) or (3.27).

Control law (3.35) in terms of variables β, β', σ looks as follows:

$$v = \begin{cases} -v_0 & \text{when} \quad \gamma(k_{11}\beta + k_{12}\beta' + k_{13}\sigma) \le -v_0 \\ \gamma(k_{11}\beta + k_{12}\beta' + k_{13}\sigma) & \text{when} \quad |\gamma(k_{11}\beta + k_{12}\beta' + k_{13}\sigma)| \le v_0 \\ v_0 & \text{when} \quad \gamma(k_{11}\beta + k_{12}\beta' + k_{13}\sigma) \ge v_0\,. \end{cases} \tag{3.36}$$

and thus the uncertain constant coefficients g_i $(i = 1, 2, 3)$ in feedback (3.31) or in (3.32) are:

$$\begin{aligned} g_1 &= \gamma k_{11} = \gamma\lambda_1(\lambda_1 + j_m e)\,, \quad g_2 = \gamma k_{12} = \gamma(\lambda_1 + j_m e)\,, \\ g_3 &= \gamma k_{13} = \gamma j_M e/\chi\,, \quad \gamma > 1/j_M\,. \end{aligned} \tag{3.37}$$

Looking closely at these expressions, it is easy to verify that all the coefficients $g_i > 0$ $(i = 1, 2, 3)$.

Therefore all coefficients g_i in feedback are defined "nearly" unambiguously. The only parameter that can be chosen arbitrarily is the total feedback gain γ, yet it must be greater that the value $1/j_M$. From expressions (3.37) it is possible to evaluate feedback coefficients, if the numerical values of system parameters are known. Prior to this, the cubic equation (3.17) must be solved to find the positive eigenvalue λ_1. To get approximate value of λ_1, formulas (3.19) or (3.20) can be used.

Using relations (3.7)–(3.11) between dimensional and nondimensional variables, the control law (3.36) can be stated in terms of original dimensional variables u, β, $\dot\beta$, ω:

$$u = \begin{cases} -u_0 & \text{when} \quad n_1\beta + n_2\dot\beta + n_3\omega \le -u_0 \\ n_1\beta + n_2\dot\beta + n_3\omega & \text{when} \quad |n_1\beta + n_2\dot\beta + n_3\omega| \le u_0 \\ u_0 & \text{when} \quad n_1\beta + n_2\dot\beta + n_3\omega \ge u_0. \end{cases} \tag{3.38}$$

Here

$$n_1 = \Gamma\lambda_1(\lambda_1 + j_m e), \quad n_2 = \Gamma(\lambda_1 + j_m e)\vartheta, \quad n_3 = \Gamma j_M e\vartheta/\chi, \quad \Gamma > c_v/(c_u\vartheta e\chi j_M)\,. \tag{3.39}$$

In study [160] a similar approach, like the one described above, is used to solve the problem of stabilizing a sensing element of a gradiometer.

The eigenvalues of open-loop (with $v = 0$) system (3.14), (3.15) are real. The eigenvalues of closed-loop system with feedback (3.33) are also real. Indeed, the positive eigenvalue of the open-loop system "turns" into a negative one as the feedback is applied, and the other two eigenvalues remain unchanged. This means that the transient in closed-loop system with feedback (3.33) (or (3.36)) is aperiodic, with any initial state from within domain of attraction B. So the described way of stabilizing equilibrium (3.13) not only realizes maximum possible domain of attraction (for linearized system), it also ensures that the transients are aperiodic.

By virtue of Lyapunov's theorem of stability on the basis of linear approximation, as control law (3.36) is applied, state (3.13) becomes asymptotically stable not only for linear system (3.12), but for nonlinear one (3.8) as well. Linear system (3.12) is close to nonlinear (3.8) if the value of angle β is small. So control law (3.36) provides that transients in nonlinear system (3.8) converge to equilibrium (3.13) if in these processes angle β remains close to zero. Domain of attraction (3.26) apparently contains such initial states that in transients starting from these states angle β deviates "far" from zero. So it would be wrong to say that from any initial state in domain (3.26) the solution to system (3.8), (3.36) converges to equilibrium (3.13). Therefore, suggested control algorithm (3.36) only provides local stabilization of the inverted pendulum. Later, in sections 5, 6, this algorithm is "upgraded" to provide global stabilization of the inverted pendulum.

Note that a flywheel with appropriate control system may be used to stabilize the vertical position of a tower crane.

3.4 Suppressing flywheel rotation

Experiments with the flywheel-fitted pendulum show that after stabilizing the pendulum in its top equilibrium, the flywheel keeps rotating with some angular velocity. The reason is, it is difficult to precisely define zero position of encoder that measures angle β. Its zero position is set up each time an experiment starts. This can be done by letting the pendulum reach its bottom equilibrium, and assuming that the encoder measurement refers to $\beta = \pi$. Let error $\Delta\beta$ appear in this calibration. This error will remain constant in all control process. The control law (3.32) will rely on this measurement, and output the following values that include this error:

$$v = \begin{cases} -v_0 & \text{with } \; g_1(\beta + \Delta\beta) + g_2\beta' + g_3\sigma \le -v_0 \\ g_1(\beta + \Delta\beta) + g_2\beta' + g_3\sigma & \text{with } \; |g_1(\beta + \Delta\beta) + g_2\beta' + g_3\sigma| \le v_0 \\ v_0 & \text{with } \; g_1(\beta + \Delta\beta) + g_2\beta' + g_3\sigma \ge v_0 \,. \end{cases} \tag{3.40}$$

Steady-state solution to nonlinear system (3.8), and to linear one (3.12), closed by control law (3.40), rather than in (3.13), is described by expressions (assuming that $\Delta\beta = \text{const}$):

$$\beta = \beta' = 0, \quad \sigma = \frac{\chi g_1 \Delta\beta}{e - \chi g_3}, \quad v = e\sigma/\chi = \frac{e g_1 \Delta\beta}{e - \chi g_3} \,. \tag{3.41}$$

This solution takes place in case when the steady-state control signal is defined by the middle row in expression (3.40). In other words, formulas (3.41) stand for the steady-state solution, if error $\Delta\beta$ is such that the voltage v, as determined by the last expression in (3.41), complies with the restriction

$$\left|\frac{eg_1\Delta\beta}{e-\chi g_3}\right| \le v_0 .$$

It is important to mention that in steady-state mode the real angle β of the pendulum deviation from vertical position is equal to zero even if the initial error in encoder calibration is present, just the same as when this error is absent. The calibration error only results in non-zero values of the steady-state flywheel velocity, and also in the steady-state voltage supplied to the motor. In experiments the flywheel indeed rotates in one direction or another, faster or slower, depending on the sign and absolute value of error in initial angle measurement.

To suppress flywheel rotation in steady-state motion mode, the linear feedback (3.40) can be supplemented by a term that is proportional to the angle that the flywheel turns, with some constant gain g_4. The angle is counted from some reference position φ^*:

$$g_4[\varphi(\tau)-\varphi^*] = g_4\int_{\tau^*}^{\tau}\frac{d\varphi}{d\xi}d\xi = g_4\int_{\tau^*}^{\tau}\sigma(\zeta)d\zeta \tag{3.42}$$

This can be reasoned theoretically, involving nonlinear system of equations (3.8), or linearized one (3.12), closed by control law (3.40). The reference value φ^* does not matter, if only the position of the flywheel is of any interest. Term (3.42) can be added into the feedback after some time τ^* lapses from the beginning of control process, and initial value φ^* should be taken as $\varphi^* = \varphi(\tau^*)$. This does not reduce the "original" domain of attraction B. Term (3.42) is proportional to the integral of flywheel angular velocity. Introducing an integral into a feedback rises the order of a system, and, as control theory postulates, this leads to possible instability of the control process. To avoid instability, feedback gain g_4 must be chosen as "not-too-large". Allowable values for this gain, from stability point of view, may be found, for example, involving Routh–Hurwitz criterion. The three coefficients g_1, g_2, g_3 may be considered as given, because they are calculated as suggested above. Experiments show that using addition (3.42) in feedback indeed reduces the flywheel velocity to zero in the steady-state mode.

3.5 Swinging and damping the pendulum

The first of equations (3.8) can be transposed to

$$\beta'' - \sin\beta = -w . \tag{3.43}$$

Here w denotes the torque produced by the electric motor (its nondimensional value)

$$w = \frac{j_M L}{(mb + Ml)g} = j_M \vartheta e L / c_v \,. \tag{3.44}$$

Torque L is considered positive if it makes the flywheel rotate counter-clockwise. At the same time, the positive torque produced by the motor tends to rotate the pendulum clockwise. This explains the fact that, unlike equation (1.4), the right-hand side of equation (3.43) has a minus sign. The value of torque L, as expression (3.4) shows, depends not only on control voltage u, but also on angular velocity of the flywheel ω ($\omega = \chi\Omega$), due to counter-electromotive forces. Neglecting counter-EMF for simplicity, consider voltage as in (3.44) as an independent control parameter. This parameter will be assumed limited in its absolute value by constant w_0,

$$|w(\tau)| \le w_0, \quad w_0 = \text{const} \,. \tag{3.45}$$

Let $w_0 < 1$. Piecewise functions $w(\tau)$ that meet condition (3.45) will be referred to as allowable control functions. The set of all such functions will be denoted as W. So, consider "separately" equation (3.43) describing the pendulum oscillations, with control functions $w(\tau)$, that satisfies (3.45).

Stable bottom equilibrium is described by relations

$$\beta = \pi, \quad \beta' = 0 \tag{3.46}$$

A new variable α will now be introduced, see Figure 3.2, that characterizes the pendulum angle of deviation from its bottom position. This angle is measured counter-clockwise,

$$\beta = \pi + \alpha \,.$$

Relations (3.46) will now look as

$$\alpha = 0, \quad \alpha' = 0 \,,$$

and equation (3.43) – as

$$\alpha'' + \sin\alpha = -w \,. \tag{3.47}$$

Consider the following problem of optimal control of the pendulum swinging. Let initial conditions be given for equation (3.47):

$$\alpha(0) \le 0, \quad \alpha'(0) = 0 \,. \tag{3.48}$$

Requirement $\alpha(0) \le 0$ does not narrow the scope of the problem. It is required to find a function $w(\tau) \in W$ that maximizes angle α, that is, $\max[\alpha(\theta)]$, at time instant θ, when angular velocity $\alpha'(\theta) = 0$. Here θ is the first (not counting zero) time instant when the angular velocity α' becomes zero. This can be stated in mathematical terms:

$$\max_{|w| \le w_0} [\alpha(\theta)], \quad \alpha'(\theta) = 0, \quad \theta > 0 \,. \tag{3.49}$$

When discussing problem (3.49), it will be considered that initial state (3.48) and value w_0 are such that with any control function $w(\tau) \in W$ there exists a nonzero time instant θ when angular velocity α' becomes zero: $\alpha'(\theta) = 0$. Otherwise, the pendulum can be spinned into a circular motion from the beginning. It will also be assumed that the optimal control function that delivers the maximum value to angle $\alpha(\theta)$ exists. If $w_0 > 1$, then from any initial value $\alpha(0)$ the pendulum can be spinned, avoiding its oscillations about its bottom equilibrium.

Time instant θ is not defined from the start. It is determined by condition $\alpha'(\theta) = 0$, and for each control function $w(\tau) \in W$ its value is different.

Along with the optimal swinging problem, the problem of optimal pendulum damping will be discussed. This problem can be stated in the following form:

$$\min_{|w| \le w_0} [\alpha(\theta)], \quad \alpha'(\theta) = 0, \quad \theta > 0 . \tag{3.50}$$

For problem (3.50), it is required to find a function $w(\tau) \in W$ that delivers a minimum value to angle α ($\min[\alpha(\theta)]$) at the first time instant θ (initial moment not included), when $\alpha'(\theta) = 0$.

If

$$\alpha(0) = 0, \quad \alpha'(0) = 0 , \tag{3.51}$$

then the pendulum is in equilibrium from the start, and the problem of pendulum damping is of no sense. At the same time, the problem of the pendulum swinging still makes sense under condition (3.51).

Introducing notation $d\alpha/d\tau = p$, equation (3.47) can be rewritten as a first order equation

$$\frac{dp}{d\alpha} p + \sin \alpha = -w . \tag{3.52}$$

With optimal control, that is supposed to exist, angle $\alpha(\tau)$ grows strictly monotonically on time interval

$$0 < \tau < \theta , \tag{3.53}$$

and thus on this interval (3.53) angular velocity $p > 0$ ($\alpha'(\tau) > 0$). At the same time, equation (3.52) can be rewritten as

$$\frac{dp}{d\alpha} = -\frac{w + \sin \alpha}{p} . \tag{3.54}$$

Investigating equation (3.54), it is easy to verify that the optimal control that provides solution to problem (3.49) must maximize derivative $dp/d\alpha$ for all values of α. Indeed, with such control function derivative p turns to zero when angle α reaches its largest possible value. The right-hand side of equation (3.54) depends linearly on control parameter w, and thus to maximize derivative $dp/d\alpha$, it must be assumed that

$$w = -w_0 \operatorname{sign} p = -w_0 .$$

Considering the next semiperiod of pendulum oscillations (when $\tau > \theta$), it can be concluded that the optimal control law that provides maximum swing to the pendulum at each oscillation semiperiod is

$$w = -w_0 \operatorname{sign} p = -w_0 \operatorname{sign} \alpha' = -w_0 \operatorname{sign} \beta' = -w_0 \operatorname{sign} \dot{\beta} \,. \tag{3.55}$$

Expression (3.55) means that at optimal control, the torque L applied to the flywheel is maximal in its absolute value, and its direction is opposite to the pendulum swing direction. At the same time, the torque $-L$ is applied to the pendulum that is directed in the same way as the pendulum swing, and this is natural from the physical point of view. Control law (3.55), as well as control law (1.33), maximizes derivative of the total energy of the pendulum (1.31) at each time instant. With control (3.55), the total energy of the pendulum increases at each semiperiod, and after several oscillations the pendulum will start to spin clockwise or counter-clockwise.

The optimal damping control law that provides solution to problem (3.50), is the "opposite" to control law (3.55)

$$w = w_0 \operatorname{sign} p = w_0 \operatorname{sign} \alpha' = w_0 \operatorname{sign} \beta' = w_0 \operatorname{sign} \dot{\beta} \,. \tag{3.56}$$

Control law (3.56) is effectively the same as adding dry friction in the pendulum pivot, its threshold equal to w_0.

The results stated above for simplified problem of optimal swinging (3.55) and damping (3.56) suggest that switching control law with voltage applied to the motor as

$$v = -v_0 \operatorname{sign} \alpha' = -v_0 \operatorname{sign} \beta' \quad \left(u = -u_0 \operatorname{sign} \frac{d\alpha}{dt} = -u_0 \operatorname{sign} \frac{d\beta}{dt} \right) \tag{3.57}$$

also makes the pendulum swing, and switching control law like

$$v = v_0 \operatorname{sign} \alpha' = v_0 \operatorname{sign} \beta' \quad \left(u = u_0 \operatorname{sign} \frac{d\alpha}{dt} = u_0 \operatorname{sign} \frac{d\beta}{dt} \right) \tag{3.58}$$

makes it stop (in the full definition of the problem), though, maybe, it will not be the optimal process in the sense of maximizing or minimizing the value of pendulum swing angle at an end each semiperiod. Numerical modeling and experiments show that control law (3.57) indeed leads to pendulum swinging, and law (3.58) – to its damping.

The application of pendulum damping problem by means of a flywheel may be to negate oscillations of a load that is hanging from a tower crane.

3.6 Translating the pendulum from the bottom equilibrium into the top one

To move the pendulum from its bottom position to the top equilibrium it must first be given a swing, and then when it reaches the topmost position, it must be "caught" and stabilized. While swinging the pendulum, it must be provided enough energy to move into the top equilibrium. Total energy E of the pendulum, not counting the energy of the flywheel, is described in terms of dimensional variables as

$$E = \frac{1}{2}\left(J_m + Ml^2 + J_r + J_M\right)\dot{\beta}^2 + (mb + Ml)g\cos\beta\,. \tag{3.59}$$

The energy of the pendulum that is resting in its top equilibrium is $E_0 = (mb + Ml)g$. Mathematical model (3.6) does not account for friction torque L_f in joint O, because it was made as small as possible ("almost zero") when the device was built. It still needs to be mentioned, however, that the friction, no matter how small, is always present in life, but it is difficult to parametrize and include into the mathematical model. Control law (3.57) swings the pendulum, pumping energy into it. If the amount of energy provided to the pendulum reaches E_0, then, in absence of friction, with small enough velocity of flywheel rotation, it will get into domain of attraction (3.27), where optimal control law like (3.38) will translate it to the top equilibrium and will stabilize it. If friction is present, then, with no extra pumping, the energy will dissipate (remember, in practice, the friction is still present), and as a result, the pendulum may not reach the domain of attraction. If the pumping is stopped when the value of energy is greater than E_0, then the pendulum may bypass the desired equilibrium. Considering that, the following way of switching to stabilization mode with control (3.38) is suggested.

While swinging the pendulum, at each time instant its energy (3.59) is evaluated. The swinging stops as the energy reaches the value of E_0 (or gets close to this value). After the desired energy level is reached, it is kept by pumping until the pendulum gets into the domain of attraction. This can be achieved by control law of the following type

$$u = k(E - E_0)\,\mathrm{sign}\,\dot{\beta}\,, \tag{3.60}$$

where k is the feedback gain. The control mode (3.60) that tracks the energy level E_0 is switched off when the system gets into domain of attraction (3.27). After that, control law (3.38) drives the pendulum to the desired top equilibrium and stabilizes it there.

Note that current study does not provide theoretical proof that such control approach as discussed above can really work. However, some parts of this approach are proved to be functional. The functionality of the control system as a whole is only shown in numerical computation and practical experiments.

3.7 Numerical experiments

The pendulum link of the device illustrated in Figure 3.1 is a homogeneous rod, so $b = l/2$ and $J_m = ml^2/3$. Its length (i.e. the length of the pendulum) $l = 0.3$ m, mass $m = 0.04$ kg, moment of inertia $J_m = 0.0012\,\mathrm{kg}\cdot\mathrm{m}^2$. No gearbox is installed between the motor and the flywheel, that means, the flywheel axle is the extension of the motor shaft; and $\chi = 1$. The flywheel is made so that its shape is close to a ring, its outside radius is $R = 0.042\,\text{m}$, and inside radius $r = 0.036\,\text{m}$. Flywheel mass is $M = 0.05\,\text{kg}$, and it is distributed evenly in the ring, and its moment of inertia is $J_M = M(r^2 + R^2)/2 = 0.0000765\,\text{kg}\cdot\text{m}^2$. The flywheel is actuated by a DC motor that has the following parameters: $J_r = 0.0000012\,\text{kg}\cdot\text{m}^2$, $c_u = 0.0069\,\text{N}\cdot\text{m/V}$, $c_v = 0.0001\,\text{N}\cdot\text{m}\cdot\text{s}$, $u_0 = 19\,\text{V}$.

With all the parameters as above, the positive eigenvalue of the system is $\lambda_1 = 0.9996$, the other two eigenvalues of the open-loop system are $\lambda_2 = -1.0004$ and $\lambda_3 = -0.2134$. Note that by applying equality (3.20), the value of λ_1 can be evaluated with high precision. Relations (3.39) to get feedback coefficients (3.38) yield: $n_1 = \Gamma \cdot 1.21\,\text{V}$, $n_2 = \Gamma \cdot 0.21\,\text{V}\cdot\text{s}$, $n_3 = \Gamma \cdot 0.000101\,\text{V}\cdot\text{s}$, $\Gamma > 141.6\,\text{V}$. The last one of expressions (3.39) limits the overall feedback gain Γ from below. However, this gain is also limited from above, if the delay in feedback loop is to be considered (see chapter 1). In numerical and practical experiments the value of this gain was taken as $\Gamma = 258\,\text{V}$, and feedback gain in control law (3.60) for energy stabilization $k = 200\,\text{V}/(\text{N}\cdot\text{m})$.

Applying the first of inequalities (3.28) (or (3.29)) gives the value for the size of the domain of attraction in terms of angle β: $|\beta(0)| < 6.34°$. If a dumbbell is used instead of the ring (and its moment of inertia is approximately a hundred times greater than that of a ring of the same mass), the domain of attraction is approximately $2.2°$ wider in terms of angle β: $|\beta(0)| < 7.43°$.

Figure 3.3 shows the transient process in angle β, its velocity $\dot{\beta}$ and the flywheel angular velocity ω for nonlinear system (3.6) with control functions (3.55), (3.60), (3.38) and initial conditions

$$\beta(0) = \pi, \quad \dot{\beta}(0) = 0, \quad \omega(0) = 0\,. \tag{3.61}$$

In Figure 3.3 time t (the abscissa) is measured in seconds, angle β in radians, $\dot{\beta}$ in radians per second, angular velocity of the flywheel – in rotations per second.

As seen in Figure 3.3, the pendulum makes a number of oscillations, getting higher and higher, and then it gets into its top equilibrium and stabilizes there. The angular velocity of the flywheel has a sawblade-shaped trajectory. Each semiperiod of oscillations it changes “almost” linearly, not reaching saturation. So the angular acceleration of the flywheel, and thus the reaction torque applied from the flywheel to the pendulum, remains “almost” constant.

3.8 Practical experiments

Control law, as described by expressions (3.57), (3.60), (3.38), and criteria to switch first from control mode (3.57) to mode (3.60) and then to mode (3.38), were programmed in *C* language and implemented into a program run on a PC, that was included into the feedback loop. Figure 3.4 shows a transient process in angle β, angular velocity $\dot{\beta}$, and the flywheel angular velocity ω, that was obtained in an experiment under initial con-

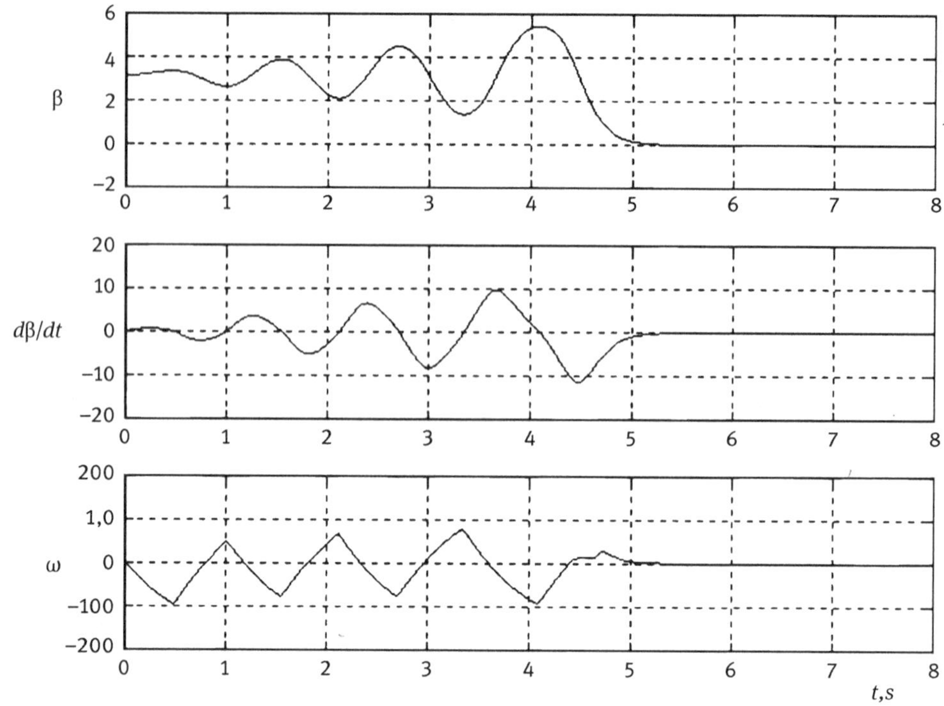

Fig. 3.3. Numerical experiment results: β is the angle of pendulum deviation from vertical (in radians), $d\beta/dt$ is the pendulum angular velocity (radians per second), ω is the angular velocity of the flywheel (in rotations per second).

ditions (3.61). This transient is similar to the transient obtained in numerical experiment and shown in Figure 3.3. In practice, the pendulum gets into its top equilibrium a little later than in numerical experiment. Video clips that illustrate these experiments are available at web site [164].

The difference in shape of transients as shown in Figures 3.3 and 3.4 are explained by drawbacks in mathematical model. In particular, it does not consider friction torque.

The experiments were successfully conducted under different initial conditions. The global stability can be characterized as follows. It is obvious that the pendulum can be stopped and put into its bottommost position, and the flywheel stopped, with any initial pendulum position and any initial angular velocities of the pendulum and the flywheel. And as law (3.57), (3.60), (3.38) can translate the pendulum from its bottom to the top equilibrium, this provides global stability of this equilibrium, that means, stability under any initial conditions.

So, mathematical modeling and experiments show that the control law described above is successful in translating the pendulum from its bottom equilibrium, as well as from any other position, into the top, unstable equilibrium, where the pendulum is

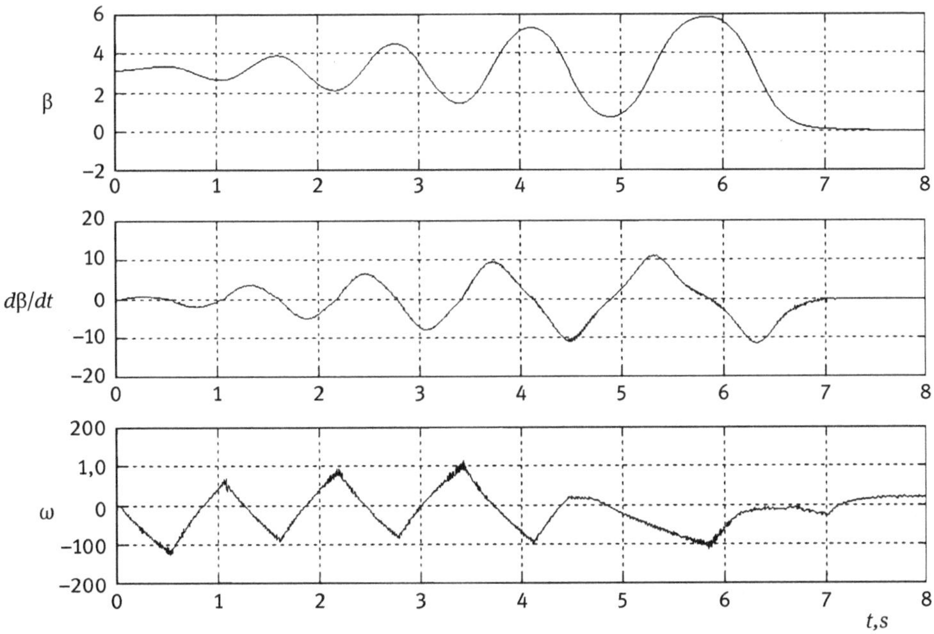

Fig. 3.4. Practical experiment results: β is the angle of pendulum deviation from vertical (in radians), $d\beta/dt$ is the pendulum angular velocity (radians per second), ω is the angular velocity of the flywheel (in rotations per second)

successfully stabilized. More control algorithms are created, that realize other motion modes. For example, such modes include pendulum rotation in one or the other direction, and then stabilizing it in its top or bottom equilibrium. Combinations of these modes may also be realized, as defined by a person controlling the process. For example, the pendulum may get to its top equilibrium, then do a series of pre-defined number of rotations, then stabilize again in the top equilibrium. After that it may do a series of rotations in the opposite direction.

The pendulum rotation in one direction is realized by the following control law

$$u = k(E - E')\,\mathrm{sgn}\,\dot{\beta}\,, \tag{3.62}$$

where E' is the value of energy E that is a little greater than E_0. Control law (3.62) tries to stabilize energy at value E', and the pendulum rotates in the same direction. The greater value E', the faster is the pendulum rotation.

4 Wheel rolling control by means of a pendulum

A single-wheel device (a monocycle) and a cart with two coaxial wheels both are of great interest from theoretical as well as from practical point of view. Web sites [162, 163] provide descriptions for monocycles equipped with a pendulum system. Longitudinal motion of such devices is realized by moving the pendulum away from its bottom equilibrium. The first device based on such principle was built in the middle of XIX century.

Studies [109, 111, 112] describe a device built in the Institute of Mechanics of Lomonosov Moscow State University, a monocycle – "gyrowheel". It is effectively a wheel equipped with a pendulum that provides longitudinal motion, and a gyroscopic stabilization system that supports its vertical position. Studies [30, 167] also describe a monocycle that is actuated and stabilized by means of a pendulum and a gyroscope. Some aspects of stability and stabilization of a single-wheel bicycle are covered in studies [87, 88].

The current chapter will discuss the plane longitudinal motion of a single-wheel device. It is considered that no slipping occurs between the wheel and the supporting surface. The device is controlled by means of a pendulum system installed on it. A mechanical model for such device is a wheel with a pendulum that is connected to its center by a joint. The system is actuated by an electric motor, its stator connected to the pendulum and its armature – to the wheel. This motor can rotate the pendulum with respect to the wheel, and at the same time, to move (roll) the wheel. Unlike in chapter 2, where the problem of stabilization of the unstable inverted pendulum on the wheel was discussed, here the question is how to organize wheel rolling by means of the pendulum. The control parameter here is the voltage supplied to the motor, which is limited in its absolute value. A feedback control law is suggested that provides maximum velocity to the device. Numerical experiments were conducted, and their results will be provided below. The mechanical system in question has two degrees of freedom and one control parameter – the voltage supplied to the motor.

When the device rolls over a rough surface it has to overcome humps and hollows. When climbing a hump or getting out of a hollow, it has to make a rise. So one of the problems to be discussed will be the device rise to a slope. The maximum slope angle will be derived that the device is capable to overcome.

Longitudinal motion of a device with *two coaxial* wheels and a pendulum-based control system can also be investigated using this model. So the results discussed here relate also to such *two-wheel* machine.

4.1 Mathematical model

Consider a wheel where in the center O a pendulum is attached (Figure 4.1).

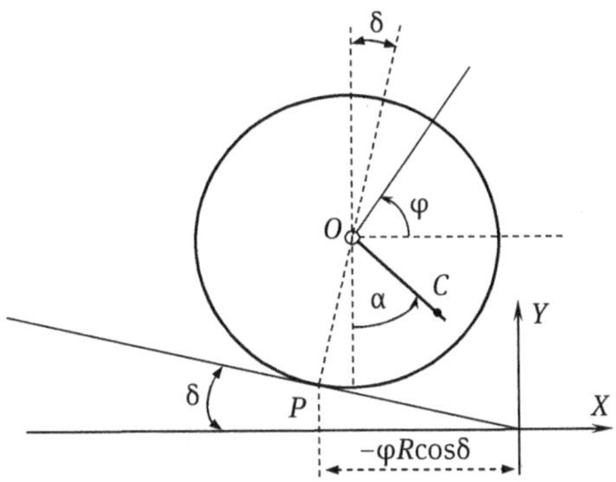

Fig. 4.1. A wheel with a pendulum on a slope.

Let the wheel be symmetric with respect to its axis O, and assume that it may roll without slipping over an even slope along a line that makes an angle δ with the horizon, and while rolling, it remains in the same vertical plane. Like above in chapter 2, M will denote the mass of the wheel, R – its geometric radius, ρ – its radius of gyration with respect to its center O. Let φ denote the angle that some particular, distinguished in the wheel radius turns counter-clockwise. This radius is selected in such a way that in the beginning of motion it is oriented in the same direction with horizontal axis X. The distance that the wheel center of mass O moves along the horizontal axis will be denoted as x, so that $\dot{x} = -\dot{\varphi}R\cos\delta$ (Figure 4.1). Let α be the angle of pendulum deviation, or, to be more exact, the angle of deviation of line OC from the bottom, stable equilibrium. The pendulum mass is denoted as m, the distance OC from the pivot point O to the pendulum center of mass C as b, and r stands for the pendulum radius of gyration with respect to its pivot point O. It will be assumed that an electric motor is mounted at the wheel axis, its stator attached to the pendulum, and its armature – to the wheel. Let L be the torque produced by this motor, and this torque (its positive value) tends to rotate the pendulum counter-clockwise, and the wheel – clockwise.

The described mechanical system has two degrees of freedom. As generalized coordinates, angles φ and α will be chosen. Variable φ is *cyclic*.

The kinetic energy of this two-body system looks like

$$T = \frac{1}{2}\left[a_{11}\dot{\varphi}^2 - 2a_{12}\cos(\alpha+\delta)\dot{\varphi}\dot{\alpha} + a_{22}\dot{\alpha}^2\right], \tag{4.1}$$

where

$$a_{11} = M(R^2+\rho^2) + mR^2, \quad a_{12} = mRb, \quad a_{22} = mr^2. \tag{4.2}$$

All coefficients in (4.2) are positive. With $\delta = 0$ and $\alpha = \beta - \pi$ expression (4.1) for kinetic energy transforms into (2.1). Expressions (4.2) and (2.2) are similar. Potential energy Π

and virtual work δW of torque L look like

$$\Pi = (M+m)gR\varphi \sin\delta - mgb\cos\alpha, \quad \delta W = L(\delta\alpha - \delta\varphi) \tag{4.3}$$

Applying Lagrangian approach of the second kind, [16, 41, 73], with expressions (4.1), (4.3), the system equations of motion can be derived:

$$\begin{aligned} a_{11}\ddot{\varphi} - a_{12}\cos(\alpha+\delta)\ddot{\alpha} + a_{12}\sin(\alpha+\delta)\dot{\alpha}^2 &= -L - (M+m)gR\sin\delta\,, \\ -a_{12}\cos(\alpha+\delta)\ddot{\varphi} + a_{22}\ddot{\alpha} + mgb\sin\alpha &= L\,. \end{aligned} \tag{4.4}$$

Neglecting inductance in the motor (the electromagnetic time constant), torque L can be presented as (see studies [42, 76, 108]):

$$L = c_u u - c_v(\dot{\alpha} - \dot{\varphi})\,. \tag{4.5}$$

Here expression $c_v(\dot{\alpha} - \dot{\varphi})$ describes counter-EMF that acts in motor, u stands for the electric voltage supplied to the motor coil. This voltage will be assumed limited in its absolute value (see (3.5)),

$$|u(t)| \le u_0 \quad (u_0 = \text{const})\,. \tag{4.6}$$

Positive constant coefficients c_u and c_v (the counter-EMF constant) can be evaluated from the technical characteristics of the stall torque, no-load torque, no-load velocity and nominal motor voltage [42, 76, 108].

Substituting expression (4.5) into (4.4) yields

$$\begin{aligned} a_{11}\dot{\omega} - a_{12}\cos(\alpha+\delta)\ddot{\alpha} + a_{12}\sin(\alpha+\delta)\dot{\alpha}^2 &= -c_u u + c_v(\dot{\alpha} - \omega) - (M+m)gR\sin\delta\,, \\ -a_{12}\cos(\alpha+\delta)\dot{\omega} + a_{22}\ddot{\alpha} + mgb\sin\alpha &= c_u u - c_v(\dot{\alpha} - \omega)\,, \end{aligned} \tag{4.7}$$

where $\omega = d\varphi/dt$.

For mathematical model of a wheel with a pendulum attached to its axis that includes a gearbox of any kind, the equations (4.7) look different, but still their structure is the same. The device investigated here belongs to the class of underactuated systems, because it has two degrees of freedom and only one control parameter. The motion control of this device is done by means of internal torques that provide relative movement of the pendulum and the wheel. The external forces that appear in this relative motion can be "organized" so that the device can move in a desired way.

The nondimensional time τ is introduced as $\tau = t\sqrt{g/R}$. Multiplying all equations in (4.7) by the common factor $R/a_{11}g$ yields a system of equations in terms of nondimensional variables:

$$\begin{aligned} \sigma' - j_1\cos(\alpha+\delta)\alpha'' + j_1\sin(\alpha+\delta)\alpha'^2 + e\sigma - e\alpha' + j_3\sin\delta + v &= 0\,, \\ -j_1\cos(\alpha+\gamma)\sigma' + j_2\alpha'' + j_1\sin\alpha - e\sigma + e\alpha' - v &= 0\,. \end{aligned} \tag{4.8}$$

The prime mark means differentiating with respect to nondimensional time τ, value $\sigma = \varphi'$, and coefficients

$$j_1 = \frac{a_{12}}{a_{11}} = \frac{mRb}{a_{11}}, \quad j_2 = \frac{a_{22}}{a_{11}}, \quad j_3 = \frac{(M+m)R^2}{a_{11}}, \quad e = \frac{c_v\sqrt{R/g}}{a_{11}} \tag{4.9}$$

are nondimensional system parameters (they are positive). Note that $j_3 < 1$, and also $j_1 < 1$, $j_2 < 1$ because $b, r < R$. Nondimensional voltage v is introduced according to the formula

$$v = \frac{c_u R}{a_{11} g} u \,. \tag{4.10}$$

Inequality (4.6) in new notation looks like

$$|v| \le v_0, \quad v_0 = \frac{c_u R}{a_{11} g} u_0 \,. \tag{4.11}$$

The wheel angle of rotation φ does not enter into motion equations (4.8), they include a new variable, the angular velocity $\sigma = \varphi'$. The total order of system (4.8) is equal to three.

4.2 Steady modes of motion

This section discussed steady (stabilized) modes of system motion that appear when the voltage supplied to the motor is constant, $v = v^* = \text{const}$ ($u = u^* = \text{const}$). Of course, this constant voltage must satisfy constraint (4.11).

Let in equations (4.8) $v = v^* = \text{const}$. A steady-state mode of motion is possible, with $\sigma = \sigma^* = \text{const}$, $\alpha = \alpha^* = \text{const}$. Steady-state values σ^*, α^* are determined by “balancing” relations that are derived from equations (4.8) (in parentheses the same “balancing” relations are provided in terms of original, dimensional variables, as derived from equations (4.7))

$$j_3 \sin\delta + v^* + e\sigma^* = 0 \quad ((M+m)gR\sin\delta + c_u u^* + c_v\omega^* = 0)\,, \tag{4.12}$$

$$j_1 \sin\alpha^* - v^* - e\sigma^* = 0 \quad (mgb\sin\alpha^* - c_u u^* - c_v\omega^* = 0)\,. \tag{4.13}$$

Relations in parentheses help better understand the balance of forces that cause the steady state. The sum $c_u u^* + c_v\omega^*$ describes the torque L^* that is produced by the electric motor in the steady-state mode.

Solving equations (4.12), (4.13) with respect to values σ^*, α^*, yields

$$\sigma^* = -\frac{v^* + j_3\sin\delta}{e}\,, \tag{4.14}$$

$$\sin\alpha^* = -\frac{(M+m)R}{mb}\sin\delta\,. \tag{4.15}$$

To ensure that inequality $\sigma^* > 0$ takes place, i.e. that when $\delta > 0$, the device would be capable of moving steadily uphill, it is necessary and sufficient, as follows from relation (4.14), that a criterion is satisfied

$$v^* < -j_3 \sin \delta \,, \tag{4.16}$$

because $e > 0$. Control voltage v is limited by expression (4.11), so inequality (4.16) is satisfied if and only if

$$v_0 \geq j_3 \sin \delta \,. \tag{4.17}$$

The value of voltage v^* influences the value σ^* of the wheel angular velocity in steady-state motion, but it does not influence the pendulum angle of deviation α^*. Let fraction $mb/(M+m)$ be denoted as r. This is the distance between the wheel center of mass O and the center of mass of the whole device, together with the pendulum. Equality (4.15) can be written as

$$r \sin \alpha^* = -R \sin \delta \,. \tag{4.18}$$

This equality has a clear geometrical sense. It means that a line drawn vertically from the center of mass of the device passes through the point where the wheel coincides with the supporting surface P (see Figure 4.1). Thus, considering (4.18), the total moment with respect to contact point P of all forces that are applied to the system is equal to zero. When condition (4.18) is satisfied, the wheel with the pendulum extended sidewards may be in equilibrium on a slope.

Condition (4.15) or (4.18) can be satisfied only in case when

$$\sin \delta \leq \frac{mb}{(M+m)R} \qquad (R \sin \delta \leq r) \,. \tag{4.19}$$

Inequality (4.19) means that when the pendulum deviates an angle $-\pi/2$ from its bottom equilibrium (and it is positioned horizontally), the line drawn vertically from its center of mass passes *through* contact point P (when (4.19) turns into equality) or *to the left* (when (4.19) is a strict inequality). If the parameters of the wheel and the pendulum are known, inequality (4.19) puts a limitation on the slope inclination angle δ. If condition (4.19) is not satisfied, then the wheel cannot climb the slope with a given angle δ; it will roll down. This statement can be put in a form of a theorem.

Theorem. *The wheel is capable of climbing a slope (in a steady-state motion) if and only if the vertical line drawn from the common center of mass of the wheel and the pendulum crosses the supporting slope in front of its contact point with the wheel.*

Figure 4.2 illustrates the case when the pendulum is horizontal, and the vertical line drawn from the common center of mass lies to the left of the vertical line that passes through point P. The common center of mass of the device is shown as a heavy point on the line that corresponds to the pendulum, and the vertical line passing through it – by a dashed line.

Under condition (4.19), equation (4.15) or (4.18) has two solutions: one of them $-\pi/2 \leq \alpha^* < 0$, the other one $\pi - \alpha^*$. In the first case, the pendulum is deviated an angle $|\alpha^*|$ to the left of the bottom (stable) equilibrium, in the second case it is deviated

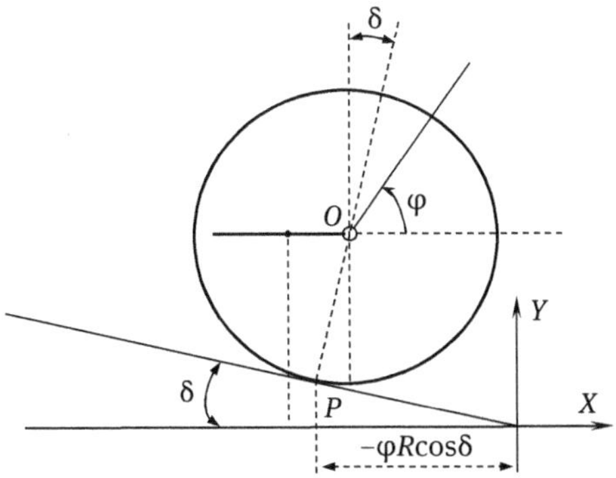

Fig. 4.2. The vertical line that passes through the common center of mass of the device lies to the left of the contact point *P*.

the same value $|\alpha^*|$ to the left of the top (unstable) equilibrium. In both cases, the pendulum is directed to the side where the device moves, that is, where the wheel rolls.

If there is a resistance force, and its moment with respect to contact point P is nonzero, then for the wheel to be capable to climb the slope it is necessary but *not sufficient* that with the pendulum positioned horizontally, the vertical line drawn from the common center of mass lies to the left of this point. To overcome an additional resistance moment, this line must be far enough away from point P.

Figure 4.3 illustrates the wheel with the pendulum that is in front of a step-shaped obstacle, contacting it at its corner B. The figure shows a straight line that passes through point B tangent to the wheel. This line forms an angle δ with the horizontal plane. The wheel with the pendulum can overcome this obstacle if and only if a *strict* inequality takes place in (4.19).

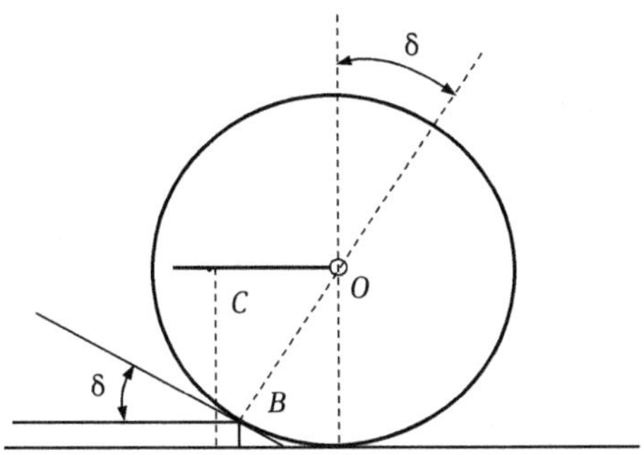

Fig. 4.3. The vertical line drawn from the common center of mass *C* of the device lies to the left of point *B* where the wheel contacts with the obstacle.

Applying the above theorem, it can be concluded that for the wheel to overcome an obstacle it is sufficient that the vertical line drawn from the common center of mass of the device (when the pendulum is positioned horizontally) lies *in front* of the contact point of the wheel and the obstacle (see Figure 4.3). If a wheel is in contact with the obstacle while being motionless, then this condition is necessary. If prior to contacting the obstacle the wheel was moving at some speed, then this requirement is not necessary.

4.3 Stability of steady-state modes

Equations that result as a deviation form with respect to steady-state mode (4.14), (4.15) of nonlinear equations (4.8) are as follows:

$$\begin{aligned} \Delta\sigma' - j_1\cos(\alpha^* + \delta)\Delta\alpha'' + e\Delta\sigma - e\Delta\alpha' &= 0 \\ -j_1\cos(\alpha^* + \gamma)\Delta\sigma' + j_2\Delta\alpha'' - e\Delta\sigma + e\Delta\alpha' + j_1\cos\alpha^*\Delta\alpha &= 0 \end{aligned} \tag{4.20}$$

Here $\Delta\sigma$ and $\Delta\alpha$ are deviations of respective variables. The characteristic equation (of third order) that corresponds to equations (4.20) looks like

$$\left[j_2 - j_1^2\cos^2(\alpha^* + \delta)\right]\lambda^3 + e\left[j_2 - 2j_1\cos(\alpha^* + \delta) + 1\right]\lambda^2 + j_1\cos\alpha^*\lambda + ej_1\cos\alpha^* = 0\,. \tag{4.21}$$

The highest order coefficient at λ^3 in equation (4.21) includes coefficients at highest order derivatives in equations (4.20). It is proportional to the determinant of kinetic energy matrix of the system taken at $\alpha = \alpha^*$ (see relations (4.9), (4.1) and (4.2)), and therefore, it is positive. Consequently,

$$\sqrt{j_2} > j_1|\cos(\alpha^* + \delta)|\,. \tag{4.22}$$

Applying inequality (4.22), it is possible to prove that the coefficient at λ^2 in equation (4.21) is also positive (considering that $j_1 < 1$):

$$j_2 - 2j_1\cos(\alpha^* + \delta) + 1 > j_1^2\cos^2(\alpha^* + \delta) - 2j_1\cos(\alpha^* + \delta) + 1 > \left[j_1\cos(\alpha^* + \delta) - 1\right]^2 > 0\,.$$

If

$$\alpha^* > -\pi/2\,, \tag{4.23}$$

then the third and the fourth coefficients are also positive. Referring again to (4.22), one can see that the only remaining Routh–Hurwitz criteria to be verified also proves to be true. Consider expression

$$ej_1\cos\alpha^*\left[j_2 - 2j_1\cos(\alpha^* + \delta) + 1\right] - ej_1\cos\alpha^*\left[j_2 - j_1^2\cos^2(\alpha^* + \delta)\right]\,. \tag{4.24}$$

If condition (4.23) is satisfied, then the sign of expression (4.24) is the same as the sign of value

$$j_1^2\cos^2(\alpha^* + \delta) - 2j_1\cos(\alpha^* + \delta) + 1 = \left[j_1\cos(\alpha^* + \delta) - 1\right]^2\,,$$

that, in turn, is always positive, since $j_1 < 1$.

So, steady-state mode (4.14), (4.15), complying with (4.23), is asymptotically stable.

The other steady state, where

$$-\pi < \alpha^* < -\pi/2\,, \tag{4.25}$$

is unstable, because under condition (4.25) the constant term, as well as the third coefficient in characteristic equation (4.21) are both negative.

Now consider a device where, unlike what was discussed above, the wheel center is fixed on a stationary horizontal axis, and the wheel can spin freely about this axis (in vertical plane), not touching any objects. Let a pendulum be mounted in the wheel center, with a motor attached to it. The stator of the motor is connected to the pendulum, and the armature – to the wheel. By moving the pendulum away from its vertical, hanging position it is possible to "organize" wheel spinning. Such system is similar to a spinning squirrel wheel.

5 Optimal swinging and damping of a swing

A pendulum can be taken as a mechanical model of a swing with a person on it. This pendulum has a point mass that can move up and down along the pendulum link, within some limits that can be described in geometric terms. The task discussed here is to find such a law of motion for this point that would drive the swing away from the vertical (that corresponds to stationary swing position) so that this deviation is maximal at the end of each oscillation semiperiod. At the same time, the problem of minimizing such deviation at each semiperiod will be considered.

The control parameter in this problem, in the same way as in studies [3, 36, 107], is considered to be the position of the point mass on the pendulum. The typical phase variables would be the angle of the pendulum deviation from the vertical, and its angular velocity. Equations of motion written in terms of such variables involve both the position and the velocity of the point mass. In other words, they include the control parameter and its derivative, that is not convenient for optimal control design. Using angular momentum of the pendulum together with point mass as one of the phase variables allows excluding the point mass velocity from motion equations. This helps solve the problem of optimal control completely.

Prior to investigating the problems of optimal swinging and damping the swing, elements of theory are provided concerning optimal control in the second-order systems. These theoretic facts will later be used in swing optimal control design.

The problems discussed in the current chapter have been discussed previously in studies [62–64, 100].

5.1 On optimal control design in second-order systems

In optimal control theory, the most important and at the same time the most difficult problem is the problem of optimal control design. By such design, deriving the optimal control law as a feedback, i.e. a function of phase variables, is understood (see [27, 130]). This problem can seldom be solved analytically. In this section, an autonomous nonlinear second-order system of general kind is concerned. The limitations imposed on control function may depend on current phase variables of the system. For such second-order systems mostly optimal control problems are explored that involve translating both phase variables to a pre-defined (desired) state. However, in the problem discussed here, only one phase variable needs to be maximized or minimized, and not at some prescribed time, but at the moment when the other variable reaches some particular value.

Let the motion of the investigated object be described by a system of two autonomous nonlinear equations that has the form of

$$\dot{x} = f_1(x, y, u), \quad \dot{y} = f_2(x, y, u) \tag{5.1}$$

As usual, dots represent differentiating with respect to time. Such differential equations describe, for example, the motion of a mechanical system with one degree of freedom. In that case, the value $f_1(x, y, u) = y$ can represent the velocity of the object, and function $f_2(x, y, u)$ – the generalized forces applied to it (divided by the mass or by the moment of inertia of the object).

A scalar control parameter u will be considered admissible if it is a piecewise continuous time function $u(t)$ that satisfies constraints

$$u_1(x, y) \leq u \leq u_2(x, y) \quad (u_1(x, y) < u_2(x, y)) \tag{5.2}$$

In three-dimensional space (x, y, u) inequalities (5.2) form a set $U(x, y)$, that is a domain of allowable values of the control parameter. The scientific studies mostly consider the case when $u_1(x, y) = \text{const}$, $u_2(x, y) = \text{const}$. In this case, set $U(x, y)$ is a stripe in the said three-dimensional space (x, y, u). If "boundary" functions $u_1(x, y)$ and $u_2(x, y)$ depend on phase coordinates, then in order to check if some piecewise continuous function $u(t)$ satisfies inequalities (5.2), one needs, generally, to solve equations (5.1) with this given control function $u = u(t)$ for $x(t)$, $y(t)$.

It will be considered that in some area of phase plane (x, y) function $f_1(x, y, u)$ does not become zero. To be more exact, let

$$f_1(x, y, u) > 0\,. \tag{5.3}$$

With constraint (5.3), the value of coordinate x can only increase with time. If (5.1) is a mathematical model of a mechanical system with one degree of freedom and $f_1(x, y, u) = y$, inequality (5.3) takes place in the top semiplane $y > 0$ of phase plane (x, y).

For now, not all of the restrictions imposed on system (5.1) and on set $U(x, y)$ (see inequality (5.2)) are provided here. It is difficult to state all assumptions at once. More limitations will be introduced as the problem is developed.

System (5.1) can be transposed to a single differential equation of the first order

$$\frac{dy}{dx} = \frac{f_2(x, y, u)}{f_1(x, y, u)} = f(x, y, u)\,. \tag{5.4}$$

Let

$$x(0) = x_0, \quad y(0) = y_0 \tag{5.5}$$

be the initial conditions for system (5.1) or for equation (5.4). To be specific, let the second coordinate at initial time instance be positive,

$$y_0 > 0\,. \tag{5.6}$$

Assume that under initial conditions (5.5), (5.6) for each piecewise-continuous function $u(t)$ the solution to system (5.1) exists and is unique. Besides, let each trajectory $y(x)$, that corresponds to an admissible control function, at some finite value of coordinate x intersect axis $Y = 0$, i.e. coordinate y becomes zero at that moment. For each admissible control function $u(t)$ coordinate y becomes zero at some respective time instance t and with some respective value of x.

Consider a set of all admissible functions $u(t)$, and a set of respective trajectories of equation (5.4), that appear with these control functions. To be more exact, consider only parts of these trajectories that begin at point (5.5) and end at abscissa $Y = 0$. The manifold of these curves sweeps a set of points that form a set of attainability [53], or a so-called "integral funnel" [34, 152]. This set of attainability D is schematically illustrated in Figure 5.1.

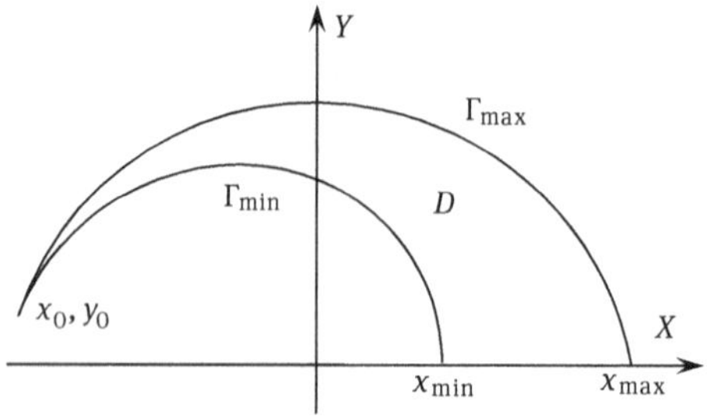

Fig. 5.1. Set of attainability *D*.

Consider a control function that maximizes derivative dy/dx with respect to its argument u at point (x, y). This control function maximizes function $f(x, y, u)$ that is included in the right-hand side of equation (5.4), and it looks like

$$u = u_{\max}(x, y) = \arg[\max_{u \in U(x,y)} f(x, y, u)] . \tag{5.7}$$

It is assumed that the maximum formulated in expression (5.7) exists, and that it is unique in the phase plane within some region that includes the set of attainability D. Let the solution to system (5.1), (5.7) under initial conditions (5.5) exist, be unique and assume that this solution yields a piecewise-continuous function $u(t)$, i.e. an allowable control function. Let $y = y_{\max}(x)$ be a trajectory that corresponds to control function (5.7), that is, the solution to equation

$$\frac{dy}{dx} = f[x, y, u_{\max}(x, y)] \tag{5.8}$$

under initial conditions (5.5), (5.6). By $\Gamma_{\max}$, a part of trajectory $y = y_{\max}(x)$ will be denoted where $x_0 \le x \le x_{\max}$, and $x_{\max}$ is the first value of argument x, that makes function $y = y_{\max}(x)$ become zero ($y_{\max}(x) = 0$). It will be shown that curve $\Gamma_{\max}$ is the upper boundary of attainability set D (see Figure 5.1).

Let with some control function $u^*(x, y)$ the trajectory of equation (5.8) that begins at some point $(x, y) \in \Gamma_{\max}$ "rise" beyond curve $\Gamma_{\max}$. Then at this point $(x, y) \in \Gamma_{\max}$ either inequality takes place

$$f[x, y, u^*(x, y)] > f[x, y, u_{\max}(x, y)] , \tag{5.9}$$

or solution $y = y_{max}(x)$ for equation (5.8) is not unique. However, inequality (5.9) is contrary to condition (5.7), and solution $y = y_{max}(x)$ for equation (5.8) that begins at point (5.5) is unique by assumption.

Now consider a control function that minimizes derivative $\frac{dy}{dx}$ with respect to its argument u at point (x, y). This control function minimizes function $f(x, y, u)$ that stands in the right-hand side of equation (5.4), and it has the following form:

$$u = u_{min}(x, y) = \arg[\min_{u \in U(x,y)} f(x, y, u)] . \tag{5.10}$$

It will be assumed that the minimum that is formulated in expression (5.10) exists, and that it is unique in phase plane within some area that includes attainability set D. Let $y = y_{min}(x)$ be a solution to equation

$$\frac{dy}{dx} = f[x, y, u_{min}(x, y)] \tag{5.11}$$

with initial conditions (5.5), (5.6). Let this solution be unique. Let Γ_{min} denote a part of trajectory $y = y_{min}(x)$ for $x_0 \le x \le x_{min}$, where x_{min} is the smalles value of argument x, that makes function $y = y_{min}(x)$ become zero ($y_{min}(x) = 0$). It can be shown that curve Γ_{min} is a lower boundary of attainability set D. The reasoning is similar to that proving that curve Γ_{max} is the upper boundary of attainability set D.

The next problem to be discussed is finding an admissible control function that yields maximum value of coordinate x at the first time instance (after the beginning of motion) when value y becomes zero. This maximization problem can be symbolically written as

$$\max_{u \subset U(x,y)}[x] \quad \text{with} \quad y = 0 . \tag{5.12}$$

To emphasize, coordinate x must be maximized not at some given (pre-defined) time instance, but at the moment when coordinate y becomes zero. A problem that is similar to (5.12) was discussed in chapter 3 (see expression (3.49)). It is assumed that the sought maximum can be achieved while applying an admissible control function. In case when equation (5.1) describes motion of a mechanical system with one degree of freedom, and $f_1(x, y, u) = y$, condition $y = 0$ means that the motion velocity becomes zero.

In parallel with the described problem, another problem will be discussed. It is, to find a control function that yields minimum value of coordinate x at the first time instance (after the beginning of motion) when the value y becomes zero. This problem can be symbolically written as

$$\min_{u \subset U(x,y)}[x] \quad \text{with} \quad y = 0 . \tag{5.13}$$

A problem similar to (5.13) was considered in chapter 3 (see expression (3.50)).

The problem of maximization – like (5.12) – is considered, for example, in study [4]. It is investigated also in later studies by the same authors, where for bilinear systems, in order to find criteria for absolute stability, a control law is built that maximally "disturbs" the system.

From the discussion above it follows that control function $u = u_{max}(x, y)$ (see expression (5.7)) solves problem (5.12). At that the maximum value of argument x is equal to x_{max} (see Figure 5.1). Control function $u = u_{min}(x, y)$ (see expression (5.10)) solves problem (5.13), and the minimum value of argument x is equal to x_{min} (see Figure 5.1).

Up to this point, the problems of maximizing and minimizing of coordinate x at the moment when coordinate y becomes zero were discussed. More general problems can also be considered in a similar way: maximizing and minimizing coordinate x at the moment when coordinate y takes on some pre-defined value $\bar{y}$. This value can be different from zero. At that, if $\bar{y} \leq y_0$, then the optimal control function will be, like before, function $u = u_{max}(x)$ (5.7), and for minimization problem – control function $u = u_{min}(x)$ (5.10). If $\bar{y} > y_0$ (see Figure 5.2), then the deductions above "swap". Maximum value of x is then reached at control function $u = u_{min}(x)$, and minimum value of x – at control function $u = u_{max}(x)$. This is explained by the fact that if $\bar{y} > y_0$, the upper boundary Γ_{max} of the attainability set crosses the line $y = \bar{y}$ at smaller value of coordinate x than the lower boundary Γ_{min}.

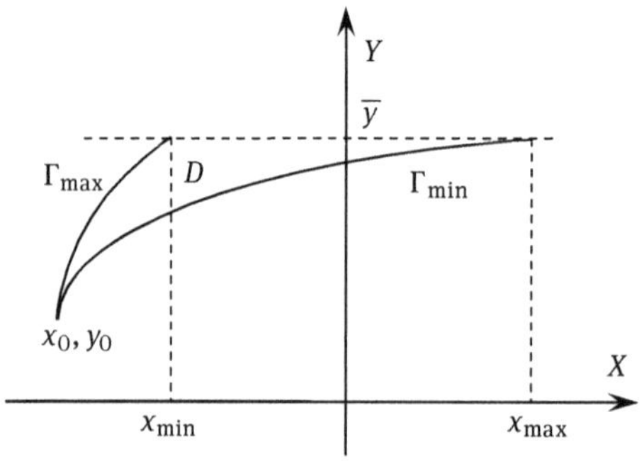

Fig. 5.2. Set of attainability *D*.

Expressions (5.7) and (5.10) for problems stated above may be referred to as the *local principle of maximum or minimum.*

Further, the obtained results are used to develop the optimal control law for swinging a swing, and for optimal damping of swing oscillations.

5.2 Mathematical model of a swing

As a model of a swing with a person on it, consider a physical pendulum with a point mass that can move along it (Figure 5.3). The mass of the pendulum is m, the mass of the point is M.

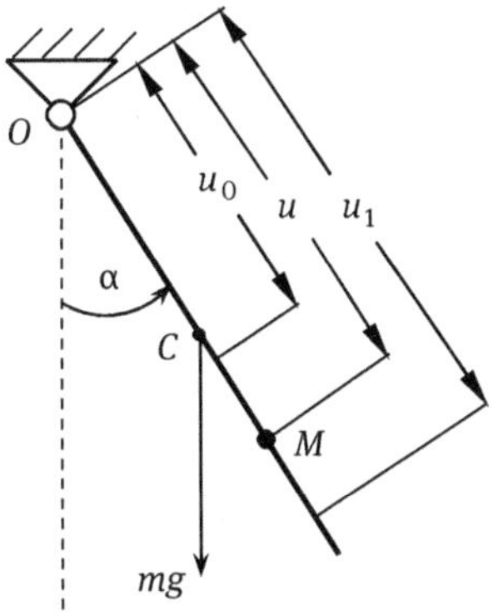

Fig. 5.3. Model of a swing.

The pendulum moment of inertia with respect to point O (the pivot) will be denoted as J, the distance between point O and its center of mass C – as b, the distance OM between point mass M and point O – as u. It will be assumed that point mass M is constrained in its motion along line OC:

$$u_0 \le u \le u_1 \tag{5.14}$$

where $u_0, u_1 = \text{const}$, $u_0 < u_1$. A nonlinear equation of swing motion looks like (see [100, 107])

$$\frac{d}{dt}\left[(J + Mu^2)\frac{d\alpha}{dt}\right] = -(Mu + mb)g \sin\alpha - c\frac{d\alpha}{dt} \tag{5.15}$$

Here α is the angle of pendulum deviation from the vertical line, it is counted counterclockwise; c is the coefficient of viscous friction that appears in the pivot O; g is the acceleration of gravity. Distance u will be considered as a control parameter.

The angular momentum $(J + Mu^2)\dot{\alpha}$ of the system with respect to its pivot point O will be denoted as K. Thus equation (5.15) of the second order can be transposed to a system of two first-order equations

$$\dot{\alpha} = \frac{K}{J + Mu^2}, \qquad \dot{K} = -\frac{cK}{J + Mu^2} - (Mu + mb)\sin\alpha . \tag{5.16}$$

5.3 Maximizing the swing oscillations magnitude

First of all, note that if

$$\alpha(0) = 0, \quad K(0) = 0 \quad (\dot{\alpha}(0) = 0) , \tag{5.17}$$

then

$$\alpha(t) \equiv 0, \quad K(t) \equiv 0 \quad (\dot{\alpha}(t) = 0) .$$

with any control function $u(t)$. That means, if the swing is motionless in its bottom equilibrium, then it can by no means be brought to motion, regardless of control function $u(t)$. It is also obvious that the swing cannot be brought to equilibrium by any control function $u(t)$, unless it was in that equilibrium from the start.

Let initial state of the system (5.16) be defined,

$$\alpha(0) < 0, \quad K(0) = 0 . \tag{5.18}$$

Consider the following problem. A law of changing distance u, within limits (5.14), is required to find, that provides maximum angle of swing deviation α at some time instant θ, when the angular momentum K (and therefore, velocity $\dot{\alpha}$) becomes zero ($K(\theta) = 0$) for the first time since the beginning of motion. In other words, it is required to maximize the swing deviation from the vertical line at the end of its oscillation semiperiod.

On time interval $0 < t < \theta$ the system moves with its angular momentum $K > 0$. So system (5.16) can be rewritten as

$$\frac{dK}{d\alpha} = -c - \frac{(Mu + mb)(J + Mu^2)\sin\alpha}{K} . \tag{5.19}$$

Search, in accordance with the results discussed in Section 1, for maximum of the right-hand side of equation (5.19) with respect to control parameter u on interval (5.14), yields the optimal law of swing control in an oscillation semiperiod where $K > 0$: $u = u_1$ when $\alpha < 0$, $u = u_0$ when $\alpha > 0$.

Doing the same for the next (the second) semiperiod of oscillations when $K < 0$, one can find out that for this semiperiod, the optimal control function is: $u = u_1$ when $\alpha > 0$, $u = u_0$ when $\alpha < 0$. To summarize, the optimal control law for a swing can be written as follows:

$$u = u(\alpha, \dot{\alpha}) = \begin{cases} u_1 & \text{when} \quad \alpha\dot{\alpha} < 0 \\ u_0 & \text{when} \quad \alpha\dot{\alpha} > 0 . \end{cases} \tag{5.20}$$

Figure 5.4 illustrates optimal control law (5.20) in phase plane $(\alpha, \dot{\alpha})$.

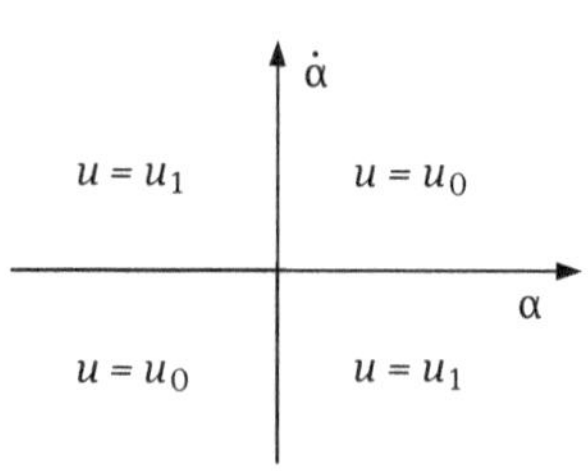

Fig. 5.4. Optimal control design for swinging a swing.

With optimal control, the point mass M must move (instantaneously) upwards to the limit when the swing passes its lowest point. Then it must move downwards to its limit when the swing reaches its maximum deviation from the vertical line, i.e. when its angular velocity becomes zero. Control law like (5.20) for a swing having no mass ($m = J = 0$) is considered in book [107], yet without discussing its optimal property.

Formulating the control law as a function of phase variables (5.20) provides the solution of the control problem not only with initial state (5.18).

Let the system have the following parameters:

$$\begin{aligned} &m = 5\,\mathrm{kg}\,, \qquad b = 4\,\mathrm{m}\,, \qquad J = 26.67\,\mathrm{kg}\cdot\mathrm{m}^2\,, \qquad M = 70\,\mathrm{kg}\,, \\ &u_0 = 3\,\mathrm{m}\,, \qquad u_1 = 3.75\,\mathrm{m}\,, \qquad c = 2\,\mathrm{N}\cdot\mathrm{m}\cdot\mathrm{s}\,. \end{aligned} \tag{5.21}$$

The pendulum is assumed to be a homogeneous rod.

Figure 5.5 shows the time plots for angle α, angular velocity $\dot{\alpha}$, control parameter u, as obtained by solving equations (5.16), (5.20) with parameters (5.21) and initial conditions $\alpha(0) = -0.1$, $K(0) = 0$.

Looking closer at Figure 5.5, one can see that with *relay* control $u(\alpha, \dot{\alpha})$ (5.20), function $u(t)$ instantaneously switches from its maximum possible value u_1 to minimum possible value u_0 when *angle* α becomes zero. Then it switches back from minimum possible value u_0 to maximum possible value u_1 when *angular velocity* $\dot{\alpha}$ becomes zero. In between the switchings, the control parameter u = const. Such control law increases the magnitude of oscillations.

Figure 5.6 shows the phase picture of the swing motion with "swinging" control law (5.20).

It follows from investigating Figures 5.5 and 5.6 that angular velocity $\dot{\alpha}$ has jump discontinuities at time instants when control function $u(t)$ switches from value u_1 to value u_0 (at $\alpha = 0$). Its absolute value at these instants rises in a step-like manner, because at these times the inertia moment $J+Mu^2$ of the swing together with the point mass becomes smaller (in a step-like manner). But angular momentum $K = (J + Mu^2)\dot{\alpha}$ at these switchings is nonzero, and it keeps its value after the switching.

When control parameter switches from value u_0 to value u_1, the moment of inertia $J + Mu^2$ changes discontinuously. But no step change of velocity takes place, because the control function switches at time instant when angular velocity $\dot{\alpha}$, and at the same time angular momentum $K = (J + Mu^2)\dot{\alpha}$, becomes zero. After the switching, angular momentum $K = (J + Mu^2)\dot{\alpha}$ keeps its value, i.e. it remains zero, and angular velocity $\dot{\alpha}$ also remains zero.

Control function (5.20) can be regarded from another point of view: it renders equilibrium (5.17) *unstable*, if damping coefficient c (see equations (5.15)) is sufficiently small.

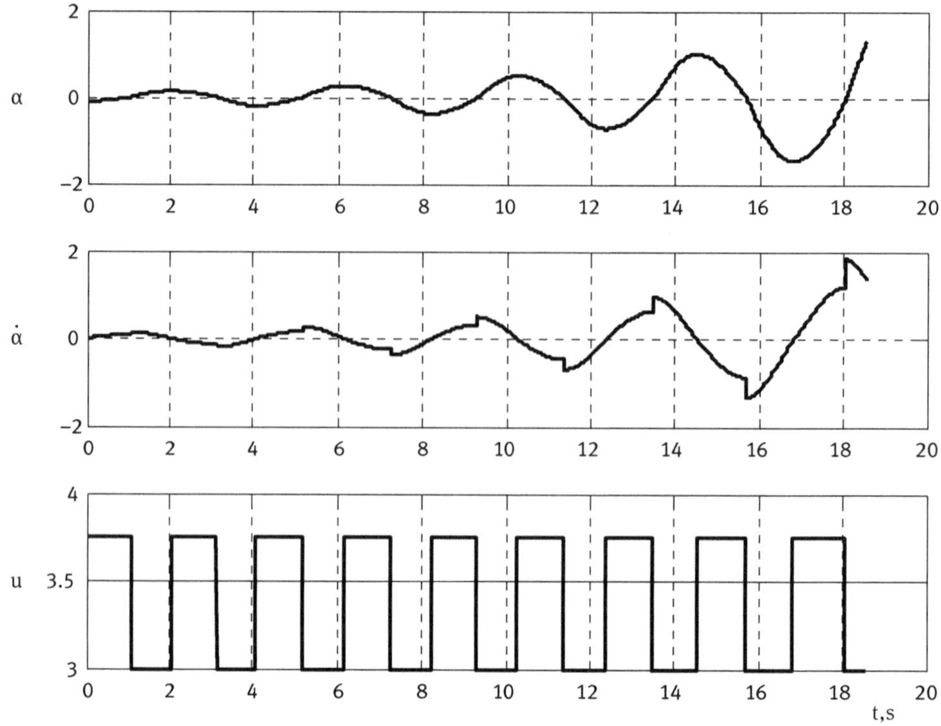

Fig. 5.5. Time plots of angle α, angular velocity $\dot{\alpha}$ and control parameter u at optimal swinging.

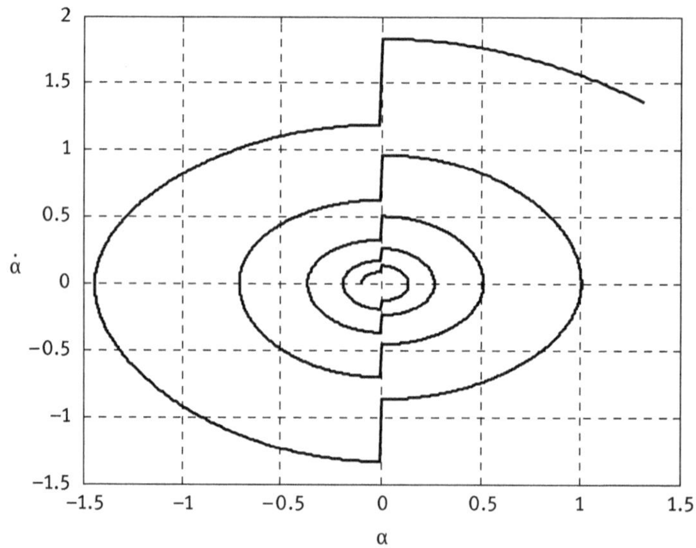

Fig. 5.6. Phase picture of the system with optimal swinging control law.

5.4 Minimizing swing oscillation magnitude

Consider now the problem of optimal swing damping, or, in other words, the problem of minimizing swing oscillation magnitude at the end of each semiperiod. To solve this task of optimal damping, the right-hand side of equation (5.19) must be minimized at each time instant with respect to parameter u.

On account of the results discussed above (in particular, expression (5.10)), minimizing the right-hand side of equation (5.19) with respect to parameter u in segment (5.14) yields the optimal control law to damp the swing when $K > 0$: $u = u_0$ when $\alpha < 0$, and $u = u_1$ when $\alpha > 0$. Considering the next semiperiod of oscillation, where $K < 0$, it can be concluded that in this semiperiod the optimal control function is: $u = u_0$ when $\alpha > 0$, and $u = u_1$ when $\alpha < 0$.

To summarize, the optimal control law can be expressed by the following notation:

$$u = u(\alpha, \dot{\alpha}) = \begin{cases} u_1, & \text{when} \quad \alpha\dot{\alpha} > 0 \\ u_0, & \text{when} \quad \alpha\dot{\alpha} < 0 \, . \end{cases} \tag{5.22}$$

Law (5.22) of optimal damping of swing oscillations, i.e. minimizing its oscillation magnitude at each semiperiod, is "the opposite" to law (5.20). To put it another way, to get the optimal swing damping law, the values u_1 and u_0 in expression (5.20) must be swapped.

In Figure 5.7, the optimal control law design (5.22) is illustrated for swing damping.

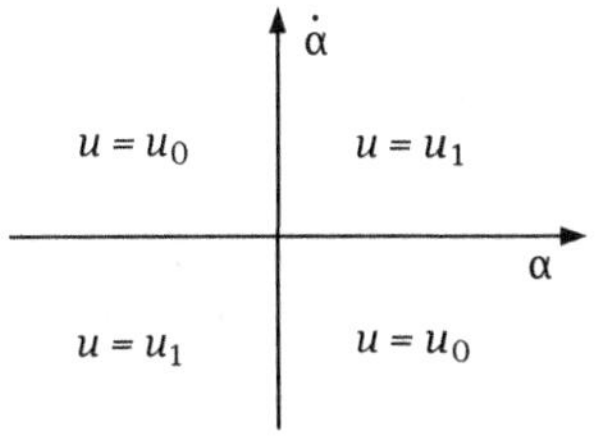

Fig. 5.7. Optimal control design for swing damping.

At optimal control law (5.22), point mass M moves instantaneously downwards to the limit at the moment when the swing passes its lowest position, and it moves upwards to the limit at the moment when the swing reaches the maximum deviation from the vertical line, that is, when its angular velocity $\dot{\alpha}$ becomes zero.

Numerical computations were done to investigate control law (5.22), while the values of parameters were taken as in (5.21).

In Figure 5.8 functions $\alpha(t)$, $\dot{\alpha}(t)$ and $u(t)$ are plotted for the optimal control law described by expression (5.22). The solution to equations (5.16), (5.22) plotted in this figure corresponds to initial state $\alpha(0) = -\pi/2 \approx -1.57$, $K(0) = 0$ ($\dot{\alpha}(0) = 0$).

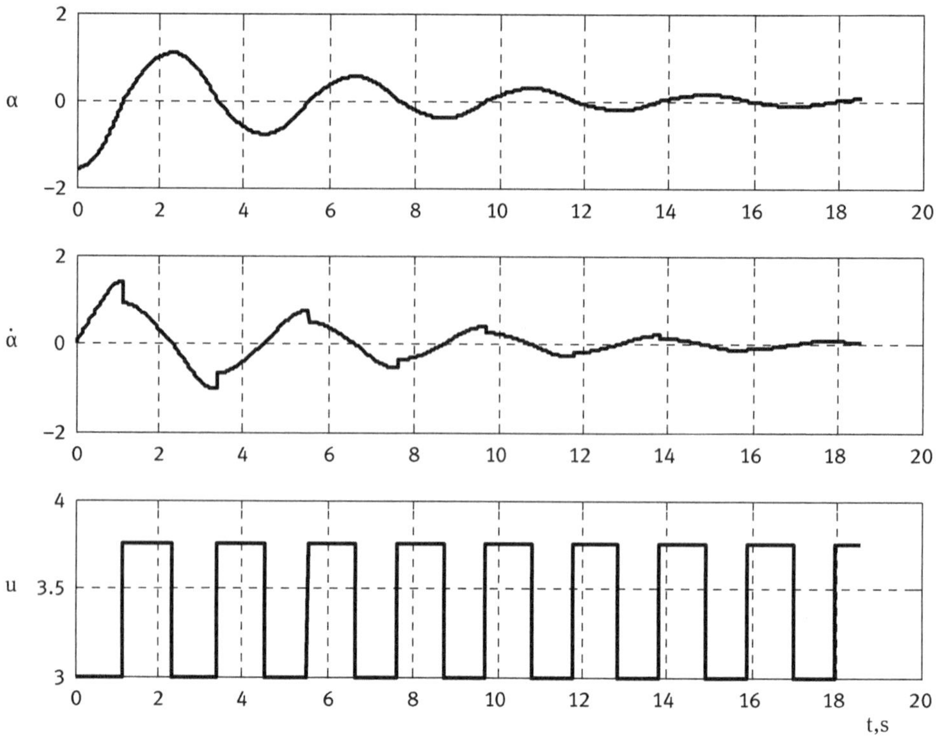

Fig. 5.8. Time plots of angle α, angular velocity $\dot{\alpha}$ and control function u at the optimal damping.

Function $u(t)$ is piecewise-constant. It switches instantaneously from its maximum possible value u_1 to the minimum possible value u_0 when *angular velocity* $\dot{\alpha}$ becomes zero. Then it switches back from the minimum possible value u_0 to the maximum possible value u_1 when *angle* α becomes zero. The magnitude of swing oscillations decreases with such control law.

Figure 5.9 demonstrates phase picture of swing motion under control law (5.22).

Investigating Figures 5.8 and 5.9, one can see that angular velocity $\dot{\alpha}$ has jump discontinuities at the time instants when control function $u(t)$ switches from its value u_0 to the value of u_1 (at $\alpha = 0$). Its absolute value at these instants decreases in a step-like manner. The reason is that the moment of inertia $J + Mu^2$ of the swing together with the point mass *M jumps up*, and the angular momentum $K = (J + Mu^2)\dot{\alpha}$ (that is non-zero) keeps its value.

When the control parameter switches from the value of u_1 to the value of u_0, the moment of inertia $J + Mu^2$ also decreases in a step-like manner. However, no step change of velocity takes place, because the switching occurs when the angular velocity

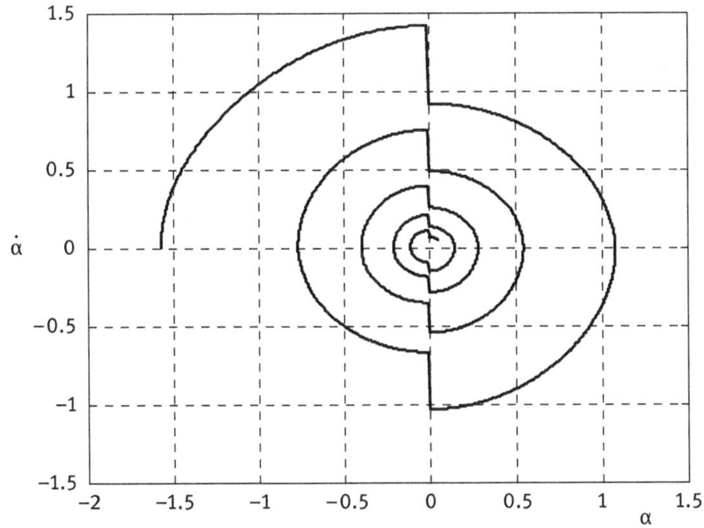

Fig. 5.9. Phase picture of the system with the optimal damping law.

together with the angular momentum $K = (J + Mu^2)\dot{\alpha}$ are both equal to zero; after the switching, both momentum $K = (J + Mu^2)\dot{\alpha}$ and velocity $\dot{\alpha}$ remain zero.

Control law (5.22) designed above, together with damping term $c(d\alpha/dt)$ (see equation (5.15)), renders equilibrium (5.17) asymptotically stable.

5.5 Controlling a swing with regard for aerodynamic resistance and dry friction

The control law for second-order systems that was discussed above is utilized in study [100] for the case when there are resistance forces that act on the system. These forces include aerodynamic drag that acts on the point mass M (or, on the person on the swing) or (and) the dry friction torque that acts in the swing pivot O. In this case, instead of (5.15), a more general equation is considered

$$\frac{d}{dt}\left[(J + Mu^2)\frac{d\alpha}{dt}\right] = -(Mu + mb)g\sin\alpha - (c + eu^2)\frac{d\alpha}{dt} + v\,.$$

Since expression $u(d\alpha/dt)$ represents the linear velocity of point mass M with constant distance u, expression $eu(d\alpha/dt)$ stands for viscous drag force of the air that is applied to point M (e = const > 0 is the damping coefficient). Expression $eu^2(d\alpha/dt)$ is the moment with respect to pivot point O of the viscous drag force. Letter v denotes the dry friction torque that is applied at the pivot point of the swing.

When these resistance forces are taken into account, the optimal control law changes. It can be seen that the switching occurs some time before the moment when the swing passes its lowest point. Besides, the control parameter may change fluently, not stepwise. The detailed discussion is not provided in frame of this study due to the awkwardness of the formulations and the results. It is also worth investigating the case when the drag force is proportional to the second power of the velocity of point M.

The swing control problem is closely related to the problem of damping of satellite oscillations in the gravity field that is performed by means of extending rods. It can also describe some athletic exercises. And, after all, this problem is of interest from the point of view of theoretical mechanics.

6 Pendulum control that minimizes energy consumption

This chapter considers a single-link physical pendulum. It discusses problems of driving the pendulum to its bottom, stable equilibrium and to the top, unstable one. Among all control laws the optimal one is proposed that minimizes consumption of mechanical energy. Such optimal control appears to be a law that includes two impulses. The problem is discussed in terms of phase coordinates that compose a cylinder.

6.1 Estimation of energy consumption

The equation that describes motion of a plane physical pendulum (see Figure 1.1) looks like (see equations (1.2) and (3.47))

$$mr^2\ddot{\alpha} + mgb\sin\alpha = L\,. \tag{6.1}$$

Here m stands for the mass of the pendulum, b – for the distance between the pivot point O and the center of mass C of the pendulum, r is the radius of gyration of the pendulum with respect to point O, α is the angle that the segment OC deviates from the position when the pendulum is in its bottom equilibrium. This angle is counted counter-clockwise. Torque L is applied at the pivot point.

Like in chapter 1, nondimensional time τ and nondimensional torque μ are introduced according to expressions (1.3)

$$\tau = t\sqrt{gb}/r, \quad \mu = L/mgb\,. \tag{6.2}$$

This transposes equation (6.1) to a simple form

$$\alpha'' + \sin\alpha = \mu\,. \tag{6.3}$$

Prime mark $'$ denotes, like above, differentiating with respect to nondimensional time τ.

With $L = 0$ ($\mu = 0$) the pendulum has equilibrium points

$$\alpha = \pi k, \quad \dot{\alpha} = 0 \quad (k = 0, \pm1, \pm2, \dots)\,. \tag{6.4}$$

Even values of k correspond to the stable (bottom) equilibrium, odd values – to the unstable (top) one.

In the current chapter, the problem of moving the pendulum by means of control torque μ to an equilibrium in a given (nondimensional) time $T > 0$.

The quality of control will be evaluated by the functional

$$J(0, T) = \int_0^T \left|\mu(\tau)\alpha'(\tau)\right| d\tau \,. \tag{6.5}$$

Expression (6.5) without absolute value marks would represent the mechanical work of control torque $\mu(\tau)$ in universal sense. Functional (6.5) characterizes mechanical energy consumption in case when the actuator that produces the control torque acts irreversibly, i.e. it consumes energy regardless of the direction of the torque – with pendulum motion or against it. Such functional is discussed in studies [19, 55, 65] regarding evaluation of energy consumption for motion control of a bipedal walking machine.

Unlike in previous chapters, no restrictions are imposed here on the control torque.

Consider time interval $[0, T]$ as broken into k segments $[\tau_0, \tau_1]$, $[\tau_1, \tau_2]$, …, $[\tau_{i-1}, \tau_i]$, …, $[\tau_{k-1}, \tau_k]$ (here $\tau_0 = 0$, $\tau_k = T$). In each segment the integration function in (6.5) maintains its sign. Then integral (6.5) can be represented as

$$\begin{aligned} J(0, T) &= \int_0^T \left|\mu(\tau)\alpha'(\tau)\right| d\tau = \sum_{i=1}^{i=k} \int_{\tau_{i-1}}^{\tau_i} \left|\mu(\tau)\alpha'(\tau)\right| d\tau = \sum_{i=1}^{i=k} \int_{\tau_{i-1}}^{\tau_i} \left|(\alpha'' + \sin\alpha)\alpha'(\tau)\right| d\tau = \\ &= \sum_{i=1}^{i=k} \int_{\tau_{i-1}}^{\tau_i} \left|\frac{d}{d\tau}\left(\frac{1}{2}\alpha'^2 - \cos\alpha\right)\right| d\tau = \int_0^T \left|\frac{dE(\alpha, \alpha')}{d\tau}\right| d\tau = \sum_{i=1}^{i=k} |\Delta E_i| \,. \end{aligned} \tag{6.6}$$

Here

$$E(\alpha, \alpha') = \frac{1}{2}\alpha'^2 - \cos\alpha \tag{6.7}$$

is the total energy of the pendulum, kinetic together with potential, and

$$\begin{aligned} \Delta E_i &= E\left(\alpha_i, \alpha_i'\right) - E\left(\alpha_{i-1}, \alpha_{i-1}'\right) = \frac{1}{2}\alpha_i'^2 - \cos\alpha_i - \frac{1}{2}\alpha i - 1'^2 + \cos\alpha_{i-1} = \\ &= \frac{1}{2}\left(\alpha_i'^2 - \alpha_{i-1}'^2\right) + \cos\alpha_{i-1} - \cos\alpha_i, \end{aligned} \tag{6.8}$$

where α_i and α_i' are the pendulum deviation angle and its angular velocity at time instant τ_i.

Integral (6.5) can be evaluated from below as follows:

$$\begin{aligned} J(0, T) &= \int_0^T |\mu(\tau)\alpha'(\tau)| d\tau \geq \int_0^T \mu(\tau)\alpha'(\tau) d\tau = \int_0^T (\alpha'' + \sin\alpha)\alpha'(\tau) d\tau = \\ &= \int_0^T \frac{d}{d\tau}\left(\frac{1}{2}\alpha'^2 - \cos\alpha\right) d\tau = \int_0^T \frac{dE(\alpha, \alpha')}{d\tau} d\tau = E[\alpha(T), \alpha'(T)] - E[\alpha(0), \alpha'(0)] \,. \end{aligned} \tag{6.9}$$

Let control law $\mu(\tau)$ be an impulse function

$$\mu(\tau) = I\delta(\tau - \theta)\,, \tag{6.10}$$

where $\delta(\tau - \theta)$ is Dirac's delta function, it is equal to zero when $\tau \neq \theta$. Constant value I represents the intensity and the direction of an impulse action (impact).

At control impulse (6.10) the pendulum deviation angle α does not change. As for angular velocity $\dot{\alpha}$, it changes in a step-like manner. The magnitude of the step $[\alpha'(\theta)]$ is evaluated in accordance with equation (6.3)

$$[\alpha'(\theta)] = \alpha'(\theta^+) - \alpha'(\theta) = I\,, \tag{6.11}$$

where $\alpha'(\theta)$ is the pendulum angular velocity prior to the control impulse (the impact), $\alpha'(\theta^+)$ is the angular velocity "immediately" after the impulse, θ and θ^+ are the instants before and after the impulse, respectively.

Let the sign of angular velocity $\alpha'(\theta^+)$ of the pendulum after impulse (6.10) be the same as the sign of angular velocity $\alpha'(\theta)$ prior to this impulse, i.e. $\alpha'(\theta^+)\alpha'(\theta) \geq 0$. Then from expressions (6.6)–(6.8) it follows that

$$J(\theta, \theta^+) = \int_{\theta}^{\theta^+} \left|\mu(\tau)\alpha'(\tau)\right| d\tau = \frac{1}{2}\left|[\alpha'(\theta^+)]^2 - [\alpha'(\theta)]^2\right|\,. \tag{6.12}$$

Equality (6.12) remains true if $\alpha'(\theta) = 0$ or $\alpha'(\theta^+) = 0$. This equality (6.12) can be proved immediately by "spreading" the control impulse (6.10) on a small time interval that begins at θ, and then shrinking this interval to zero [54, 55].

Now let $\alpha'(\theta^+)\alpha'(\theta) < 0$, that means, velocities $\alpha'(\theta)$ and $\alpha'(\theta^+)$ have different signs. Then

$$J(\theta, \theta^+) = \int_{\theta}^{\theta^+} \left|\mu(\tau)\alpha'(\tau)\right| d\tau = \frac{1}{2}\left\{[\alpha'(\theta^+)]^2 + [\alpha'(\theta)]^2\right\}\,. \tag{6.13}$$

Indeed, break the impulse applied to the pendulum into two impulses that are applied one after another immediately. After the first impulse, velocity α' takes a step from its value $\alpha'(\theta)$ to become zero. After the second impulse it changes from zero to value $\alpha'(\theta^+)$. Then, according to relations (6.6)–(6.8), at the first impulse energy $[\alpha'(\theta)]^2/2$ is consumed, and at the second one – energy $[\alpha'(\theta^+)]^2/2$. In total, the energy consumption is expressed as (6.13).

Combining (6.12) and (6.13), one can get the following expression for energy consumption at control impulse (6.10):

$$J(\theta, \theta^+) = \frac{1}{2}\begin{cases} \left|[\alpha'(\theta^+)]^2 - [\alpha'(\theta)]^2\right| & \text{when} \quad \alpha'(\theta^+)\alpha'(\theta) \geq 0 \\ [\alpha'(\theta^+)]^2 + [\alpha'(\theta)]^2 & \text{when} \quad \alpha'(\theta^+)\alpha'(\theta) < 0 \end{cases} \tag{6.14}$$

Further, the problem of translating the pendulum to its equilibrium in time T will be considered. The control law will be discussed that minimizes functional (6.5). Actually, two problems will be investigated: one of moving the pendulum into the unstable equilibrium

$$\alpha = \pm\pi, \quad \dot{\alpha} = 0 \tag{6.15}$$

and the other one – of moving it to the stable equilibrium

$$\alpha = 0, 2\pi, \quad \dot{\alpha} = 0\,. \tag{6.16}$$

6.2 Translating the pendulum to the unstable equilibrium

Let the pendulum motion be described by nonlinear equation (6.3). Consider a task of translating this pendulum to equilibrium (6.15) in a given time T. The control algorithm that solves this task must minimize functional (6.5).

The trajectories of motion will be studied within stripe

$$-\pi \leq \alpha \leq \pi \tag{6.17}$$

Superposing the two straight lines

$$\alpha = -\pi\,, \tag{6.18}$$

$$\alpha = \pi\,, \tag{6.19}$$

gives a phase cylinder. Equilibriums (6.15)

$$\alpha = \pi, \quad \dot{\alpha} = 0 \tag{6.20}$$

and

$$\alpha = -\pi, \quad \dot{\alpha} = 0 \tag{6.21}$$

being placed on this phase cylinder, coincide. Trajectories of equation (6.3) in stripe (6.17) that appear when $\mu = 0$, i.e. the "ballistic" trajectories, are shown in Figure 6.1 [136]

These trajectories can be described by the energy integral (see expression (6.7))

$$E(\alpha, \alpha') = \frac{1}{2}\alpha'^2 - \cos\alpha = \frac{1}{2}[\alpha'(0)]^2 - \cos\alpha(0) = h = \text{const}\,. \tag{6.22}$$

In Figure 6.1, the so-called separating curve $R(T)$ is shown; its description and equation will be provided further.

Let

$$\alpha' = f(\alpha, T) \tag{6.23}$$

be the equation of curve $K(\pi, T)$ in plane α, α'. A representative point that starts from this curve and moves along a ballistic trajectory (with $\mu = 0$) will get onto line (6.19)

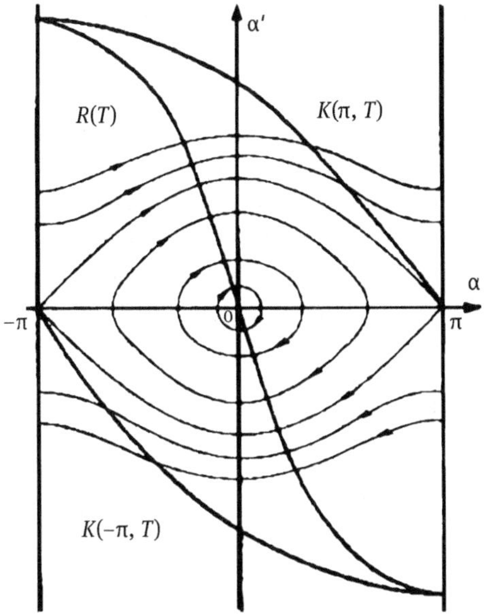

Fig. 6.1. Trajectories of natural (with $\mu = 0$) oscillations of the pendulum.

exactly in time T. Of course, this equation is considered only in stripe (6.17). Curve $K(\pi, T)$ is shown in Figure 6.1. It follows from expression (6.22) for the energy integral that equation (6.23) can be written implicitly using elliptic integral

$$\int_{\alpha}^{\pi} \frac{d\xi}{(\alpha'^2 - 2\cos\alpha + 2\cos\xi)^{1/2}} = T\,. \tag{6.24}$$

Curve $K(-\pi, T)$ determines a set of points, so that a representative point that starts from any of these points gets onto line (6.18) in time T. This curve is symmetric to curve $K(\pi, T)$ with respect to the origin (see Figure 6.1), and thus it can be represented by equation

$$\alpha' = -f(-\alpha, T)\,. \tag{6.25}$$

Curve $K(\pi, T)$ intersects each of the integral trajectories (6.22) (that result for each value of h) not more than once. Curves $K(\pi, T)$ that correspond to larger time T, are apparently located below the curves that correspond to smaller values of T. With all values of T they are located above the separatrix

$$\alpha' = 2\cos\frac{\alpha}{2}\,. \tag{6.26}$$

As for separatrix (6.26), each curve $K(\pi, T)$ has only one common point with it – point (6.20), because, when a representative point moves along curve (6.26), it reaches state (6.20) in infinite time. As $T \to \infty$, curve $K(\pi, T)$ infinitesimally approaches integral curve (6.26). Curves $K(\pi, T)$ for different values of T can be built using value tables for the elliptic integral (6.24). However, it is easier to build them by means of numerical integration of a homogeneous (with $\mu = 0$) equation (6.3), starting from points locates at straight line $\alpha = \pi$ (6.19) and using reverse time from $\tau = 0$ to $\tau = -T$. In other words, curve $K(\pi, T)$ for a particular value of T can be built by reverse motion that starts at line (6.19). Curves $K(\pi, T)$ (6.23) and $K(-\pi, T)$ (6.25) are shown in Figure 6.1, together with the ballistic trajectories of the pendulum. As curve $K(\pi, T)$ is located above separatrix (6.26), similarly curve $K(-\pi, T)$ is located below separatrix $\alpha' = -2\cos(\alpha/2)$.

Assume that the initial point $(\alpha(0), \alpha'(0))$ is located above curve $K(\pi, T)$ (6.23), i.e. $\alpha'(0) > f[\alpha(0), T]$. Then in ballistic motion the representative point at some time instant $\theta < T$ reaches line (6.19). The angular velocity $\alpha'(\theta)$ at this instant θ equals

$$\alpha'(\theta) = \left\{[\alpha'(0)]^2 - 4\cos^2\frac{\alpha(0)}{2}\right\}^{1/2} . \tag{6.27}$$

Applying at this time instant $\tau = \theta$ a damping impulse

$$\mu(\tau) = -\alpha'(\theta)\delta(\tau - \theta) \tag{6.28}$$

diminishes this velocity, and brings the representative point to equilibrium (6.20). The energy consumption in this control process (6.28), (6.27) is estimated as

$$J(0, \theta^+) = \frac{1}{2}[\alpha'(\theta)]^2 = \frac{1}{2}[\alpha'(0)]^2 - 2\cos^2\frac{\alpha(0)}{2} . \tag{6.29}$$

The difference (6.29) is equal to the difference in total energy of the pendulum taken at the initial moment and at the end of motion. Letting $\mu(\tau) = 0$ at $\tau > \theta$ will render the pendulum motionless in its equilibrium.

If the pendulum is required to reach equilibrium (6.20) *exactly* in time T, and not earlier, this can be done by control law

$$\mu(\tau) = I_0\delta(\tau) + I_T\delta(\tau - T) , \tag{6.30}$$

where

$$I_0 = f[\alpha(0), T] - \alpha'(0), \quad I_T = -\left\{f^2[\alpha(0), T] - 4\cos^2\frac{\alpha(0)}{2}\right\}^{1/2} . \tag{6.31}$$

Control function (6.30) includes two impulses. The first one ($I_0 < 0$), applied at time instant $\tau = 0$, decreases the pendulum energy in a step-like manner, thus moving the representative point onto curve $K(\pi, T)$. Then in ballistic motion (with $\mu = 0$) the representative point reaches line (6.19) at $\tau = T$. The second impulse ($I_T < 0$), applied at time instant $\tau = T$, provides one more step to decrease the pendulum energy, thus moving the representative point to equilibrium (6.20).

The first line in relation (6.14) helps find the energy consumed in the first impulse in control law (6.30), (6.31)

$$\frac{1}{2}[\alpha'(0)]^2 - \frac{1}{2}f^2[\alpha(0), T] . \tag{6.32}$$

Energy consumption in the second impulse in control law (6.30), (6.31) is

$$\frac{1}{2}f^2[\alpha(0), T] - 2\cos^2\frac{\alpha(0)}{2}. \tag{6.33}$$

Adding together values (6.32) and (6.33) yields the total energy consumption for the impulse control law (6.30), (6.31)

$$J(0, T) = \int_0^T |\mu(\tau)\alpha'(\tau)|d\tau = \frac{1}{2}[\alpha'(0)]^2 - 2\cos^2\frac{\alpha(0)}{2}. \tag{6.34}$$

The value of (6.34), as well as of (6.29), is equal to the absolute value of the difference between the pendulum total energy (6.7) at initial time $\tau = 0$ and at the final time instant $\tau = T$. It is clear from inequality (6.9) that it is impossible to achieve less energy consumption.

Thus if initial point $(\alpha(0), \alpha'(0))$ is located above curve $K(\pi, T)$ (6.23), then the "double-impulse" control law (6.30), (6.31), and the "single-impulse" one (6.28), is optimal in regard for energy consumption. Any control function $\mu(\tau)$ will be optimal if the total energy (6.7) monotonically decreases in process of motion. Such control function may be continuous on some intervals. That means, under initial conditions that are located *above* curve $K(\pi, T)$ the optimal control function is not unique. However, there is only one, unique optimal control function for initial points located exactly on curve $K(\pi, T)$.

For initial conditions that are located below curve $K(-\pi, T)$ (6.25) (see Figure 6.1) and on this curve, the picture is similar to above. Yet for the points located on curve $K(-\pi, T)$ and below it, the target equilibrium is state (6.21).

Now consider initial states that are located in between curves $K(\pi, T)$ (6.23) and $K(-\pi, T)$ (6.25).

Control law (6.30), (6.31) that consists of two impulses apparently translates the pendulum to equilibrium (6.20). If the initial state is in between curves $K(\pi, T)$ and $K(-\pi, T)$ then value $I_0 > 0$, and value $I_T < 0$. The corresponding trajectory $ABC\pi$ is shown in Figure 6.2

Point B lies on curve $K(\pi, T)$. Together with trajectory $ABC\pi$ that corresponds to double-impulse control law (6.30), (6.31), Figure 6.2 shows trajectory $AD\pi$ that connects the same points A and π. Point D of this trajectory lies above curvilinear segment BC, and energy (6.7) of the pendulum at point D is greater than at points of segment BC. Thus integral (6.5) calculated for trajectory $AD\pi$, is, in accordance with relation (6.6), greater than the same integral calculated for trajectory $ABC\pi$. The time period in which the pendulum is translated to position (6.20) along trajectory $ABC\pi$ equals T. The time period in which the pendulum moves to the same position (6.20) along any other trajectory that lies below trajectory $ABC\pi$ ("inside" quadrangle $ABC\pi$), is greater than T, because in some of its segments velocity α' is less than the velocity on trajectory $ABC\pi$. Hence the double-impulse control law (6.30), (6.31) translates the system to position (6.20) in a *given* time T with *minimum* energy consumption.

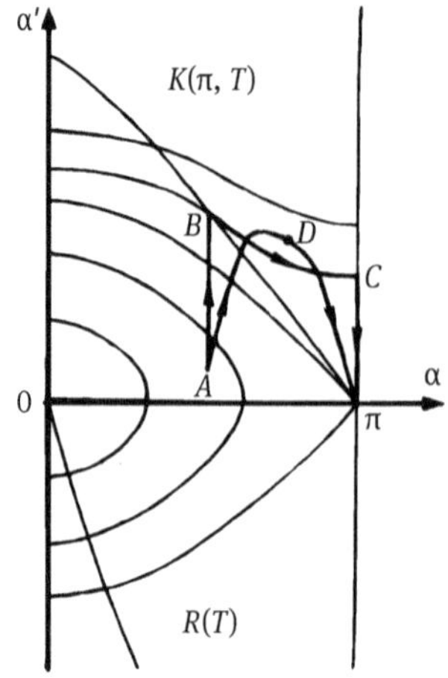

Fig. 6.2. Trajectory $ABC\pi$ corresponds to double-impulse control law, $\alpha'(0) > 0$.

For initial states that lie in between curves $K(\pi, T)$ and $K(-\pi, T)$, the energy used in the first impulse in control law (6.30), (6.31) is evaluated according to expression (6.14), as

$$\frac{1}{2}f^2[\alpha(0), T] - \frac{1}{2}[\alpha'(0)]^2 , \tag{6.35}$$

in case when $\alpha'(0) \geq 0$, and as

$$\frac{1}{2}f^2[\alpha(0), T] + \frac{1}{2}[\alpha'(0)]^2 , \tag{6.36}$$

when $\alpha'(0) < 0$.

The energy used in the second impulse is evaluated by expression (6.33) in both cases.

Adding together values (6.33) and (6.35) yields the expression to evaluate energy consumption in double-impulse control law (6.30), (6.31) with $\alpha'(0) \geq 0$

$$f^2[\alpha(0), T] - \frac{1}{2}[\alpha'(0)]^2 - 2\cos^2\frac{\alpha(0)}{2} . \tag{6.37}$$

Adding (6.33) and (6.36) yields the expression for evaluating energy consumption in double-impulse control law (6.30), (6.31) with $\alpha'(0) < 0$

$$f^2[\alpha(0), T] + \frac{1}{2}[\alpha'(0)]^2 - 2\cos^2\frac{\alpha(0)}{2} . \tag{6.38}$$

Double-impulse control law (6.30) where

$$I_0 = -f[-\alpha(0), T] - \alpha'(0), \quad I_T = \left\{f^2[-\alpha(0), T] - 4\cos^2\frac{\alpha(0)}{2}\right\}^{1/2} , \tag{6.39}$$

translates the pendulum by time instant $\tau = T$ into position (6.21), that corresponds, as well as position (6.20), to the top, unstable equilibrium. Indeed, the first impulse

($I_0 < 0$) moves the representative point onto curve $K(-\pi, T)$ (6.25), then in ballistic motion at time instant $\tau = T$ this point reaches line (6.18), where the second impulse ($I_T > 0$) moves it to equilibrium (6.21).

Energy consumption in control law (6.30), (6.39) in case when $\alpha'(0) \geq 0$ is

$$f^2[-\alpha(0), T] + \frac{1}{2}[\alpha'(0)]^2 - 2\cos^2\frac{\alpha(0)}{2}\,. \tag{6.40}$$

When $\alpha'(0) < 0$, the total energy consumed by control impulses (6.30), (6.39) is

$$f^2[-\alpha(0), T] - \frac{1}{2}[\alpha'(0)]^2 - 2\cos^2\frac{\alpha(0)}{2}\,. \tag{6.41}$$

The next step is to find initial states such that the energy consumption for double-impulse control law (6.30), (6.31), that moves the system to state (6.20), and the energy consumption for double-impulse control law (6.30), (6.39), that moves the system to state (6.21), are the same. To find such initial states, set equal expressions (6.37) and (6.40), and also expressions (6.38) and (6.41). Doing some transposition, one can find the equation of the curve $R(T)$ where the sought states are located:

$$\alpha' = -\left|f^2[\alpha, T] - f^2[-\alpha, T]\right|^{1/2} \operatorname{sign}\alpha\,. \tag{6.42}$$

Curve $R(T)$ (6.42) is shown in Figure 6.1, and also it is partially present in Figure 6.2. This curve passes the origin $\alpha = \alpha' = 0$, and it is symmetric with respect to the origin. The ends of curve $R(T)$ that lie on straight lines (6.18) and (6.19) coincide with the ends of curves $K(\pi, T)$ and $K(-\pi, T)$. This fact follows from inspection of equations (6.42), and also of (6.23) and (6.25). From these equations it follows that curves $R(T)$ and $K(\pi, T)$, and also $R(T)$ and $K(-\pi, T)$ are tangent in coinciding points. Curve $R(T)$ can be easily plotted from equation (6.42), if the curves $K(\pi, T)$ and $K(-\pi, T)$ are already built.

From initial states that belong to curve $R(T)$ both control laws (6.30), (6.31), and (6.30), (6.39) translate the pendulum to the top, unstable equilibrium in one and the same time interval T, consuming the same energy. Yet the trajectories that correspond to these control processes are different. In one case, the pendulum swings counterclockwise, in the other case – clockwise. For initial states that are located above curve $R(T)$, control law (6.30), (6.31) is less energy-consuming than law (6.30), (6.39), and for initial states that are below curve $R(T)$ it is vice versa. Curve $R(T)$ is a *separating* line. Such separating lines typically exist in systems that are investigated in phase cylinder [18, 50, 94].

To summarize the discussion that was presented above, the following theorem is proposed.

Theorem. *An energy-optimal control law for pendulum translation to the top, unstable equilibrium can be built for any initial state as a double-impulse function, where the first impulse is applied at the initial time instant, and the second one – at the final time instant.*

The next section will consider the problem of optimal control of the pendulum translation to the bottom, stable equilibrium.

6.3 Translating the pendulum to the stable equilibrium

Let the pendulum motion be described by nonlinear equation (6.3). Consider the problem of translating this pendulum to equilibrium (6.16) on a given time interval T. The control law that will be proposed in this section minimizes functional (6.5).

Let the trajectories of motion be located within phase stripe

$$0 \leq \alpha \leq 2\pi \,. \tag{6.43}$$

Combining the two straight lines

$$\alpha = 0 \,, \tag{6.44}$$

$$\alpha = 2\pi \,, \tag{6.45}$$

yields a phase cylinder. Equilibriums (6.16)

$$\alpha = 2\pi, \quad \dot{\alpha} = 0 \tag{6.46}$$

and

$$\alpha = 0, \quad \dot{\alpha} = 0 \tag{6.47}$$

being placed on the phase cylinder, coincide. Trajectories of equation (6.3) in stripe (6.43), that result when $\mu = 0$, i.e. the "ballistic" trajectories, are shown in Figure 6.3 [136].

These trajectories are described by energy integral (6.22). Equation

$$\alpha' = f(\alpha, T) \tag{6.48}$$

describes a curve $K(2\pi, T)$ in plane α, α'. A representative point that starts from this curve will reach line (6.45) exactly on time interval T. Of course, this equation is considered only within stripe (6.43).

Curve $K(0, T)$ is a set of starting representative points that get onto line (6.44) exactly in time T. It is symmetric to curve $K(2\pi, T)$ with respect to point $(\pi, 0)$, and hence it is described by equation

$$\alpha' = -f(2\pi - \alpha, T) \,. \tag{6.49}$$

With any value of $0 < T \leq \pi$ curves $K(2\pi, T)$ (6.48) end at point (6.46) (see Figure 6.3). With $0 < T \leq \pi/2$ these curves are located above the abscissa $\alpha' = 0$. With $T = \pi/2$ curve $K(2\pi, T)$ ends at point (6.46), being tangent with the abscissa $\alpha' = 0$. Each curve $K(2\pi, T)$ (6.48) with $T > \pi/2$ intersects the abscissa at point

$$\alpha = \alpha_T \quad (\pi < \alpha_T < 2\pi) \tag{6.50}$$

i.e. $f(\alpha_T, T) = 0$. At that, $f(\alpha, T) > 0$ if $0 \leq \alpha < \alpha_T$, and $f(\alpha, T) < 0$ if $\alpha_T < \alpha < 2\pi$.

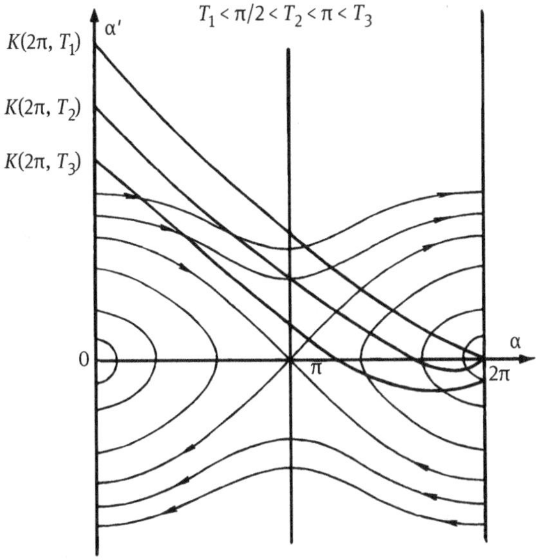

Fig. 6.3. Trajectories of "ballistic" (with $\mu = 0$) motion modes of the pendulum.

With $\pi/2 < T \leq \pi$ curves $K(2\pi, T)$ intersect the abscissa and end at point (6.46), approaching it from below. With $T = \pi$ curve $K(2\pi, T)$ approaches point (6.46) tangent to half line $\alpha = 2\pi$, $\alpha' \leq 0$. When $T > \pi$ curves $K(2\pi, T)$ end on line (6.45) with $\alpha' < 0$. As the value of T grows, curve $K(2\pi, T)$ lowers (see Figure 6.3). However, with all values of T it is located above separatrix (6.26). Figure 6.3 demonstrates curves $K(2\pi, T)$ with three values T_1, T_2, T_3, that satisfy inequalities

$$T_1 < \frac{\pi}{2} < T_2 < \pi < T_3 .$$

As $T \to \infty$, curve $K(2\pi, T)$ infinitesimally approaches curve (6.26).

Consider initial states $(\alpha(0), \alpha'(0))$ located above curve $K(2\pi, T)$. Following the same reasoning that was proposed when considering the problem of pendulum translation to the unstable equilibrium, it is easy to show that the optimal control law for these states can be built as a single impulse like in (6.28). This impulse is applied to the pendulum at time instant $\tau = \theta < T$, when it reaches vertical position (the representative point, moving along a ballistic trajectory, gets onto line (6.45)). At that instant, the pendulum velocity is equal to

$$\alpha'(\theta) = \left\{[\alpha'(0)]^2 + 4\sin^2\frac{\alpha(0)}{2}\right\}^{1/2} . \tag{6.51}$$

The damping impulse (6.28), (6.51) moves the pendulum directly to equilibrium state (6.46). The energy consumption at such optimal single-impulse control law is

$$J(0, \theta) = \frac{1}{2}[\alpha'(\theta)]^2 = \frac{1}{2}[\alpha'(0)]^2 + 2\sin^2\frac{\alpha(0)}{2} . \tag{6.52}$$

It is possible to translate the pendulum to equilibrium (6.46) *exactly* in time T (and not earlier), using a double-impulse control law (6.30), where

$$I_0 = f[\alpha(0), T] - \alpha'(0), \quad I_T = -\left\{f^2[\alpha(0), T] + 4\sin^2\frac{\alpha(0)}{2}\right\}^{1/2}. \tag{6.53}$$

If time interval T and initial angle $\alpha(0)$ are such that

$$f[\alpha(0), T] \geq 0\,, \tag{6.54}$$

then the amount of energy consumed in the double-impulse control process (6.30), (6.53), as one can easily verify, are equal to value (6.52) and thus it is minimal. Conversely, if

$$f[\alpha(0), T] < 0 \quad (\alpha_T < \alpha(0) < 2\pi)\,, \tag{6.55}$$

then the energy consumption in process controlled by law (6.30), (6.53) occurs to exceed the minimum possible value of (6.52). In this case, the first impulse in control law (6.30), (6.53) provides the pendulum with a velocity that is directed away from equilibrium (6.46).

If condition (6.55) takes place, the single-impulse law that translates the pendulum to equilibrium (6.46) on time interval that is less than T is not unique. This translation can be done in many ways, including a double-impulse control law. At that the energy (6.52) consumed in process of motion is minimal. If afterwards the control function is assigned as $\mu(\tau) = 0$, then by time T (and later) the pendulum will remain in this equilibrium.

For initial states that are located below curve $K(0, T)$ the picture is similar to what was discussed above.

Consider now the initial states located *in between* curves $K(2\pi, T)$ and $K(0, T)$. To remind, these curves are symmetrical to each other with respect to point $(\pi, 0)$.

Double-impulse control law (6.30), (6.53) can move the representative point to position (6.46) by time instant $\tau = T$. The first impulse in this law moves the point onto curve $K(2\pi, T)$. Then in ballistic motion it reaches straight line (6.45), where the second impulse moves it to the equilibrium.

The energy consumption for law (6.30), (6.53), evaluated in accordance with formulas (6.6), (6.14), are represented by expression

$$J(0, T) = f^2[\alpha(0), T] - \frac{1}{2}[\alpha'(0)]^2 + 2\sin^2\frac{\alpha(0)}{2}\,, \tag{6.56}$$

if

$$\alpha'(0) \geq 0, \quad \alpha(0) \leq \alpha_T\,, \tag{6.57}$$

by expression

$$J(0, T) = f^2[\alpha(0), T] + \frac{1}{2}[\alpha'(0)]^2 + 2\sin^2\frac{\alpha(0)}{2}\,, \tag{6.58}$$

if

$$\alpha'(0) \leq 0, \quad \alpha(0) \leq \alpha_T\,, \tag{6.59}$$

and by expression

$$J(0, T) = \frac{1}{2}[\alpha'(0)]^2 + 2\sin^2\frac{\alpha(0)}{2}, \tag{6.60}$$

if

$$\alpha'(0) \le 0, \quad \alpha(0) \ge \alpha_T. \tag{6.61}$$

Control law (6.30) where

$$I_0 = -f[2\pi - \alpha(0), T] - \alpha'(0), \quad I_T = \left\{f^2[2\pi - \alpha(0), T] + 4\sin^2\frac{\alpha(0)}{2}\right\}^{1/2}, \tag{6.62}$$

moves the representative point to position (6.47) by time instant $\tau = T$. The first impulse in this law moves the point onto curve $K(0, T)$. Then in ballistic motion it reaches line (6.44), and then the second impulse moves it to equilibrium (6.47).

The energy consumed in motion process ruled by control law (6.30), (6.62) is described by expression

$$J(0, T) = f^2(2\pi - \alpha(0), T) - \frac{1}{2}[\alpha'(0)]^2 + 2\sin^2\frac{\alpha(0)}{2}, \tag{6.63}$$

if

$$\alpha'(0) \le 0, \quad \alpha(0) \ge 2\pi - \alpha_T, \tag{6.64}$$

by expression

$$J(0, T) = f^2(2\pi - \alpha(0), T) + \frac{1}{2}[\alpha'(0)]^2 + 2\sin^2\frac{\alpha(0)}{2}, \tag{6.65}$$

if

$$\alpha'(0) \ge 0, \quad \alpha(0) \ge 2\pi - \alpha_T, \tag{6.66}$$

and by expression

$$J(0, T) = \frac{1}{2}[\alpha'(0)]^2 + 2\sin^2\frac{\alpha(0)}{2}, \tag{6.67}$$

if

$$\alpha'(0) \ge 0, \quad \alpha(0) \le \alpha_T. \tag{6.68}$$

Note that expressions (6.60), (6.67) coincide with each other and with expression (6.52).

Like in unstable equilibrium case, the next step is to find initial states such that the energy consumption for control process ruled by law (6.30), (6.53) is equal to the amount of energy used by control law (6.30), (6.62). These initial states lie on the separating curve $R(T)$, that is symmetric with respect to point $(\pi, 0)$, as one can easily check by himself.

From expressions (6.57), (6.59), (6.64), (6.66) it follows that evaluations (6.56), (6.58), (6.63), (6.65) of energy consumption are true on interval

$$2\pi - \alpha_T \le \alpha(0) \le \alpha_T.$$

At that, equalities (6.56), (6.65) are true in area $\alpha'(0) \geq 0$, and equalities (6.58), (6.63) in area $\alpha'(0) \leq 0$. Setting values (6.56) and (6.65) equal to each other, and value (6.58) to (6.63), yields the equation that describes the separating curve $R(T)$ on interval

$$2\pi - \alpha_T \leq \alpha \leq \alpha_T\,. \tag{6.69}$$

This equation looks as follows

$$\alpha' = |f^2[\alpha, T] - f^2[2\pi - \alpha, T]|^{1/2}\,\mathrm{sign}(\pi - \alpha)\,. \tag{6.70}$$

Equation (6.70) is "similar" to equation (6.42).

From investigation of inequalities (6.57), (6.68) it turns out that both expressions (6.56), (6.67) are true in area (6.57). Setting equal values (6.56) and (6.67) shows that on interval

$$0 \leq \alpha \leq 2\pi - \alpha_T \tag{6.71}$$

the separating curve $R(T)$ coincides with curve $K(2\pi, T)$. Curves $K(2\pi, T)$ and $K(0, T)$ are symmetric to each other with respect to point $(\pi, 0)$, and curve $R(T)$ is symmetric with respect to the same point. It follows from these facts that on interval

$$\alpha_T \leq \alpha \leq 2\pi \tag{6.72}$$

the separating curve $R(T)$ coincides with curve $K(0, T)$. Figure 6.4 shows curve $R(T)$, and also curves $K(2\pi, T)$ and $K(0, T)$. The vertical dashed lines pass through points $\alpha'(0) = 0$, $\alpha(0) = 2\pi - \alpha_T$ and $\alpha'(0) = 0$, $\alpha(0) = \alpha_T$. Equation (6.70) describes curve $R(T)$ that is located between these dashed lines.

If $0 < T \leq \pi/2$, then curve $K(2\pi, T)$ is located in the upper half of phase plane (α, α'), and curve $K(0, T)$ in the lower semiplane. Note that separating curve $R(T)$ does not coincide anywhere with curve $K(2\pi, T)$, or with $K(0, T)$, and it is at all points described by equation (6.70). The separating curve can be easily built from curves $K(2\pi, T)$ and $K(0, T)$.

It will now be shown that the double-impulse control law (6.30), (6.53) is optimal for all initial states that are located in between curves $R(T)$ and $K(2\pi, T)$.

First let $\alpha'(0) \geq 0$, then $\alpha(0) \leq \alpha_T$, because the initial point is below curve $K(2\pi, T)$. In this case, the fact that control law (6.30), (6.53) is optimal is proved in the same manner as above (see Figure 6.2), where the fact of optimality of double-impulse control law is proved in the problem of pendulum translation to the unstable equilibrium.

Now let $\alpha'(0) < 0$. This case is divided into two: $\alpha(0) \geq \alpha_T$ and $\alpha(0) \leq \alpha_T$. First consider $\alpha(0) \geq \alpha_T$ (see Figure 6.5).

The amount of energy (6.5) consumed at trajectory $ABCE$ is minimal. This follows from relation (6.6) together with the fact that the total energy (6.7) of the pendulum is monotonically decreasing while the pendulum moves along this trajectory.

Now let $\alpha'(0) < 0$ and $\alpha(0) \leq \alpha_T$. Figure 6.6 shows trajectory $ADBCE$ that corresponds to double-impulse control (6.30), (6.53) and, for comparison, trajectory $AD'B'C'E$.

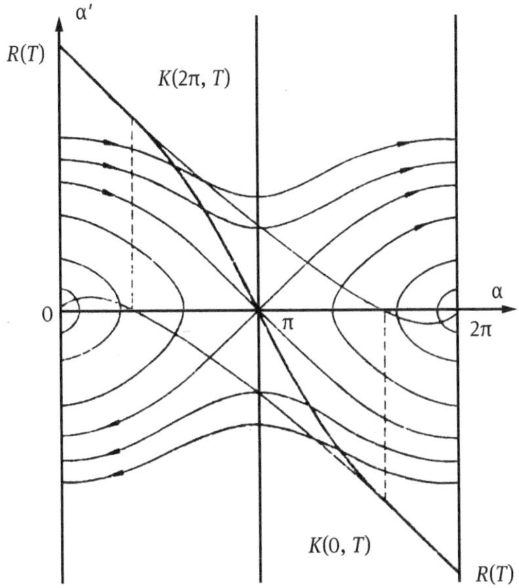

Fig. 6.4. Curves $K(2\pi, T)$, $K(0, T)$ and separating curve $R(T)$.

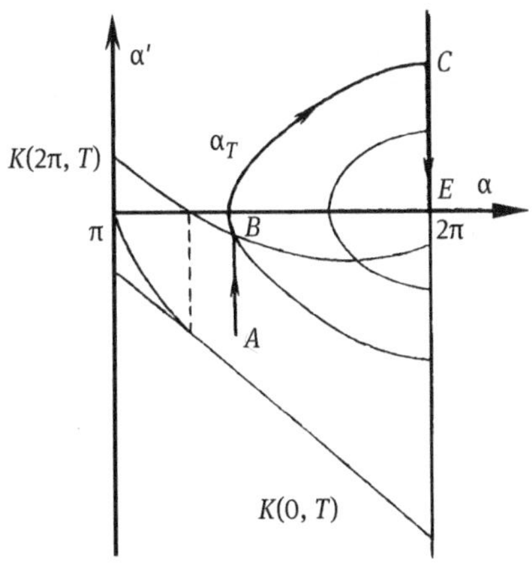

Fig. 6.5. Trajectory $ABCE$ corresponds to double-impulse control law, $\alpha'(0) < 0$, $\alpha(0) > \alpha_T$.

Energy consumption in process ruled by control law (6.30), (6.53) that realizes the pendulum motion along trajectory $ADBCE$ is described by expression (6.58). Compare it to the amount of energy consumed in motion along the "reference" trajectory $AD'B'C'E$. A part of this trajectory, namely, $D'B'C'E$, corresponds to the double-impulse control law because, as it was shown above, such law is optimal when $\alpha'(0) = 0$, $\alpha(0) \leq \alpha_T$.

In part AD' of trajectory $AD'B'C'E$ angle α decreases strictly monotonically, because in this part $\alpha' < 0$. For integral (6.5) evaluated at trajectory AD' that starts at point A and ends at a given point D' to be minimal, it is necessary that energy (6.7) changes monotonically as the representative point moves along this trajectory. First,

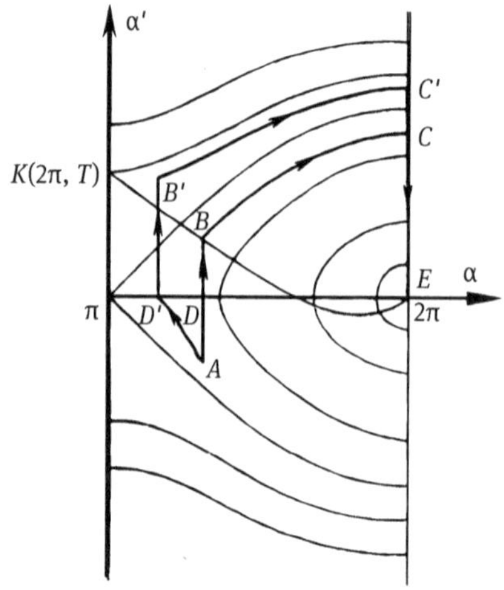

Fig. 6.6. Trajectory *ADBCE* corresponds to double-impulse control, $\alpha'(0) < 0$, $\alpha(0) \le \alpha_T$.

it will be assumed that energy (6.7) monotonically decreases, or remains unchanged, i.e. at point D' its value (6.7) is no greater than at point A. Then integral (6.5) evaluated for trajectory AD' is equal to

$$J(0, \theta) = \frac{1}{2}[\alpha'(0)]^2 - \cos\alpha(0) + \cos\alpha^*(\theta) . \tag{6.73}$$

Here $\theta > 0$ is the time interval of motion along trajectory part AD', and $\alpha^*(\theta) < \alpha(0)$ is the value of angle α at point D'. Evaluating integral (6.5) at part $D'B'C'E$ of the trajectory is equal to

$$\begin{aligned} J(\theta, T) &= \frac{1}{2}f^2[\alpha^*(\theta), T-\theta] + \frac{1}{2}f^2[\alpha^*(\theta), T-\theta] - \cos\alpha^*(\theta) + \cos 2\pi = \\ &= f^2[\alpha^*(\theta), T-\theta] - \cos\alpha^*(\theta) + 1 \end{aligned} \tag{6.74}$$

Adding values (6.73) and (6.74) together yields

$$J(0, T) = \frac{1}{2}[\alpha'(0)]^2 + f^2[\alpha^*(\theta), T-\theta] + 1 - \cos\alpha(0) \tag{6.75}$$

The value of (6.75) is greater than of (6.58), because $f[\alpha^*(\theta), T-\theta] > f[\alpha(0), T]$. The latter inequality is true since the value of $f(\alpha, T)$ increases as time T and/or angle α decreases. Note that trajectory $ADBCE$ is more energy-efficient than $AD'B'C'E$ also in case when on its part AD' control function $\mu = 0$, i.e. when AD' is a trajectory of ballistic motion.

Now let the value of energy (6.7) at point D' be greater than at point A. Then the value of integral (6.5) calculated for trajectory AD' equals

$$J(0, \theta) = -\cos\alpha^*(\theta) - \frac{1}{2}[\alpha'(0)]^2 + \cos\alpha(0) > 0 . \tag{6.76}$$

The value of integral (6.5) calculated for part $D'B'C'E$ of trajectory $AD'B'C'E$ is still in accordance with expression (6.74). Adding together expressions (6.76) and (6.74) yields

$$\begin{aligned} J(0, T) &= -\cos\alpha^*(\theta) - \frac{1}{2}[\alpha'(0)]^2 + \cos\alpha(0) + f^2[\alpha^*(\theta), T-\theta] - \cos\alpha^*(\theta) + 1 > \\ &> f^2[\alpha^*(\theta), T-\theta] - \cos\alpha^*(\theta) + 1 > \\ &> f^2[\alpha^*(\theta), T-\theta] + \frac{1}{2}[\alpha'(0)]^2 - \cos\alpha(0) + 1 = \\ &= \frac{1}{2}[\alpha'(0)]^2 + f^2[\alpha^*(\theta), T-\theta] + 1 - \cos\alpha(0)\,. \end{aligned} \tag{6.77}$$

Transpositions in (6.77) rely on inequality (6.76).

Expression (6.77) is the same as (6.75), and this value is greater than (6.58).

Therefore, the energy consumed on the "reference" trajectory $AD'B'C'E$ is greater than in motion along trajectory $ADBCE$ that results from the double-impulse control.

To summarize the results obtained for the problem of pendulum translation to the bottom, stable equilibrium, a theorem may be given.

Theorem. *The energy-optimal control law for pendulum translation to the bottom, stable equilibrium from any initial state can be built as a double-impulse function. The first impulse is applied at the initial time instant* $\tau = 0$ *or at time instant* $0 < \tau < T$*, and the second one – at the final time instant* $\tau = T$.

The previous section, Section 2, considered the problem of translating the pendulum to the top, unstable equilibrium. The theorem stated in that section is a little different from the theorem stated here.

Part II: **Double physical pendulum**

A plane double physical pendulum is a system with two degrees of freedom. Controlling such system is a complicated task, if there is only one control input in it. This control input can be a torque applied in the pivot point – the "shoulder" joint, or in the joint between the two links – the "elbow". A system where the number of control parameters is less than the number of degrees of freedom is often referred to as underactuated. The problem gets even more complicated if the control parameter is limited in its value. In fact, the real control resources are always limited [53]. Limitation imposed on control input is significant when the desired mode of system motion without control is unstable. In this case, as a result, a set of initial states from which the system can be put into a desired mode is limited. An important part is the problem of building a control law as a feedback that provides maximum possible domain of attraction of this desired mode.

Controlling an inverted double pendulum is rather popular among researchers. The studies that concern such mechanical system include, in particular, those where a pendulum with stationary pivot is considered, like in the current study. Some works discuss not only the task of local stabilization of a desired unstable equilibrium, but also the problem of its global stabilization. For example, a pendulum with a control torque applied in its inter-link joint is investigated in studies [28, 101, 124–126, 145], where a substitution of variables transposes the equations to "cascade" structure when the inter-link angle may be regarded as an intermediate control variable. In case when the control torque is applied at the pivot, the "cascade" form of equations appears to be more complicated [124–126]. In said studies no limitation on control parameters are considered. One idea of pendulum control design that has much potential in it is to consider the system energy to control a pendulum, as it is proposed in [17, 68] for a single-link pendulum. Studies [38, 131] discuss the problem of moving a double pendulum to a neighbourhood of the top, unstable equilibrium by means of a torque applied in the pivot. This torque is limited in its absolute value, and the system is subject to external disturbance forces. The control algorithm is designed with the help of decomposition method developed in work [132]. Energy based control of the double pendulum by the torque applied in the pivot is designed in [48].

The current part considers the problems of local stabilization, and also the global stabilization, of an unstable equilibrium of a double inverted pendulum. Only one control parameter is present in the system – the torque in the "elbow" or in the "shoulder" joint. This torque is assumed limited in its absolute value. The problems of optimal swinging and damping of the double pendulum are also discussed here.

7 Local stabilization of an inverted pendulum by means of a single control torque

The current chapter develops a control law that stabilizes a double inverted pendulum in case of given (limited) control resources, i.e. the problem of local stabilization. Two cases are considered: first, when the motion is controlled by a torque applied at the pivot point of the pendulum; second, when the control torque is applied at the inter-link joint. The mathematical model of double pendulum without control, linearized about its top equilibrium, has two corresponding eigenvalues in the right semiplane and two – in the left semiplane. When the control torque is limited, the domain of controllability is limited with respect to two "unstable" Jordan variables. For each of the two cases, the domains of controllability are found and discussed below. Building a domain of controllability provides for estimation of a domain of attraction that can be realized when a feedback control law is designed. In each case, it is possible to create a linear feedback control law (with saturation) such that its domain of attraction is close to the domain of controllability, and thus to the maximum possible domain of attraction.

Despite many studies dedicated to problems of double pendulum control design, the author never had a chance to find any works that consider the problems of local stabilization and maximizing domain of attraction like those discussed in this chapter.

7.1 Mathematical model of the pendulum

Figure 7.1 shows a plane double physical pendulum with a stationary pivot point O. The pivot is assumed an ideal (frictionless) cylindrical joint. A similar ideal joint connects at point D the two links of the pendulum. Both links are assumed perfectly rigid. The axes of the joints in points O and D are perpendicular to the drawing plane. The center of mass of the first link is located within segment OD. Notation φ_1 and φ_2 represents the angles that the first (segment OD) and the second links, respectively, deviate from the vertical line, counted counter-clockwise.

Kinetic T and potential Π energy of the double pendulum can be presented as

$$T = \frac{1}{2}\left[a_{11}\dot{\varphi}_1^2 + 2a_{12}\dot{\varphi}_1\dot{\varphi}_2\cos(\varphi_1 - \varphi_2) + a_{22}\dot{\varphi}_2^2\right], \quad \Pi = b_1\cos\varphi_1 + b_2\cos\varphi_2 . \quad (7.1)$$

The virtual work δW of torque L that is applied at joint O or D can be written, respectively, as

$$\delta W = L\delta\varphi_1 \quad \text{or} \quad \delta W = L(\delta\varphi_2 - \delta\varphi_1) . \quad (7.2)$$

In expressions (7.1) $a_{11} = I_1 + m_2 l^2$, $a_{22} = I_2$, $a_{12} = m_2 r_2 l$, $b_1 = (m_1 r_1 + m_2 l)g$, $b_2 = m_2 r_2 g$, I_1 and I_2 are the inertia moments of the first and the second links with respect to their corresponding joints O and D, m_1 and m_2 denote the mass of the first

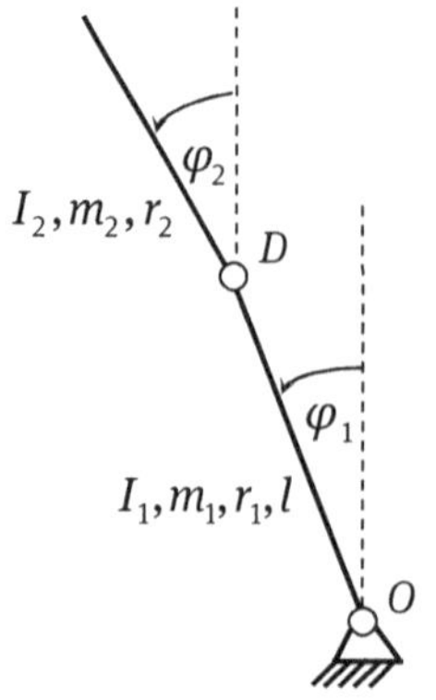

Fig. 7.1. A double inverted pendulum.

and the second links, r_1 and r_2 – the distances between joints O and D and the centers of mass of the first and second links, respectively, l is the length of the first link OD, g represents the gravity acceleration. The length of the second link is not present in expression (7.1), and hence it does not enter in the equations of motion; only the distance r_2 between joint D and the center of mass of the second link is present in motion equations. It will be assumed that $r_1, r_2 > 0$, i.e. the center of mass of the first link OD is not coincident with joint O, and the center of mass of the second link – with joint D.

Taking expressions (7.1) and (7.2) and using Lagrangian approach of the second kind [16, 41, 73] yields the motion equations of the double pendulum in matrix form

$$A\ddot{\varphi} + F\dot{\varphi}^2 + B\sin\varphi = c^{(i)}L \quad (i = 1 \quad \text{or} \quad i = 2)\,. \tag{7.3}$$

Here

$$\varphi = \begin{Vmatrix}\varphi_1\\ \varphi_2\end{Vmatrix}, \quad \dot{\varphi}^2 = \begin{Vmatrix}\dot{\varphi}_1^2\\ \dot{\varphi}_2^2\end{Vmatrix}, \quad \sin\varphi = \begin{Vmatrix}\sin\varphi_1\\ \sin\varphi_2\end{Vmatrix},$$

$$A = \begin{Vmatrix} a_{11} & a_{12}\cos(\varphi_2 - \varphi_1)\\ a_{12}\cos(\varphi_2 - \varphi_1) & a_{22}\end{Vmatrix},$$

$$F = \begin{Vmatrix} 0 & -a_{12}\sin(\varphi_2 - \varphi_1)\\ a_{12}\sin(\varphi_2 - \varphi_1) & 0\end{Vmatrix},$$

$$B = \begin{Vmatrix} -b_1 & 0\\ 0 & -b_2\end{Vmatrix}, \quad c^{(1)} = \begin{Vmatrix}1\\ 0\end{Vmatrix}, \quad c^{(2)} = \begin{Vmatrix}-1\\ 1\end{Vmatrix}.$$

If the control torque is applied at the pivot point O, then in equations (7.3) it must be set $i = 1$, and if the torque L is applied at the inter-link joint D, then $i = 2$.

As admissible control input the piecewise-continuous functions $L(t)$ will be taken, that are limited in absolute value by constant L_0

$$|L| \le L_0 \quad (L_0 = \text{const})\,. \tag{7.4}$$

The set of such functions will be denoted as W.

When $L \equiv 0$, system (7.3) has a trivial solution

$$\varphi_1 = \varphi_2 = 0, \quad \dot{\varphi}_1 = \dot{\varphi}_2 = 0\,, \tag{7.5}$$

that corresponds to the unstable equilibrium of the pendulum without control when both links are directed upwards.

The goal of this chapter is to design a control law that will stabilize state (7.5) and provide a “large” domain of attraction.

When considering the problem of pendulum stabilization in the top, unstable equilibrium (7.5), it will be assumed that at the beginning of the process (and actually during the whole process) of stabilization the pendulum is within some small neighbourhood of this state. The circular motions of the whole pendulum or some of its links will not be considered. In other words, only the *local* stabilization problem will be investigated.

7.2 Linearized model

Linearization of equations (7.3) about equilibrium (7.5) yields a system

$$A_0\ddot{\varphi} + B\varphi = c^{(i)}L, \quad A_0 = \begin{Vmatrix} a_{11} & a_{12} \\ a_{12} & a_{22} \end{Vmatrix} \quad (i = 1 \quad \text{or} \quad i = 2)\,. \tag{7.6}$$

With $L = 0$ nonlinear system (7.3), as well as linearized one (7.6), are conservative. The characteristic equation of a homogeneous system (7.6) is as follows:

$$(a_{11}a_{22} - a_{12}^2)\lambda^2 - (a_{11}b_2 + a_{22}b_1)\lambda + b_1b_2 = 0 \tag{7.7}$$

This equation has two positive roots. If its discriminant is nonzero, then the roots are different

$$\lambda_1 > \lambda_2 > 0\,. \tag{7.8}$$

Note. *If $r_1 > 0$, and $r_2 = 0$ ($b_2 = 0$), i.e. the center of mass of the second link is located in joint D, then one of the roots of equation (7.7) is equal to zero, and the other one remains positive. In this case, with $i = 2$, when the control torque is also applied at joint D, the considered double pendulum becomes similar to a single-link pendulum with a flywheel at its end. Such a pendulum with a controlled flywheel was discussed in the first part – in chapter 3.*

Uncontrolled system (7.6) (with $L = 0$) is conservative, so a nondegenerate transformation

$$\varphi = Kx\,, \tag{7.9}$$

where K is a constant matrix, can reduce it to *normal* coordinates [40, 41] x_1, x_2, so that it will look like

$$\ddot{x} - \Lambda x = d^{(i)}L, \quad x = \begin{Vmatrix} x_1 \\ x_2 \end{Vmatrix}, \quad \Lambda = \begin{Vmatrix} \lambda_1 & 0 \\ 0 & \lambda_2 \end{Vmatrix}\,. \tag{7.10}$$

Transformation matrix K consists of eigenvectors of matrix $A_0^{-1}B$. It can be chosen, for example, as

$$K = \left\| \begin{matrix} \dfrac{b_2 - a_{22}\lambda_1}{a_{12}\lambda_1} & \dfrac{b_2 - a_{22}\lambda_2}{a_{12}\lambda_2} \\ 1 & 1 \end{matrix} \right\| . \tag{7.11}$$

Here it is assumed that $r_2 \neq 0$, and therefore, $a_{12} \neq 0$.

A matrix that transforms the system to normal coordinates may, of course, be different from (7.11). Any other transformation matrix can be derived from (7.11) by multiplying its first or/and second column by some factor. It the transformation matrix is chosen as (7.11), then

$$d^{(i)} = \left\| \begin{matrix} d_1^{(i)} \\ d_2^{(i)} \end{matrix} \right\| = K^{-1}A_0^{-1}c^{(i)} = \frac{1}{(\lambda_2 - \lambda_1)(a_{11}a_{22} - a_{12}^2)} \left\| \begin{matrix} a_{12}\lambda_1 & \dfrac{\lambda_1 b_1}{\lambda_2 b_2}(a_{22}\lambda_2 - b_2) \\ -a_{12}\lambda_2 & -\dfrac{\lambda_2 b_1}{\lambda_1 b_2}(a_{22}\lambda_1 - b_2) \end{matrix} \right\| c^{(i)}$$

$$(i = 1 \quad \text{or} \quad i = 2) . \tag{7.12}$$

System (7.10) is equivalent to two scalar differential equations written in terms of normal coordinates x_1 and x_2

$$\ddot{x}_1 - \lambda_1 x_1 = d_1^{(i)} L, \quad \ddot{x}_2 - \lambda_2 x_2 = d_2^{(i)} L . \tag{7.13}$$

Here

$$d^{(1)} = \left\| \begin{matrix} d_1^{(1)} \\ d_2^{(1)} \end{matrix} \right\| = K^{-1}A_0^{-1}c^{(1)} = \frac{1}{(\lambda_2 - \lambda_1)(a_{11}a_{22} - a_{12}^2)} \left\| \begin{matrix} a_{12}\lambda_1 \\ -a_{12}\lambda_2 \end{matrix} \right\| ,$$

$$d^{(2)} = \left\| \begin{matrix} d_1^{(2)} \\ d_2^{(2)} \end{matrix} \right\| = K^{-1}A_0^{-1}c^{(2)} = \frac{1}{(\lambda_2 - \lambda_1)(a_{11}a_{22} - a_{12}^2)} \left\| \begin{matrix} b_1 - (a_{11} + a_{12})\lambda_1 \\ -b_1 + (a_{11} + a_{12})\lambda_2 \end{matrix} \right\| .$$

Columns $d^{(1)}$ and $d^{(2)}$ are obtained by means of expression (7.12). Besides, the expression for column $d^{(2)}$ can be simplified by applying expression for characteristic polynomial (7.7).

From inequalities (7.8) it follows that the spectrum of system (7.13) has two positive eigenvalues

$$\mu_1 = \sqrt{\lambda_1}, \quad \mu_2 = \sqrt{\lambda_2} \quad (\mu_1 > \mu_2 > 0) \tag{7.14}$$

and two negative eigenvalues that are equal to the first two in absolute values: $-\mu_1, -\mu_2$.

7.3 Domains of controllability

Values λ_1, λ_2 are not the same (see relations (7.8)), so the left-hand sides of equations (7.13) are different. Besides, none of the eigenvalues is equal to zero if $b_1 b_2 \neq 0$. Therefore both elements of column $d^{(1)}$ are nonzero. This means that [69, 98] mechanical system (7.13), and thus the original one (7.6), is completely controllable in Kalman's sense [88, 89–92] if the control torque is applied only at pivot point O. In [69], among other things, it is proved that a *multi-link* inverted pendulum (with any number of links) is a completely controllable system whenever a control torque is applied at the pivot point.

Let the following equality take place: $b_1-(a_{11}+a_{12})\lambda_1=0$, or $b_1-(a_{11}+a_{12})\lambda_2=0$. Substituting $\lambda_1=b_1/(a_{11}+a_{12})$ or $\lambda_2=b_1/(a_{11}+a_{12})$ in equation (7.7) shows that these equalities can be true if and only if $(a_{12}+a_{22})b_1=(a_{12}+a_{11})b_2$. If the latter equality does not apply, then $b_1-(a_{11}+a_{12})\lambda_1\neq 0$ and $b_1-(a_{11}+a_{12})\lambda_2\neq 0$. At that $d_1^{(2)}\neq 0$, $d_2^{(2)}\neq 0$, and mechanical system (7.13), together with the original one (7.6), is completely controllable in Kalman's sense [88, 89–92] when a control torque is present only in inter-link joint D.

Controllable linear system (7.6) has two positive eigenvalues (7.14): $\mu_1>\mu_2>0$; and to speak of control resource, it is limited (see inequality (7.4)). Thus the set of initial states from which the system (7.6) can be translated to a desired equilibrium (7.5) by applying a control function $L(t)\in W$ is limited [53]. This set will be denoted as $Q^{(i)}$, and it is usually called domain of controllability. To build domain $Q^{(i)}$ for system (7.6) or, equivalently, for system (7.13), the fourth-order system (7.13) is represented in Jordan form. Then two differential equations that correspond to the positive eigenvalues (7.14) can be separated

$$\begin{aligned}\dot y_1&=\mu_1y_1+\frac{d_1^{(i)}}{\mu_1}L\,, \quad \text{where}\quad y_1=x_1+\frac{1}{\mu_1}\dot x_1\,,\\ \dot y_2&=\mu_2y_2+\frac{d_2^{(i)}}{\mu_2}L\,, \quad \text{where}\quad y_2=x_2+\frac{1}{\mu_2}\dot x_2\,.\end{aligned}\tag{7.15}$$

The first equation in (7.15) is derived from the first equation in (7.13), the second one – from the second equation in system (7.13).

As Jordan variables, in place of y_1, y_2 any variables may be taken, that result from multiplying each of y_1, y_2, by an arbitrary positive or negative number. In other words, the eigenvalues do not depend on the way a system is transposed to Jordan form, but this cannot be said about the Jordan variables themselves.

Note that with control function $L(t)\equiv L_0$ system (7.15) has a stationary point (that is unstable)

$$y_1=-\frac{d_1^{(i)}}{\mu_1^2}L_0,\quad y_2=-\frac{d_2^{(i)}}{\mu_2^2}L_0\,,\tag{7.16}$$

and with control function $L(t)\equiv -L_0$ – stationary point (also unstable)

$$y_1=\frac{d_1^{(i)}}{\mu_1^2}L_0,\quad y_2=\frac{d_2^{(i)}}{\mu_2^2}L_0\,.\tag{7.17}$$

Domain $Q^{(i)}$ is limited in four-dimensional space of variables y_1, y_2, y_3, y_4 with respect to "unstable" variables y_1 and y_2. It is not limited with respect to the other two Jordan variables y_3, y_4 that correspond to the negative eigenvalues $-\mu_1, -\mu_2$, i.e. it is a cylindric area [53]. Slice $S^{(i)}$ of this area by planes $y_3 = 0$, and $y_4 = 0$ is an open set that is symmetric with respect to the origin

$$y_1 = 0, \quad y_2 = 0\,. \tag{7.18}$$

Set $S^{(i)}$ is the domain of controllability of system (7.15), i.e. it is the set of initial states $y_1(0)$, $y_2(0)$ from which system (7.15) can be translated to the origin (7.18) by means of a control function $L(t) \in W$. The boundary of this set is formed by two integral trajectories of system (7.15) that are symmetric with respect to the origin. These trajectories result from assigning $L(t) \equiv L_0$ and $L(t) \equiv -L_0$ [27, 53, 158]. One of these trajectories that result from putting $L(t) \equiv L_0$ begins at $t = 0$ at point (7.17), and as $t \to -\infty$ it converges to point (7.16). Its equation is as follows:

$$y_1(t) = \frac{d_1^{(i)} L_0}{\mu_1^2}(2e^{\mu_1 t} - 1), \quad y_2(t) = \frac{d_2^{(i)} L_0}{\mu_2^2}(2e^{\mu_2 t} - 1) \quad (-\infty < t \le 0)\,. \tag{7.19}$$

To put it in other words, trajectory (7.19) begins at $t = -\infty$ at point (7.16) and ends at $t = 0$ in point (7.17).

The other trajectory built for $L(t) \equiv -L_0$ begins as $t = 0$ at point (7.16), and at $t \to -\infty$ it ends in point (7.17). Its equation looks like

$$y_1(t) = -\frac{d_1^{(i)} L_0}{\mu_1^2}(2e^{\mu_1 t} - 1), \quad y_2(t) = -\frac{d_2^{(i)} L_0}{\mu_2^2}(2e^{\mu_2 t} - 1) \quad (-\infty < t \le 0)\,. \tag{7.20}$$

In other words, trajectory (7.20) begins at $t = -\infty$ at point (7.17) and ends at $t = 0$ in point (7.16).

At points (7.16) and (7.17) the bounding trajectories (7.19), (7.20) coincide, and they (the points) are the corner points of domain $S^{(i)}$. These points are also the corners of rectangle

$$|y_1| < \left|d_1^{(i)}\right| \frac{L_0}{\mu_1^2}, \quad |y_2| < \left|d_2^{(i)}\right| \frac{L_0}{\mu_2^2}\,,$$

that encloses domain of controllability $S^{(i)}$. The other two vertices of this rectangle are symmetric to points (7.16), (7.17) with respect to abscissa (or ordinate axis). The question of existence of corner points in domains of attainability, domains of controllability is regarded in study [56]. The presence of such points is related to the fact that Bellman's function may be not smooth (in our case, the Bellman's function shows the relationship between the minimal time of bringing the system to the origin and its initial state in the phase space).

Consider a double pendulum with identical homogeneous links. The task is to build sets $S^{(1)}$ and $S^{(2)}$ for such pendulum. Let

$$m_1 = m_2 = m, \quad r_1 = r_2 = l/2, \quad I_1 = I_2 = \frac{1}{3} m l^2\,. \tag{7.21}$$

It will be assumed that the maximum possible value of L_0 is the same for both torques – in pivot point O as well as in inter-link joint D.

The characteristic equation (7.7) for a double pendulum with identical homogeneous links has roots

$$\lambda_1, \lambda_2 = 3\left(1 \pm \frac{2}{\sqrt{7}}\right)\frac{g}{l}\,. \tag{7.22}$$

Column matrix $d^{(i)}$ can be represented as

$$d^{(i)} = \begin{Vmatrix} d_1^{(i)} \\ d_2^{(i)} \end{Vmatrix} = -\frac{9}{14ml^2}\begin{Vmatrix} 2+\sqrt{7} - \dfrac{16+5\sqrt{7}}{3} \\ 2-\sqrt{7} - \dfrac{16-5\sqrt{7}}{3} \end{Vmatrix} c^{(i)}\,. \tag{7.23}$$

Considering expressions for matrices $c^{(i)}$ with $i = 1$ and $i = 2$ (see equations (7.3)), column vector $d^{(1)}$ that corresponds to the case when the control torque is applied in pivot point O is

$$d^{(1)} = \begin{Vmatrix} d_1^{(1)} \\ d_2^{(1)} \end{Vmatrix} = -\frac{9}{14ml^2}\begin{Vmatrix} 2+\sqrt{7} \\ 2-\sqrt{7} \end{Vmatrix}\,, \tag{7.24}$$

and column vector $d^{(2)}$ that corresponds to control torque in inter-link joint D is

$$d^{(2)} = \begin{Vmatrix} d_1^{(2)} \\ d_2^{(2)} \end{Vmatrix} = \frac{3}{7ml^2}\begin{Vmatrix} 11+4\sqrt{7} \\ 11-4\sqrt{7} \end{Vmatrix}\,. \tag{7.25}$$

Note that relations $d_1^{(2)}/d_1^{(1)}$ and $d_2^{(2)}/d_2^{(1)}$ do not depend on pendulum parameters, namely, the mass and length of the pendulum links.

Figure 7.2 illustrates domains $S^{(1)}$ and $S^{(2)}$ built for the two-link pendulum with identical homogeneous links. The pendulum parameters that satisfy conditions (7.21) are

$$m = 0.2\,\text{kg}, \quad l = 0.15\,\text{m}, \quad L_0 = 0.15\,\text{N}\cdot\text{m}\,. \tag{7.26}$$

Domain $S^{(2)}$, as it can be seen in Figure 7.2, is more spread along the abscissa than domain $S^{(1)}$. In turn, domain $S^{(1)}$ is more spread along the ordinate axis. The two domains have an intersection. But none of domains $S^{(1)}$, $S^{(2)}$ contains another one. So it cannot be said that to apply torque in one of the joints is more efficient than in the other one. There exist such initial deviations from desired equilibrium (7.5), that can be countered by a torque applied at pivot point O, and not by a torque applied at inter-link joint D. And vice versa, there are such deviations that can be countered by a torque in the inter-link joint D and not by a torque applied at pivot point O.

Inspecting expressions (7.24), (7.25), and also (7.22) for the roots of the characteristic equation shows that values $d_1^{(i)}L_0/\mu_1^2$ and $d_2^{(i)}L_0/\mu_2^2$ are proportional to a nondimensional constant combination $L_0/(mgl)$, thus its variation expands or shrinks domains $S^{(1)}$, $S^{(2)}$, keeping them homothetic to themselves. That means, with any value of $L_0/(mgl)$ these domains look the same as in Figure 7.2.

Note. *If a control torque (limited in absolute value) is applied at each of the joints, then the domain of controllability becomes larger, though remains limited. In this case, domain of controllability S (of the second-order system with control torques applied in both*

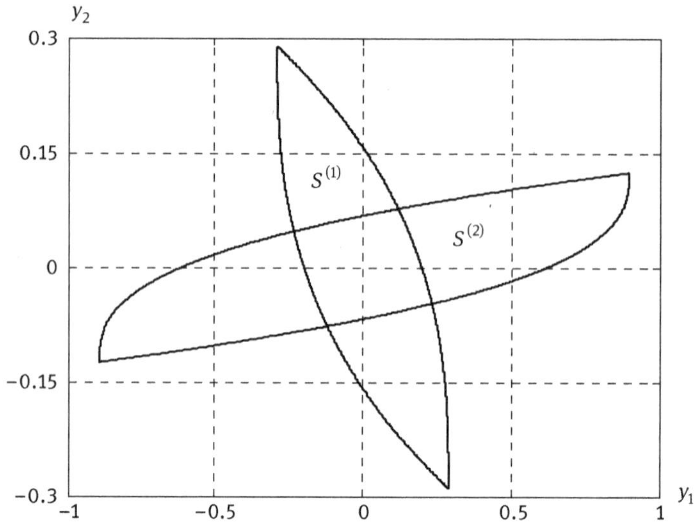

Fig. 7.2. Domains of controllability $S^{(1)}$ and $S^{(2)}$.

joints) is equal to geometric sum of domains $S^{(1)}$ and $S^{(2)}$, i.e. $S = S^{(1)} \oplus S^{(2)}$. In other words, each vector $(y_1, y_2) \in S$ is a sum of two vectors belonging to domains $S^{(1)}$ and $S^{(2)}$. At that, $Q^{(1)} \oplus Q^{(2)} = Q$, where Q is the controllability domain of the fourth-order linear system with control torques applied in both joints. This statement is valid by virtue of Cauchy's formula for solution to a system of linear differential equations where a sum of two Duhamel's integrals is involved, one for each control parameter.

Knowing domains $S^{(1)}$ and $S^{(2)}$, one can evaluate domains of attraction from above, for any kind of feedback control that is constrained by inequality (7.4). The point is that the domain of *attraction* of a system with constrained feedback is contained in *controllability* domain of such constrained system.

7.4 Feedback design. Maximizing domain of attraction

Now consider a feedback design that makes the desired equilibrium (7.5) asymptotically stable. This equilibrium corresponds to the top, unstable position of the double pendulum. The feedback will be designed in such a way that its domain of attraction will be largest possible. A domain of attraction, like previously, is a set of initial states starting from which the close-loop system asymptotically approaches the equilibrium.

Consider a linearized system (7.15). The control (stabilization) law will be designed as a linear feedback that involves the unstable variables y_1 and y_2

$$L = g_1 y_1 + g_2 y_2 \,. \tag{7.27}$$

Applying linear feedback (7.27) will apparently "suppress" the instability of variables y_1 and y_2. Since the control torque L is limited in absolute value (see inequality (7.4)), a *saturated* linear feedback should be taken instead of (7.27):

$$L = \begin{cases} -L_0 & \text{when} \quad g_1y_1 + g_2y_2 \le -L_0 \\ g_1y_1 + g_2y_2 & \text{when} \quad |g_1y_1 + g_2y_2| \le L_0 \\ L_0 & \text{when} \quad g_1y_1 + g_2y_2 \ge L_0 \end{cases} . \tag{7.28}$$

Note that variables y_1, y_2 are expressed in terms of variables $x_1, x_2, \dot{x}_1, \dot{x}_2$, like in relations (7.15). These variables, in turn, are expressed by means of relations (7.9), (7.11) in terms of the original variables $\varphi_1, \varphi_2, \dot{\varphi}_1, \dot{\varphi}_2$. Thus expression (7.28) describes a feedback that actually involves all phase coordinates of linear system (7.6) or the original nonlinear one (7.3).

The feedback coefficients g_1, g_2 will be taken so that the domain of attraction of system (7.15), (7.28) will be maximum possible.

Suppose that constant coefficients g_1, g_2 are chosen so that the solution to linear second-order system (7.15), (7.28) is asymptotically stable. This means, there exists a neighbourhood of the origin (7.18), such that when the initial state $y_1(0)$, $y_2(0)$ lies within this neighbourhood, the solution $y_1(t), y_2(t) \to 0$ as $t \to \infty$. Then, by virtue of relation (7.28) $L(t) \to 0$ as $t \to \infty$. Equations (7.15) with $L = 0$ (i.e. the open-loop system) correspond to positive eigenvalues μ_1, μ_2, the other two linear equations in Jordan coordinates with $L = 0$ correspond to negative eigenvalues $-\mu_1$, $-\mu_2$. Nonhomogeneous members in these equations are proportional to torque L, therefore, as $L(t) \to 0$, they asymptotically (as $t \to \infty$) converge to zero. So the solution of the two linear nonhomogeneous equations with corresponding negative eigenvalues also converges to zero under *any* initial conditions [45]. Consequently, solution (7.5) of the fourth-order system (7.6), (7.28), (and therefore of the original nonlinear system (7.3)), (7.28) is also asymptotically stable. Speaking of the attraction domain of linear system (7.6), (7.28), it is limited only with respect to unstable variables y_1, y_2, i.e. it is a cylindric area in the four-dimensional state space. Note that the eigenvalues of system (7.6) that are positive when $L = 0$ shift to the left semiplane when the linear feedback (7.27) is applied. The negative eigenvalues of system (7.6) with control law (7.27) remain unchanged.

In studies [156, 157] it is demonstrated that coefficients g_1, g_2 in control law (7.28) may be chosen in such a way that domain of attraction $B^{(i)}$ of system (7.15), (7.28) is arbitrarily close to domain of controllability $S^{(i)}$. A method of choosing such coefficients is provided below. It is slightly different from the way proposed in [156, 157].

A straight line that passes through points (7.16), (7.17) that are symmetric with respect to the origin is described by equation

$$k_1y_1 + k_2y_2 = 0 . \tag{7.29}$$

Coefficients k_1, k_2 cannot be determined uniquely, only their ratio can. These coefficients may be taken, for example, as

$$k_1 = d_2^{(i)}/\mu_2^2, \quad k_2 = -d_1^{(i)}/\mu_1^2 . \tag{7.30}$$

Then the equation of line (7.29) transposes to

$$\frac{d_2^{(i)}}{\mu_2^2} y_1 - \frac{d_1^{(i)}}{\mu_1^2} y_2 = 0 . \tag{7.31}$$

In feedback law (7.28) coefficients g_1, g_2, may be chosen proportional to values (7.30)

$$g_1 = \frac{\gamma d_2^{(i)}}{\mu_2^2}, \quad g_2 = -\frac{\gamma d_1^{(i)}}{\mu_1^2} , \tag{7.32}$$

where γ is an “overall” feedback gain that needs to be determined. The linear feedback law (7.27) with coefficients (7.32) then gets the following form:

$$L = \frac{\gamma d_2^{(i)}}{\mu_2^2} y_1 - \frac{\gamma d_1^{(i)}}{\mu_1^2} y_2 \quad , \tag{7.33}$$

and the linear saturated feelback law (7.28) is as follows:

$$L = \begin{cases} -L_0 & \text{when} \quad \gamma \left[\dfrac{d_2^{(i)}}{\mu_2^2} y_1 - \dfrac{d_1^{(i)}}{\mu_1^2} y_2 \right] \le -L_0 \\ \gamma \left[\dfrac{d_2^{(i)}}{\mu_2^2} y_1 - \dfrac{d_1^{(i)}}{\mu_1^2} y_2 \right] & \text{when} \quad \left| \gamma \left[\dfrac{d_2^{(i)}}{\mu_2^2} y_1 - \dfrac{d_1^{(i)}}{\mu_1^2} y_2 \right] \right| \le L_0 \\ L_0 & \text{when} \quad \gamma \left[\dfrac{d_2^{(i)}}{\mu_2^2} y_1 - \dfrac{d_1^{(i)}}{\mu_1^2} y_2 \right] \ge L_0 . \end{cases} \tag{7.34}$$

The characteristic equation that corresponds to the second-order system (7.15) with linear saturated feedback (7.33) is (λ stands for the spectral parameter):

$$\lambda^2 + \lambda \left[\frac{\gamma d_1^{(i)} d_2^{(i)}}{\mu_1 \mu_2} \left(\frac{1}{\mu_1} - \frac{1}{\mu_2} \right) - (\mu_1 + \mu_2) \right] + \mu_1 \mu_2 = 0 . \tag{7.35}$$

For both roots of equation (7.35) to be in the left semiplane of complex numbers, it is necessary and sufficient that the second coefficient is positive, i.e.

$$\gamma d_1^{(i)} d_2^{(i)} < -\frac{\mu_1^2 \mu_2^2 (\mu_1 + \mu_2)}{\mu_1 - \mu_2} . \tag{7.36}$$

Remember that $\mu_1 > \mu_2 > 0$. Condition (7.36) provides asymptotic stability of the trivial solution (7.18) to system (7.15) looped by linear feedback (7.33), as well as to the same system looped by nonlinear feedback (7.34). The latter control law involves limitation (7.4).

The right-hand side of inequality (7.36) is negative because $\mu_1 > \mu_2 > 0$. So for asymptotic stability it is necessary that

$$\gamma d_1^{(i)} d_2^{(i)} < 0 \quad \left(\operatorname{sign}\gamma = -\operatorname{sign}\left[d_1^{(i)} d_2^{(i)}\right]\right) . \tag{7.37}$$

If

$$|\gamma| > \frac{\mu_1^2\mu_2^2(\mu_1+\mu_2)}{\left|d_1^{(i)} d_2^{(i)}\right|(\mu_1-\mu_2)} , \tag{7.38}$$

then inequality (7.37) becomes a necessary and sufficient condition for asymptotic stability. In other words, condition (7.37) is necessary and sufficient for asymptotic stability of the trivial solution (7.18) when the absolute value of feedback gain value γ is large enough. Note that when this absolute value of gain γ is large enough, the roots of equation (7.35) are real.

Assume that criterion of asymptotic stability (7.36) is satisfied, i.e. there exists such a neighbourhood of the origin (7.18) that the solution to system (7.15), (7.34) that starts from initial state $y_1(0)$, $y_2(0)$ belonging to this neighbourhood will asymptotically converge to the origin (7.18). This neighbourhood itself should be contained in domain of controllability $S^{(i)}$. Otherwise, system (7.15) could be translated to the origin by means of a limited in absolute value control input from point $y_1, y_2 \notin S^{(i)}$. This would be in a contradiction with the definition of domain of controllability $S^{(i)}$.

When the time is reversed, $t \to -t$, system (7.15) transposes to

$$\dot{y}_1 = -\mu_1 y_1 - \frac{d_1^{(i)}}{\mu_1} L, \quad \dot{y}_2 = -\mu_2 y_2 - \frac{d_2^{(i)}}{\mu_2} L . \tag{7.39}$$

With condition (7.36) in effect, origin (7.18) is an unstable equilibrium of system (7.39), (7.34). Let initial state $y_1(0)$, $y_2(0)$ of system (7.39), (7.34) be close enough to the origin, within the attraction domain of system (7.15), (7.34). Remember that this domain of attraction is assumed to exist and be contained in domain of controllability $S^{(i)}$. The corresponding solution to system (7.39), (7.34) at $0 \le t < \infty$ will remain inside domain of controllability $S^{(i)}$, that itself is limited in plane of variables y_1, y_2. Control law (7.34) with a finite value of γ is a linear function of coordinates y_1, y_2 in stripe

$$\left|\frac{\gamma d_2^{(i)}}{\mu_2^2} y_1 - \frac{\gamma d_1^{(i)}}{\mu_1^2} y_2\right| \le L_0 . \tag{7.40}$$

The control torque takes its terminal values $L = \pm L_0$ outside stripe (7.40). The boundary (corner) points (7.16) and (7.17) of controllability domain $S^{(i)}$ that correspond to $L = \pm L_0$ lie inside stripe (7.40), and thus they are not the equilibrium points for system (7.15), (7.34), nor for system (7.39), (7.34). Consequently, inside domain of controllability $S^{(i)}$ system (7.15), (7.34), as well as system (7.39), (7.34), has no steady states except the origin $y_1 = y_2 = 0$. Since no other equilibriums are present, the solution to system (7.39), (7.34) that begins in a small neighbourhood of the origin converges to a periodic solution as $t \to \infty$ [121]. Such periodic solution is a stable limit cycle of system (7.39), (7.34). A the same time, it is an unstable solution to system (7.15), (7.34), and thus it is the boundary of attraction domain $B^{(i)}$ of the origin (7.18). Domain of attraction $B^{(i)}$ is certainly contained in domain of controllability $S^{(i)}$: $B^{(i)} \in S^{(i)}$.

The next step is to show that as $|\gamma| \to \infty$, with condition (7.37) taking effect, domain of attraction $B^{(i)}$ approaches domain of controllability from within. In other words, the boundary of attraction domain $B^{(i)}$ (the limit cycle mentioned above) converges to the boundary of domain $S^{(i)}$ as $|\gamma| \to \infty$.

As the value of $|\gamma|$ increases, stripe (7.40) becomes narrower, and as $|\gamma| \to \infty$, it "shrinks" to a straight line (7.31). In turn, control law (7.34) becomes a relay law as $|\gamma| \to \infty$:

$$L = -L_0 \operatorname{sign}\left[d_1^{(i)} d_2^{(i)} \left(\frac{d_2^{(i)}}{\mu_2^2} y_1 - \frac{d_1^{(i)}}{\mu_1^2} y_2 \right) \right] . \tag{7.41}$$

The interval of line (7.31) that lies between points (7.16) and (7.17) is described by relations

$$\frac{d_2^{(i)}}{\mu_2^2} y_1 - \frac{d_1^{(i)}}{\mu_1^2} y_2 = 0, \qquad -\frac{\left|d_1^{(i)}\right|}{\mu_1^2} L_0 < y_1 < \frac{\left|d_1^{(i)}\right|}{\mu_1^2} L_0 . \tag{7.42}$$

It will be shown that interval (7.42) is an attractor of system (7.15) with relay control law (7.41) [8, 51, 52, 142, 159]. Equivalently, the phase velocities of system (7.15), (7.41) from both sides of interval (7.42) are directed towards this interval. In particular, at points of interval (7.42), where

$$\frac{d_2^{(i)}}{\mu_2^2} y_1 - \frac{d_1^{(i)}}{\mu_1^2} y_2 = +0 , \tag{7.43}$$

the time derivative of expression $y_1 d_2^{(i)}/\mu_2^2 - y_2 d_1^{(i)}/\mu_1^2$ with respect to system (7.15), (7.41) is negative

$$\frac{d}{dt}\left[\frac{d_2^{(i)}}{\mu_2^2} y_1 - \frac{d_1^{(i)}}{\mu_1^2} y_2 \right] < 0 , \tag{7.44}$$

and at points of interval (7.42), where

$$\frac{d_2^{(i)}}{\mu_2^2} y_1 - \frac{d_1^{(i)}}{\mu_1^2} y_2 = -0 , \tag{7.45}$$

this derivative is positive.

$$\frac{d}{dt}\left[\frac{d_2^{(i)}}{\mu_2^2} y_1 - \frac{d_1^{(i)}}{\mu_1^2} y_2 \right] > 0 . \tag{7.46}$$

These assertions can be proved as follows. Consider equations (7.15), (7.41) and the first of relations (7.42). The said derivative can be expressed as (the derivation process is omitted)

$$\frac{d_2^{(i)}}{\mu_2^2} \dot{y}_1 - \frac{d_1^{(i)}}{\mu_1^2} \dot{y}_2 = \frac{d_2^{(i)}}{\mu_2^2} y_1 (\mu_1 - \mu_2) - \frac{\left|d_1^{(i)} d_2^{(i)}\right|}{\mu_1^2 \mu_2^2} (\mu_1 - \mu_2) L_0 \operatorname{sign}\left(\frac{d_2^{(i)}}{\mu_2^2} y_1 - \frac{d_1^{(i)}}{\mu_1^2} y_2 \right) . \tag{7.47}$$

On interval (7.42) the first term in the right-hand side of expressions (7.47) satisfies inequalities

$$-\frac{\left|d_1^{(i)}d_2^{(i)}\right|}{\mu_1^2\mu_2^2}L_0(\mu_1-\mu_2) < \frac{d_2^{(i)}}{\mu_2^2}y_1(\mu_1-\mu_2) < \frac{\left|d_1^{(i)}d_2^{(i)}\right|}{\mu_1^2\mu_2^2}L_0(\mu_1-\mu_2)\,. \tag{7.48}$$

From equality (7.47) and inequalities (7.48) it follows that on interval (7.42) the following relation is valid

$$\operatorname{sign}\left[\frac{d_2^{(i)}}{\mu_2^2}\dot{y}_1 - \frac{d_1^{(i)}}{\mu_1^2}\dot{y}_2\right] = -\operatorname{sign}\left[\frac{d_2^{(i)}}{\mu_2^2}y_1 - \frac{d_1^{(i)}}{\mu_1^2}y_2\right]. \tag{7.49}$$

Relation (7.49) shows that under condition (7.43) (to one side of interval (7.42)) inequality (7.44) takes place, and under (7.45) (to the other side of that interval) – inequality (7.46). Thus all trajectories that begin at points in some neighbourhood of interval (7.42) reach that interval in some finite time.

Interval (7.42), being an attractor for system (7.15), (7.41), is a "deflector" for system (7.39), (7.41) – a system that describes reverse motion. Therefore, a solution to system (7.39), (7.41) that begins outside interval (7.42) will never reach that interval.

Any trajectory of system (7.39), (7.41) that begins at a point within domain of controllability $S^{(i)}$ does not leave that domain, and as $t \to \infty$ it converges to one of the states (7.16) or (7.17), remaining at all times to the same side of interval (7.42) where the initial point of that trajectory is. For example, if initial state $y_1(0)$, $y_2(0)$ of the system lies in the area where

$$d_1^{(i)}d_2^{(i)}\left(\frac{d_2^{(i)}}{\mu_2^2}y_1 - \frac{d_1^{(i)}}{\mu_1^2}y_2\right) < 0\,,$$

then entire trajectory will be located in this area. At the same time, $L = L_0$ and the corresponding solution asymptotically converges to point (7.16). If additionally initial state $y_1(0)$, $y_2(0)$ is "close" to point (7.17), then the trajectory of the system will be "close" to trajectory (7.19), which is on one of the boundaries of controllability domain $S^{(i)}$. The value of $|\gamma|$ may be chosen so large that this initial state $y_1(0)$, $y_2(0)$ would be outside stripe (7.40). Thus with the same initial state and large enough value of $|\gamma|$ the initial part of trajectory of system (7.39), (7.34) that lies outside stripe (7.40) will be close to trajectory (7.19). When the value of $|\gamma|$ is "large", stripe (7.40) is "narrow". The part of trajectory that is located in that stripe is also small. Consequently, when the value of $|\gamma|$ is large enough, domain of attraction $B^{(i)}$ is close to domain of controllability $S^{(i)}$.

So to conclude, in order to maximize the domain of attraction of system (7.6), *all* control resources must be used to suppress the *unstable* modes of motion.

It is worth investigating the trajectories of system (7.15) with relay control law (7.41) after these trajectories reach interval (7.42). Once the system gets onto that interval, it cannot leave it. It remains on that interval and moves in it in a sliding mode [8, 51, 52, 142, 159], or, alternatively, it stays in the point where it hit that interval. When the system is on interval (7.42), the following identity is valid:

$$\frac{d_2^{(i)}}{\mu_2^2}y_1 - \frac{d_1^{(i)}}{\mu_1^2}y_2 \equiv 0 \, .$$

From that identity and equations (7.15) it follows:

$$\frac{d_2^{(i)}}{\mu_2^2}\dot{y}_1 - \frac{d_1^{(i)}}{\mu_1^2}\dot{y}_2 = \frac{d_2^{(i)}}{\mu_2^2}y_1(\mu_1 - \mu_2) + \frac{d_1^{(i)}d_2^{(i)}}{\mu_1^2\mu_2^2}(\mu_1 - \mu_2)L \equiv 0 \, . \tag{7.50}$$

Solving equation (7.50) for torque L yields the equivalent continuous control law [8, 51, 52, 142, 159], that can be substituted for relay law (7.41) on interval (7.42)

$$L = -\frac{\mu_1^2}{d_1^{(i)}}y_1 \, . \tag{7.51}$$

Making substitution (7.51) in the first equation in (7.15) yields $\dot{y}_1 = 0$. This means, when system (7.15), (7.41) gets onto interval (7.42), all motion stops, and the system "gets stuck" in the point where it arrived. This "phenomenon" is related to the fact that as $|\gamma| \to \infty$, one of the roots of characteristic equation (7.35), while remaining to the left of the imaginary axis, converges to zero (it is easy to show that the other one converges to $-\infty$). Therefore, equilibrium (7.18) of system (7.15), (7.34) can be asymptotically stable only when the value of the "overall" gain γ is finite. Larger values of $|\gamma|$, on the one hand, increase the domain of attraction, but, on the other hand, one of the eigenvalues gets closer to zero, reducing the robustness of the system. Besides, when one of the eigenvalues is close to zero (while still remaining negative), the transients in the system take longer time to settle. The system then moves to the desired state – the origin – "slowly".

7.5 Numerical experiments

In Figure 7.3, domain of controllability $S^{(1)}$ is shown for a double pendulum with control torque applied at the pivot point – the same that was illustrated in Figure 7.2; its boundary is represented by a dashed line. The pendulum links are considered similar, they are homogeneous, and their parameters satisfy conditions (7.21), (7.26). Inside domain $S^{(1)}$ domain of attraction $B^{(1)}$ is located. This domain of attraction corresponds to control law (7.34) with gain value $\gamma = -2$. The boundary of domain $B^{(1)}$ (the limit cycle), represented by a solid line, has been determined numerically by solving equations (7.39), that is, by reverse-solving original equations (7.15). To find the limit cycle, the initial state can be taken both inside domain of controllability $S^{(1)}$ and outside that domain.

It follows from inspection of Figure 7.3 that the boundary of domain $B^{(1)}$ is close to the boundary of domain $S^{(1)}$. As the value of $|\gamma|$ grows, the boundary of domain

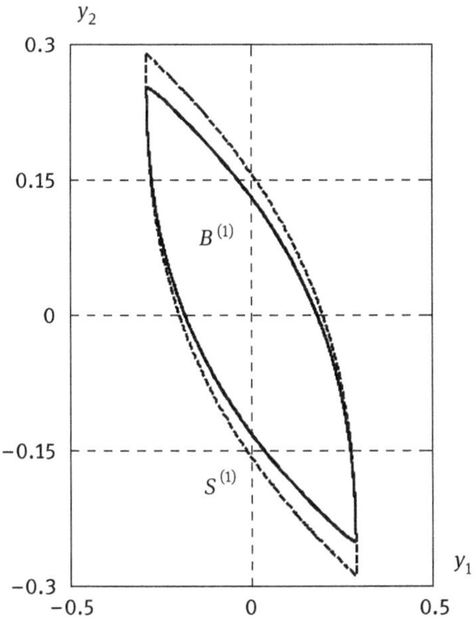

Fig. 7.3. Domain of controllability $S^{(1)}$ (its boundary is shown in a dashed line) and domain of attraction $B^{(1)}$.

$B^{(1)}$ infinitely approaches the boundary of domain $S^{(1)}$. In practice, when $\gamma = -20$, the boundaries of the two domains cannot be distinguished visually. The boundary of domain of controllability $S^{(1)}$, as mentioned above, has corner points (7.16) and (7.17). The boundary of attraction domain $B^{(1)}$ does not have corner points, though visually it looks like that such points are present in Figure 7.3. The boundary points of domain $B^{(1)}$ that seem to be corners lie on straight lines that enclose stripe (7.40). In these points the angle between the tangent line to the limit cycle changes abruptly, but not in a step-like manner. Hence in the seeming corner points the curvature of the trajectory is large but not infinite.

Figure 7.4 illustrates domain of controllability $S^{(2)}$ of a double pendulum with control torque applied in the inter-link joint, the same as in Figure 7.2. The boundary of this domain is shown in a dashed line. Numerical values of system parameters are provided in relations (7.21), (7.26). Inside domain $S^{(2)}$ attraction domain $B^{(2)}$ is located, determined for control law (7.34) with gain value $\gamma = 2$. The boundary of domain $B^{(2)}$ (the limit cycle) represented by a solid line has been found numerically by solving equations (7.39).

It follows from inspection of Figure 7.4 that the boundary of domain $B^{(2)}$ is close to the boundary of domain $S^{(2)}$.

Linearized system (7.6) and nonlinear one (7.3) were investigated numerically. The values of pendulum parameters were taken as in (7.21), (7.26). The control law was taken as in (7.34). In particular, the initial conditions were taken so that

$$\varphi_1(0) = \varphi_2(0), \quad \dot{\varphi}_1(0) = \dot{\varphi}_2(0) = 0\,. \tag{7.52}$$

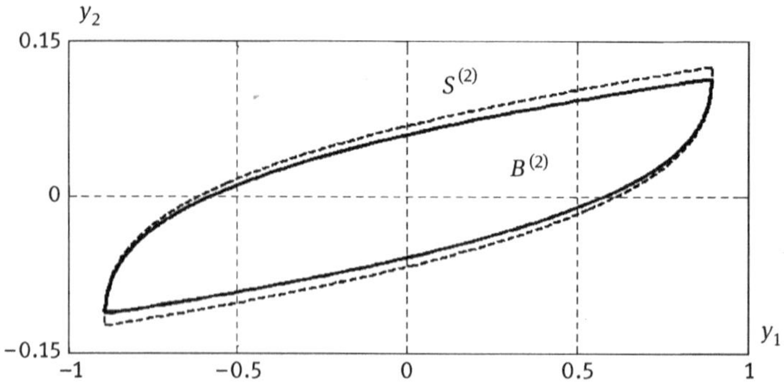

Fig. 7.4. Domain of controllability $S^{(2)}$ (its boundary is shown in a dashed line) and domain of attraction $B^{(2)}$.

These conditions correspond to a "stretched" pendulum, where both links are positioned along a common straight line. The initial velocities of the links are equal to zero.

First, the results obtained for the pendulum with torque (7.34) (feedback gain $\gamma = -2$), applied at the pivot point. Investigation of the linear model (7.6) shows that the pendulum can be driven (asymptotically) to its unstable equilibrium (7.5) when its deviation from the vertical that satisfies condition (7.52), is within range

$$|\varphi_1(0), \varphi_2(0)| \leq 0.116 \quad (6.70°) .$$

The initial nonlinear system (7.3) has a "slightly" smaller range of admissible deviations

$$|\varphi_1(0), \varphi_2(0)| \leq 0.1155 \quad (6.62°) . \tag{7.53}$$

Certainly, the numbers provided here are approximate.

Figure 7.5 illustrates the transients in full nonlinear system (7.3) with control law (7.34) ($\gamma = -2$) when $\varphi_1(0) = \varphi_2(0) = 0.1155$ (6.62°), $\dot{\varphi}_1(0) = \dot{\varphi}_2(0) = 0$ (refer to relations (7.52), (7.53))

The abscissa represents time in seconds.

Now the results obtained for the pendulum with torque (7.34) (feedback gain $\gamma = 2$) applied in the inter-link joint will be provided. It appears for linear system (7.6) that the double pendulum can be translated to equilibrium (7.5), when its deviation from the vertical that satisfies condition (7.52) is within range

$$|\varphi_1(0), \varphi_2(0)| \leq 0.046 \quad (2.65°) .$$

The original nonlinear system (7.3) has a "significantly" smaller range

$$|\varphi_1(0), \varphi_2(0)| \leq 0.0255 \quad (1.46°) . \tag{7.54}$$

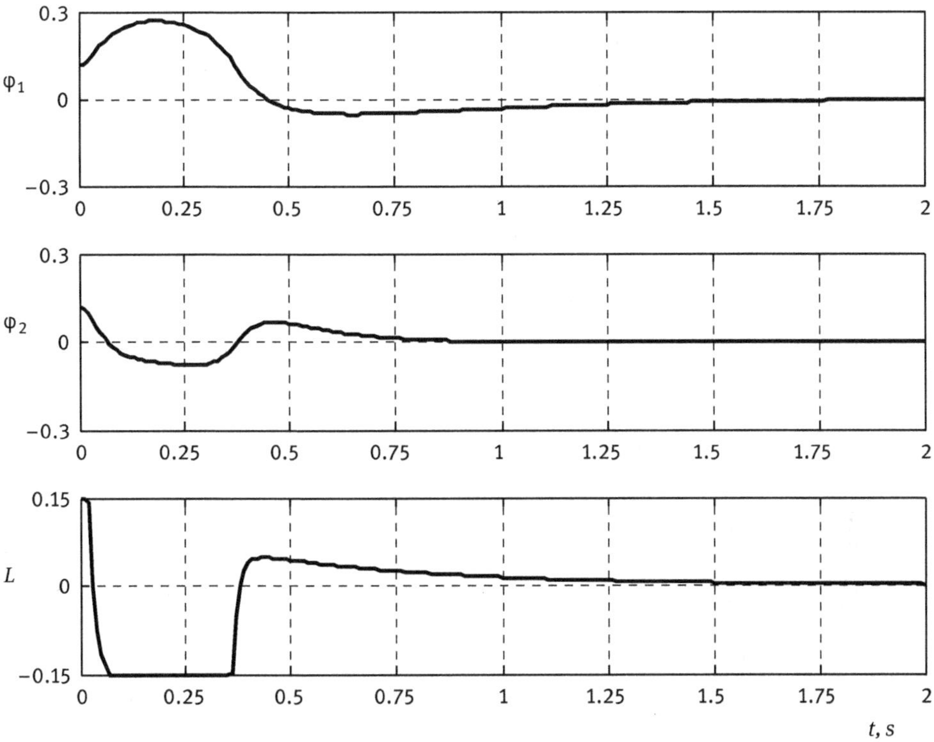

Fig. 7.5. Transients with control torque applied at the pivot point.

In Figure 7.6, the transients in nonlinear system (7.3) are plotted, with control law (7.34) and initial conditions $\varphi_1(0) = \varphi_2(0) = 0.0255\ (1.46°)$, $\dot{\varphi}_1(0) = \dot{\varphi}_2(0) = 0$ (refer to relations (7.52), (7.54))

Inspection of Figures 7.5 and 7.6 shows that in both cases control torque L only in the beginning takes its terminal values $L = \pm L_0 = \pm 0.15 N \cdot m$. Later its values are within allowable range (7.4). At that, the feedback is linear (7.33), and the system itself is inside stripe (7.40). Variables φ_1 and φ_2 in the beginning of the transients do a couple of oscillations, but later they monotonically converge to equilibrium (7.5) without any oscillations. Such behavior of these variables on time interval where the feedback is linear (7.33) can be explained as follows. When the deviation from equilibrium (7.5) is small, the nonlinear system is "close" to the linear one. Speaking of linear system (7.15), (7.33) its characteristic equation (7.35) with large values of $|\gamma|$ has real roots. The other two eigenvalues of fourth-order linear system (7.6) do not depend on control function, and they are also real: $-\mu_1$, $-\mu_2$. And with all eigenvalues real, the rest of the transient has no oscillations.

Consider an initial state that satisfies equalities

$$\varphi_1(0) = \varphi_2(0) = 0, \quad \dot{\varphi}_1(0) = \dot{\varphi}_2(0)\,. \tag{7.55}$$

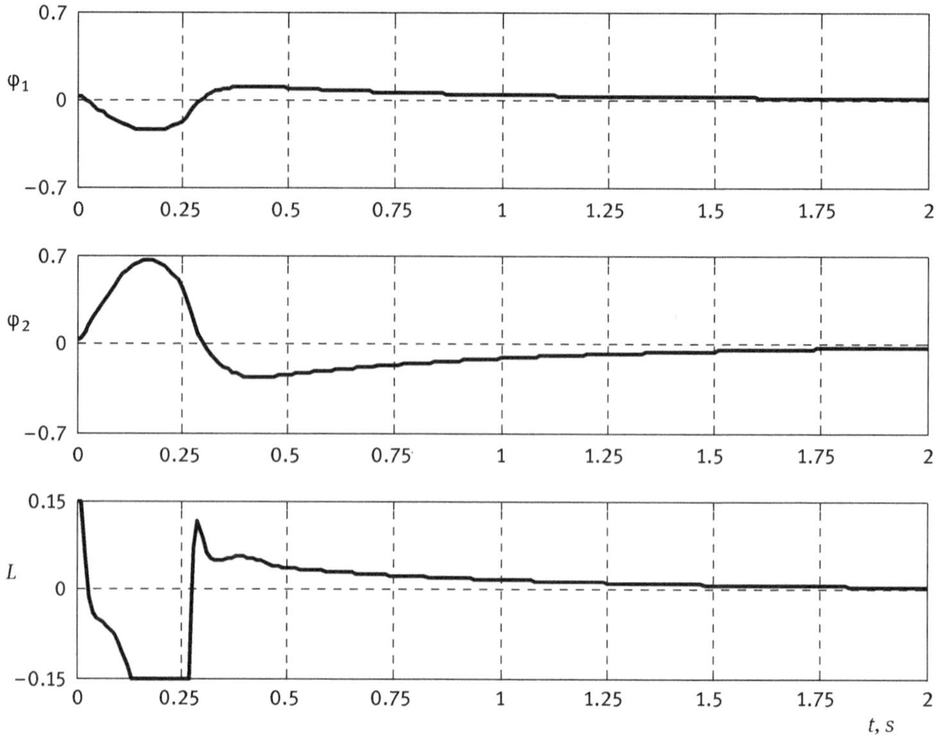

Fig. 7.6. Transients with control torque in the inter-link joint.

First, consider the numbers obtained for the pendulum with torque (7.34) ($\gamma = -2$) applies in the pivot. Investigation of linear model (7.6) shows that in order to be able to translate the pendulum to equilibrium (7.5), its angular velocities that satisfy condition (7.55) must be kept within range

$$|\dot{\varphi}_1(0), \dot{\varphi}_2(0)| \leq 0.754 \frac{1}{\mathrm{s}} \quad \left(43.20 \frac{\circ}{\mathrm{s}}\right).$$

The initial nonlinear system (7.3) has a smaller range of allowable velocities, yet not much smaller

$$|\dot{\varphi}_1(0), \dot{\varphi}_2(0)| \leq 0.738 \frac{1}{\mathrm{s}} \quad \left(42.28 \frac{\circ}{\mathrm{s}}\right).$$

The results for the pendulum with torque (7.34) ($\gamma = 2$) applied in the inter-link joint are as follows. Investigation of linear model (7.6) with such torque shows that the pendulum can be translated to equilibrium (7.5) when its angular velocities that satisfy condition (7.56) are within range

$$|\dot{\varphi}_1(0), \dot{\varphi}_2(0)| < 0.32 \frac{1}{\mathrm{s}} \quad \left(18.33 \frac{\circ}{\mathrm{s}}\right).$$

The full nonlinear system (7.3) has the range of allowable velocities that is almost twice as small

$$|\dot{\varphi}_1(0), \dot{\varphi}_2(0)| < 0.173\,\frac{1}{\mathrm{s}} \quad \left(9.91\,\frac{^\circ}{\mathrm{s}}\right).$$

Note that in all investigated cases, and also in some others, whenever the control torque is applied in the pivot point, the range of initial disturbances is wider than when the control torque is applied in the inter-link joint. This may create an impression that the torque applied in the pivot is more effective.

Note. *A double pendulum has two more unstable equilibriums, in addition to one with both links upside down. These states are not discussed here. In one of these states, the first link hangs down, and the second one is inverted. In the other state, the first link is inverted, and the second one hangs down. In both cases the motion equations that are linearized about the equilibrium have only one corresponding eigenvalue in the right complex semiplane (a positive eigenvalue), that is, the degree of instability of the system is equal to one. This significantly simplifies the investigation. To remind, a domain of controllability of a system with the degree of instability equal to one is an area in the phase space that is bounded by two parallel hyperplanes. When linear (with saturation) feedback involves the only unstable variable, the domain of attraction coincides with the entire domain of controllability, and thus it is maximal. The pendulum discussed in the current chapter has two inverted links. Its motion equations, being linearized about the equilibrium, have two corresponding eigenvalues in the right semiplane. It is said that the degree of instability of a double inverted pendulum is equal to two. The domain of controllability in this case is a cylindric area that is limited in two unstable Jordan variables. Unlike the system with only one unstable variable, in this case the domain of attraction of the system with a linear (saturated) feedback that involves the two unstable variables does not coincide with the domain of controllability. However, the domain of attraction may be set arbitrarily close to the domain of controllability. Thus the problem of stabilizing a double pendulum with both links inverted is more complicated. The systems discussed above, in chapters 1–3, namely, a single inverted pendulum with stationary pivot, a pendulum with a wheel-based pivot, a pendulum with a flywheel, all have one eigenvalue in the right semiplane.*

The method of stabilizing a double pendulum that maximizes the domain of attraction may be applied for other systems that have degree of instability equal to two. The third part utilizes this method for stabilizing a ball on a curvilinear rod.

8 Optimal control design for swinging and damping a double pendulum

Problems of controlling oscillating systems draw interest of many researchers. For example, such systems are investigated in studies [2, 3, 36, 37, 107]. Usually, the most complicated task is the design of a control law, especially an optimal one, for systems that have fewer control parameters than degrees of freedom. These systems are called "underactuated". Such systems are, in particular, swings, pendulum systems, walking machines that have drives installed only in the inter-link joints. A person can control swing motion about its pivot by moving his body in an appropriate way (see chapter 5). No control torque is applied in the pivot. A gymnast swings on a bar mostly by controlling the angle in his hip joint. The torque in his wrist joint is negligible. In the last two cases, the person makes proper use of the gravity force.

The current chapter discusses the problems of optimal swinging and damping of a double pendulum. The control parameter is the inter-link joint angle – the "elbow" angle. It is assumed that this angle has a limited range, and it may change arbitrarily fast. The task is to find a law of variating this angle, as a feedback, so that at each semiperiod of pendulum oscillations the magnitude of oscillations increases. The optimal control law can be designed without applying Pontryagin's principle of maximum. In study [75], two problems of optimal control of swinging a double physical pendulum are investigated. In one of them, the inter-link joint angle is also taken as a control parameter. The authors, however, do regard the derivative of the inter-link joint angle in motion equations. The necessary conditions of optimality are formulated. This makes possible to do numerical experiments. Designing a control law as a feedback (the synthesis problem) is not possible without excluding the angle derivative.

The considered problem is interesting from the point of view of theoretical mechanics. It also helps modeling motion modes of a gymnast on a bar. Such modes are investigated in a range of studies (see, for example, article [119] and its reference list).

8.1 Mathematical model

The system of interest – the double physical pendulum with stationary pivot point O – is illustrated in Figure 8.1. The pendulum is attached by means of a cylindric joint, and the friction in this joint is neglected. A similar ideal cylindric joint connects the pendulum links at point D. The links are assumed to be perfectly rigid bodies. The resistance of environment to the pendulum motion is not considered. The axes of both joints are perpendicular to the drawing plane. The center of mass of the first (upper) link is located in segment OD. The center of mass of the second link is located in the second segment shown in Figure 8.1 that begins at point D.

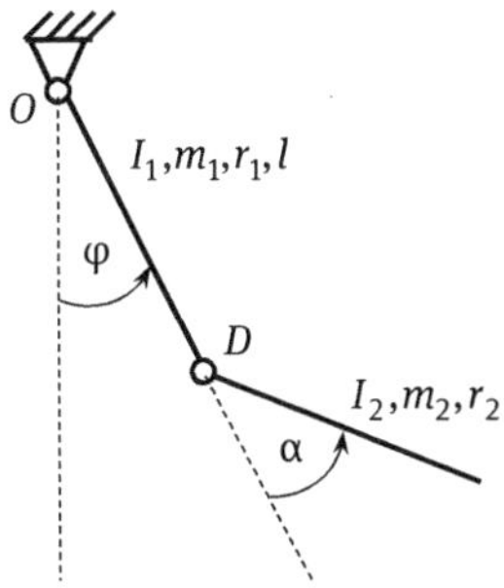

Fig. 8.1. Double pendulum.

Let φ be the angle that the first link (segment OD) deviates from the vertical, counted counter-clockwise, and α – the angle that the second link deviates from the line that continues segment OD (see Figure 8.1).

Kinetic energy T and potential energy Π of the pendulum, and also the virtual work of torque L that is applied in the inter-link joint D, are expressed as

$$T = \frac{1}{2}\left[a_{11}\dot{\varphi}^2 + 2a_{12}\dot{\varphi}\,(\dot{\varphi}+\dot{\alpha})\cos\alpha + a_{22}\,(\dot{\varphi}+\dot{\alpha})^2\right], \qquad \Pi = -b_1\cos\varphi - b_2\cos(\varphi+\alpha), \quad \delta W = L\delta\alpha\,. \tag{8.1}$$

Like in chapter 7, here $a_{11} = I_1 + m_2 l^2$, $a_{22} = I_2$, $a_{12} = m_2 r_2 l$, $b_1 = (m_1 r_1 + m_2 l)g$, $b_2 = m_2 r_2 g$; I_1 and I_2 are the moments of inertia of the first and the second links about joints O and D, respectively; m_1 and m_2 are the masses of the first and the second link; r_1 and r_2 are the distances from joints O and D to the centers of mass of the first and the second link, respectively; l is the length of the first link OD; g is the gravity acceleration.

In chapter 7, for describing a double pendulum (see Figure 7.1) as generalized coordinates the angles φ_1 and φ_2 were taken. These angles represent deviations of the first and the second link from the vertical. In contrast, in the current chapter the generalized coordinates are the angle φ of deviation of the first link from the vertical and the angle α of deviation of the second link from the line that continues the first one. The reason for such choice is that the inter-link angle α in this problem plays the role of a control parameter, and it is convenient to take it as one of the coordinates.

Applying relations (8.1), one can derive equations of a double pendulum with control torque L in the inter-link joint D as follows:

$$j_1(\alpha)\ddot{\varphi} + j_2(\alpha)\ddot{\alpha} - 2a_{12}\dot{\varphi}\dot{\alpha}\sin\alpha - a_{12}\dot{\alpha}^2\sin\alpha = -b_1\sin\varphi - b_2\sin(\varphi+\alpha)\,, \tag{8.2}$$

$$j_2(\alpha)\ddot{\varphi} + a_{22}\ddot{\alpha} + a_{12}\dot{\varphi}^2\sin\alpha = -b_2\sin(\varphi+\alpha) + L\,. \tag{8.3}$$

Here expression $j_1(\alpha) = a_{11} + a_{22} + 2a_{12}\cos\alpha$ represents the moment of inertia with respect to pivot point O, and hence $j_1(\alpha) > 0$ for all values α; $j_2(\alpha) = a_{22} + a_{12}\cos\alpha$. Equation (8.2) results from adding together the two scalar equations in system (7.3) with $i = 2$, and replacing angle φ_2 with α in accordance with relation $\varphi_2 = \alpha + \varphi_1$, and then renaming angle φ_1 as φ.

8.2 Reduced angle

Equation (8.2) may be transposed as

$$\dot{K} = -b_1 \sin\varphi - b_2 \sin(\varphi + \alpha)\,. \tag{8.4}$$

Here K is the angular momentum of the system with respect to pivot point O

$$K = \frac{\partial T}{\partial \dot{\varphi}} = j_1(\alpha)\dot{\varphi} + j_2(\alpha)\dot{\alpha}\,. \tag{8.5}$$

Relation (8.4) results from the Theorem of angular momentum that is written with respect to pivot point O. The right-hand side of equation (8.4) represents the moment with respect to point O generated by gravity force applied to pendulum links. It is the derivative of expression (8.5) that enters in the left-hand side of equation (8.2).

Dividing both sides of relation (8.5) by $j_1(\alpha)$ yields

$$\dot{\varphi} + \frac{j_2(\alpha)}{j_1(\alpha)}\dot{\alpha} = \frac{K}{j_1(\alpha)}\,. \tag{8.6}$$

Equations (8.4) and (8.6) may be taken instead of (8.2) and (8.3) as a new system of motion equations. The new phase variables are K, φ, and the control parameter is α. However, these equations (8.4), (8.6) involve derivative $\dot{\alpha}$ together with angle α itself. This makes angle α difficult to use as a control parameter and to solve the problems of optimal control design that are formulated below. In order to exclude derivative $\dot{\alpha}$ from consideration a new variable is introduced – the reduced angle p.

First, note that the left-hand side of relation (8.6) may be written as

$$\frac{d\varphi}{dt} + \frac{j_2(\alpha)}{j_1(\alpha)}\frac{d\alpha}{dt} = \frac{d}{dt}[\varphi + F(\alpha)]\,, \tag{8.7}$$

where

$$F(\alpha) = \int \frac{j_2(\alpha)}{j_1(\alpha)}d\alpha = \int \frac{a_{22} + a_{12}\cos\alpha}{a_{11} + a_{22} + 2a_{12}\cos\alpha}d\alpha = \frac{\alpha}{2} - A\,\mathrm{arctg}\left[B\,\mathrm{tg}\,\frac{\alpha}{2}\right]\,. \tag{8.8}$$

Constant values A and B are determined by expressions

$$A = \frac{a_{11} - a_{22}}{\sqrt{(a_{11} + a_{22})^2 - 4a_{12}^2}}\,, \qquad B = \sqrt{\frac{a_{11} + a_{22} - 2a_{12}}{a_{11} + a_{22} + 2a_{12}}}\,.$$

Note that, as it follows from equality (8.8),

$$F(0) = 0\,. \tag{8.9}$$

Instead of angle φ, a new variable p is introduced according to expression (see equality (8.7))

$$p = \varphi + F(\alpha)\,. \tag{8.10}$$

The left-hand side of equation (8.6) is the time derivative of variable p (see equality (8.7)), so this equation can be written as

$$\dot{p} = \frac{K}{j_1(\alpha)} . \tag{8.11}$$

Variable p is also introduced for investigating motion of a double pendulum in studies [100, 124–126].

Using relation (8.10), angle φ can be expressed in terms of variables p and α

$$\varphi = p - F(\alpha) . \tag{8.12}$$

After substituting expression (8.12) into equation (8.4), the latter transposes to

$$\dot{K} = f(p, \alpha), \quad f(p, \alpha) = -b_1 \sin[p - F(\alpha)] - b_2 \sin[p - F(\alpha) + \alpha] . \tag{8.13}$$

Relations (8.11), (8.13) may be considered as a second-order system of equations where the variables are reduced angle p and angular momentum K. Angle α in equations (8.11), (8.13) is an external variable that may be interpreted as a control parameter, assuming that applying torque L in the inter-link joint it is possible to realize any behavior of this angle.

Since $j_1(\alpha) > 0$ for all values of angle α, the value of p monotonically increases on time intervals where $K > 0$, and it monotonically decreases on intervals where $K < 0$. On each of these intervals system (8.11), (8.13) can be rewritten as a single equation of the first order

$$\frac{dK}{dp} = \frac{f(p, \alpha) j_1(\alpha)}{K} . \tag{8.14}$$

So an idealized control model will be investigated. This model is described by equations (8.11), (8.13), or by equation (8.14). As a control parameter in this model the inter-link angle α is taken. It is assumed that this angle can vary in a given range

$$\alpha_{\min} \le \alpha \le \alpha_{\max}, \quad \alpha_{\min}, \alpha_{\max} = \text{const}, \quad \alpha_{\min}, \alpha_{\max} \in (-\pi, \pi) . \tag{8.15}$$

The allowable control functions are piecewise continuous functions $\alpha(t)$ belonging to segment (8.15). The set of admissible control functions will be denoted as U.

8.3 Optimal control that swings the pendulum

Let the initial state of the system be given

$$-\pi < p(0) < 0, \quad K(0) = 0 . \tag{8.16}$$

If angle $\alpha(0) = 0$, then both links at $t = 0$ are stretched along the same line, and $\varphi(0) = p(0)$, as it follows from relations (8.9), (8.10). Then under initial conditions (8.16) $-\pi < \varphi(0) < 0$. If $-\pi < p(0) < 0$ then there exists such value of $\alpha(0)$ (for example, $\alpha(0) =$

0), that the moment of the gravity force is positive, i.e. $f[p(0), \alpha(0)] > 0$ (see (8.13)). From differential equation (8.13) it follows that at such value of $\alpha(0)$ derivative $\dot{K}$ is positive. So the angular momentum K that was equal to zero at $t = 0$, at the beginning of motion (when $t > 0$) will start to increase, and it will become positive. Let initial condition $p(0)$ and the values of $\alpha_{\min}, \alpha_{\max}$ be such that with each control function $\alpha(t) \in U$ the angular momentum K at some time instant $t_1 > 0$ becomes zero: $K(t_1) = 0$. Each control function $\alpha(t) \in U$ has a different corresponding time t_1. From equation (8.11) it follows that on entire time interval $(0, t_1)$ the value of angle p increases strictly monotonically, because on this interval the angular momentum $K > 0$.

The problem of optimal pendulum swinging is stated as follows. It is required to find such law of varying the control parameter α that angle p reaches its maximum possible value at some time $t_1 > 0$, when the angular momentum K becomes zero (again) *for the first time* since the beginning of motion: $K(t_1) = 0$. This can be symbolically written as:

$$\max_{\alpha_{\min} \le \alpha \le \alpha_{\max}} [p(t_1)], \quad K(t_1) = 0, \quad t_1 > 0\,. \tag{8.17}$$

The stated problem is the problem of maximizing the reduced angle p at the end of a "half-cycle" of pendulum oscillations.

A similar problem to that written in (8.17) was discussed in Part 1, and it was solved in general in chapter 5 (see section 1 of that chapter).

To solve problem (8.17) (of maximizing $p(t_1)$) it is necessary and sufficient to maximize derivative dK/dp during the whole motion. In other words, the right-hand side of equation (8.14) must be maximal (refer to chapter 5). So at each time instant angle α must be chosen from within range (8.15) so that product $f(p, \alpha) j_1(\alpha)$ would be maximal. Therefore the sought optimal control function $\alpha(p)$ that depends only on variable p can be written as

$$\alpha(p) = \arg \max_{\alpha_{\min} \le \alpha \le \alpha_{\max}} [f(p, \alpha) j_1(\alpha)] \tag{8.18}$$

The maximum of function $f(p, \alpha) j_1(\alpha)$ with respect to variable α is attained on the boundary of segment (8.15) or inside this segment. If $\alpha \equiv 0$, then the double pendulum will move like if it were a single pendulum. At that, $p(t_1) = -p(0)$. Function (8.18) is not identically equal to zero, so under control law (8.18) $p(t_1) > -p(0)$. Thus at an end of an oscillation semiperiod (the end of half-cycle) the deviation of $|p|$ from zero will be greater than at the beginning of this semiperiod, i.e. $|p(t_1)| > |p(0)|$.

Note. *If at time instant $t = t_1$, when $K = 0$, angle α changes instantaneously (stepwise) then, as it follows from equation (8.4), angular momentum K remains zero. Reduced angle p also remains unchanged, as it follows from equation (8.11). As for angle φ, it changes stepwise, according to expression (8.12). If the value of α is taken from within interval (8.15) so that function $F(\alpha)$ takes its maximum value, the value of angle φ at time instant t_1 is maximized (see relation (8.12)). So to maximize the deviation angle of the first (upper) link at time instant when the angular momentum becomes zero, i.e. at the end of a half-cycle, it is required, first, to find control function (8.18), and second, to assign at time instant $t = t_1$ value α that minimizes function $F(\alpha)$.*

Now consider the task of maximizing oscillation magnitude of the pendulum at the next semiperiod, when $K < 0$, i.e. the task of maximizing angle $p(t_2)$. Here $t_2 > t_1$ is the first (after t_1) time instant when angular momentum K becomes zero: $K(t_2) = 0$. Since on time interval (t_1, t_2) angular momentum $K < 0$, optimal control law $\alpha(p)$ that delivers minimum to the value of $p(t_2)$ is obviously determined by expression

$$\alpha(p) = \arg \min_{\alpha_{\min} \leq \alpha \leq \alpha_{\max}} [f(p, \alpha) j_1(\alpha)] . \tag{8.19}$$

If at time instant t_2 the value of α is chosen so that $F(\alpha)$ reaches its maximum, then angle $\varphi(t_2)$ will be minimal.

So, by interleaving control laws (8.18) and (8.19), it is possible to swing the double pendulum, providing it maximal oscillation magnitude at the end of each semiperiod. When $K > 0$, control law (8.18) should be used, and when $K < 0$ – control law (8.19). This means, optimal control described by expressions (8.18), (8.19) depends on angle p as well as on angular momentum K. Combining expressions (8.18), (8.19), the control law may be symbolically written as

$$\alpha = \alpha^* (p, K) . \tag{8.20}$$

Below the results of some numerical experiments are provided. These experiments were carried out with the following "anthropomorphic" values of parameters [161, 168]:

$$\begin{aligned} &m_1 = 38.4\,\text{kg}\,, \quad &&l_1 = 1.19\,\text{m}\,, \quad &&r_1 = 0.77\,\text{m}\,, \quad &&I_1 = 28.72\,\text{kg}\cdot\text{m}^2\,, \\ &m_2 = 26\,\text{kg}\,, &&l_2 = 1\,\text{m}\,, &&r_2 = 0.415\,\text{m}\,, &&I_2 = 6.3\,\text{kg}\cdot\text{m}^2\,. \end{aligned} \tag{8.21}$$

Index 1 refers to the first (upper) link OD, index 2 – to the second (lower) link (see Figure 8.1). The lengths of the first and the second links are denoted respectively as l_1 and l_2, though in Figure 8.1 the length of the first link is designated as l. The length l_2 of the second link does not enter in the equations of motion. These equations involve only the distance between point D and the center of mass of the second link. This distance is denoted as r_2. Parameters (8.21) are calculated supposing that the upper link is a perfectly rigid body that consists of combined torso and stretched arms of a gymnast. The lower link is considered to be the legs.

Let $\alpha_{\min} = -2\pi/3$, $\alpha_{\max} = 0$. Such terminal values of control parameter α describe a case when the gymnast cannot bend backwards, and he also cannot completely fold his body – he lacks 60°. So for the purpose of numerical experiment it is considered that

$$-2\pi/3 \leq \alpha \leq 0 . \tag{8.22}$$

Figure 8.2 illustrates control function $\alpha(t)$ that provides optimal swinging of the double pendulum in six semicycles. Also, corresponding functions $K(t)$, $p(t)$ and $\varphi(t)$ are shown. The solution is found for initial conditions

$$p(0) = -\pi/30 \ \ (6^\circ), \quad K(0) = 0 .$$

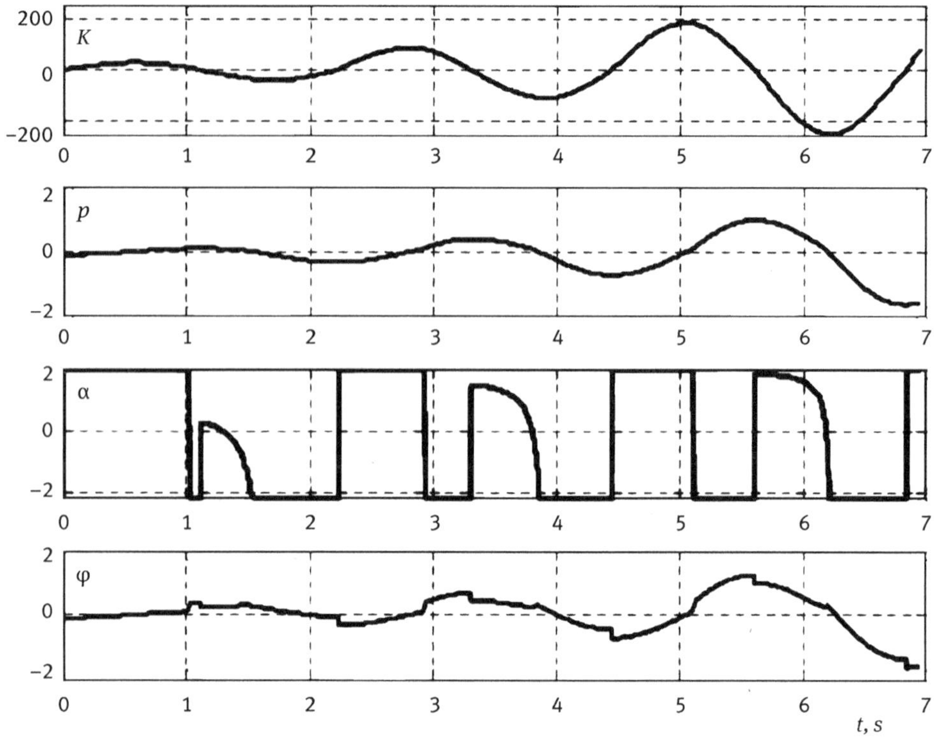

Fig. 8.2. Dynamics of angular momentum K and angles p, α, φ in process of swinging of the double pendulum.

The optimal control law is built in accordance with expressions (8.18), (8.19) or, equivalently, in accordance with expression (8.20). For the purpose of maximizing and minimizing function $f(p, \alpha)j_1(\alpha)$ the numerical experiments involve a discrete set of points from interval (8.22).

Inspecting Figure 8.2, one can see that the magnitude of oscillations of variables K, p and φ increases as time passes. Angle α mostly takes its terminal values $\alpha_{\min} = -2\pi/3$ and $\alpha_{\max} = 0$. Function $\alpha(t)$ has discontinuities at instants when angular momentum K passes its zero value. It also has discontinuities in other points; some of them are close to extremum points of the angular momentum. However, on some intervals function $\alpha(t)$, not being constant, changes continuously. Functions $K(t)$ and $p(t)$ are continuous. The way the pendulum moves from left to right is different from the way it moves from right to left, because the limitations imposed on control function are non-symmetric. Angle $\varphi(t)$ changes in phase with reduced angle $p(t)$. On time intervals when $\alpha(t) = 0$, equality $\varphi(t) = p(t)$ takes place, that is in accordance with equalities (8.9), (8.10). On other intervals angle φ is different from angle p, but these angles are "close". Whenever control function $\alpha(t)$ has a discontinuity, angle φ takes a stepwise change, because function $F(\alpha)$ (8.8) has a discontinuity.

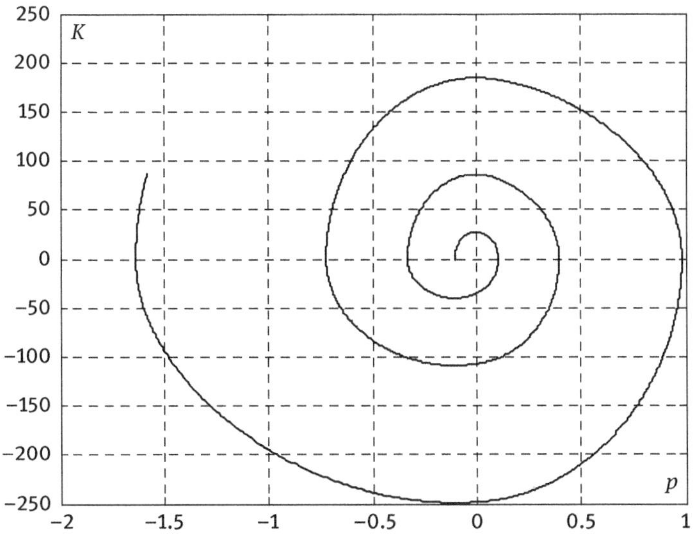

Fig. 8.3. Dynamics of phase variables p and K during pendulum swinging.

Figure 8.3 shows the trajectory of optimal pendulum swinging in phase plane (p, K). The same trajectory in terms of variables φ and K is shown in Figure 8.4.

Figures 8.3 and 8.4 also show that proposed control function swings the pendulum. Figure 8.4 better than Figure 8.2 shows the discontinuity points of function $\varphi(t)$. These points are represented in Figure 8.4 as horizontal segments. These segments are located on the abscissa, where angular momentum K changes its sign, and also in some other points of plane (φ, K).

Figure 8.5 illustrates functions $K(t)$, $p(t)$, $\alpha(t)$ and $\varphi(t)$ that correspond to the process of swinging and consequent rotation of the pendulum with limitation (8.22) imposed on control parameter, and under initial conditions

$$p(0) = 0, \quad K(0) = 0\,. \tag{8.23}$$

If $\alpha(0) = 0$, the first of conditions (8.23) results in equality $\varphi(0) = 0$, that means that both links of the pendulum at initial time instant $t = 0$ hang down motionless. As it is seen in Figure 8.5, angle α in the beginning of motion changes stepwise, and at time $t = +0$ its value is $\alpha_{\min} = -2\pi/3$. At the same time, angle φ also changes stepwise. After that the pendulum swings are controlled by law (8.20), and it oscillates back and forth. After several oscillations the pendulum goes on rotating non-stop about pivot point O, like a swinging gymnast. This roundabout motion is clockwise, and in this motion $K < 0$, and the control function is determined according to expression (8.19). Figure 8.6 shows the same motion of the double pendulum as in Figure 8.5, but in phase plane (p, K). When the pendulum rotates non-stop in one direction, derivative dK/dp at each value of p takes the least possible value.

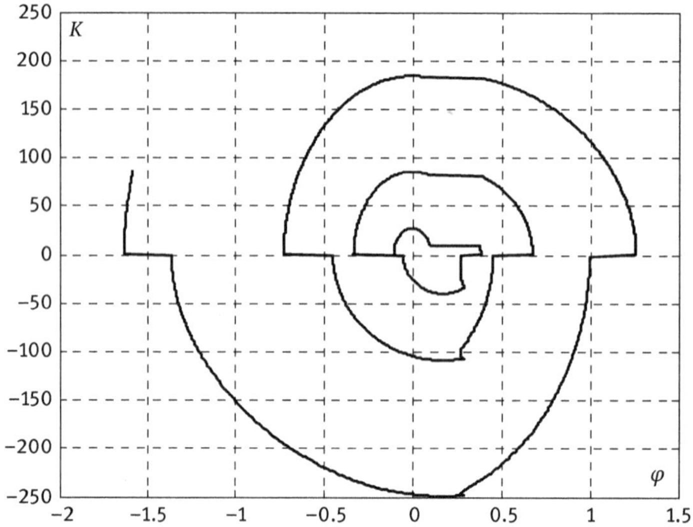

Fig. 8.4. Dynamics of variables φ and K during pendulum swinging.

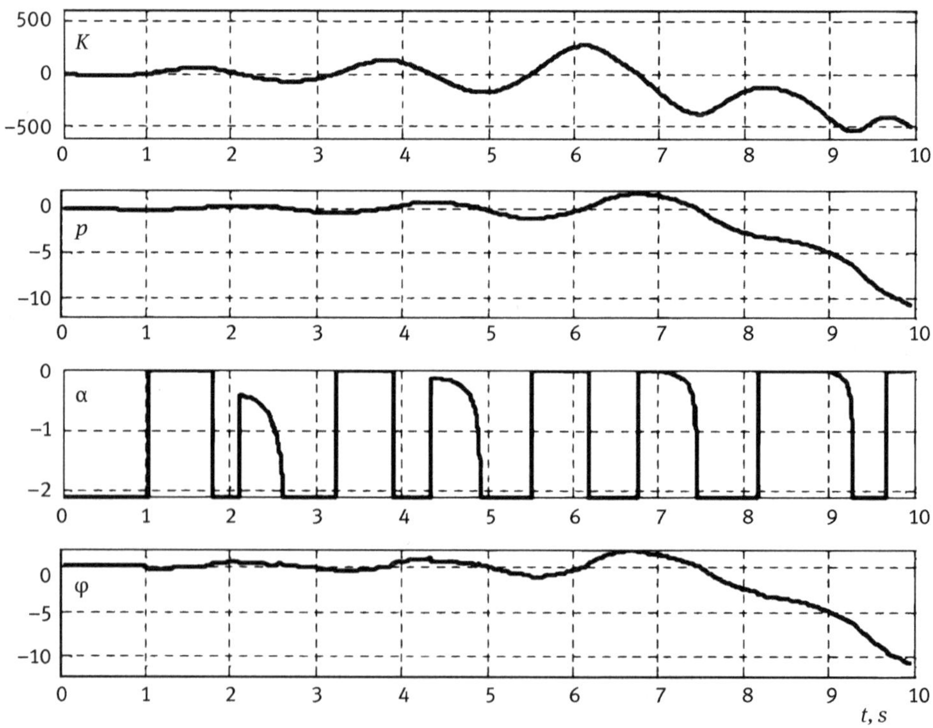

Fig. 8.5. Dynamics of variables K, p, α, φ in process of swinging and consequent rotation of the pendulum.

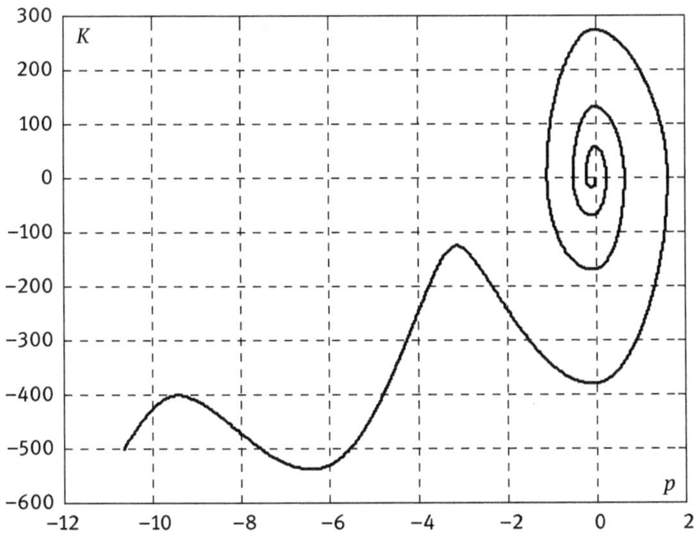

Fig. 8.6. Dynamics of phase variables p and K during pendulum swinging and rotation.

8.4 Optimal control law for pendulum damping

Let the initial conditions for system of equations (8.11), (8.13) or equation (8.14) be described by relations (8.16). Like in previous section, assume that the initial condition $p(0)$ and the values of $\alpha_{\min}$, $\alpha_{\max}$ are such that with any control function $\alpha(t) \in U$ angular momentum K first increases, and then at some time instant $t_1 > 0$ it becomes zero: $K(t_1) = 0$. For each control function $\alpha(t) \in U$ there will be a different corresponding time t_1. It follows from equation (8.11) that on entire time interval $(0, t_1)$ angle p increases strictly monotonically, because $K > 0$. Let for each control function $\alpha(t) \in U$ be $p(t_1) > 0$.

The task of optimal pendulum damping is as follows. It is required to determine a law of varying control parameter α such that angle p reaches its minimum possible value at time instant $t_1 > 0$. This is the instant when angular momentum K becomes zero for the *first* time after the beginning of motion. This can be stated as follows:

$$\min_{\alpha_{\min} \leq \alpha \leq \alpha_{\max}} [p(t_1)], \quad K(t_1) = 0, \quad t_1 > 0\,. \tag{8.24}$$

To solve problem stated in (8.24) (to minimize the value of $p(t_1)$), it is necessary and sufficient that during all motion derivative dK/dp is the least possible. In other words, the right-hand side of equation (8.14) must be minimal (see chapter 5). Or, consequently, at each time instant angle α should be chosen from within interval (8.15) in order to minimize product $f(p, \alpha) j_1(\alpha)$. So when $K > 0$ the sought optimal control law

$\alpha(p)$ as a function of variable p, can be presented as

$$\alpha(p) = \arg \min_{\alpha_{min} \le \alpha \le \alpha_{max}} [f(p, \alpha) j_1(\alpha)] . \tag{8.25}$$

The value of α can be changed stepwise at time instant t_1, in order to minimize angle $\varphi(t_1)$ that represents deviation of the first link from the vertical.

Expression (8.25) coincides with expression (8.19). However, control law (8.19) is used when $K < 0$, and it maximizes the pendulum oscillation magnitude. Contrary, control law (8.25) is used when $K > 0$, and this law minimizes the magnitude of oscillations.

Next, consider the task of minimizing the pendulum oscillation magnitude at the next semiperiod. Namely, the task is to maximize angle $p(t_2)$. Here $t_2 > t_1$ is the first (after t_1) time instant when angular momentum K becomes zero: $K(t_2) = 0$. It is assumed that with any control function $\alpha(t) \in U$ taken on interval (t_1, t_2) angle $p(t_2) < 0$. On time interval (t_1, t_2) angular momentum $K < 0$, so the optimal control function $\alpha(p)$ that delivers maximum value to $p(t_2)$, is determined by expression

$$\alpha(p) = \arg \max_{\alpha_{min} \le \alpha \le \alpha_{max}} [f(p, \alpha) j_1(\alpha)] . \tag{8.26}$$

Expression (8.26) coincides with expression (8.18). However, control law (8.18) is used when $K > 0$, and it maximizes the pendulum oscillation magnitude, while control law (8.26) is used when $K < 0$ to minimize oscillation magnitude. Thus the optimal damping laws are actually the "reversed" laws (8.18) and (8.19) of pendulum swinging.

So, interleaving control laws (8.25) and (8.26), it is possible to damp pendulum oscillations, making their magnitude minimal at the end of each oscillation semiperiod. When $K > 0$, control law (8.25) should be used, and when $K < 0$ – control law (8.26). The optimal control law that is represented by expressions (8.25), (8.26) thus depends on reduced angle p as well as on angular momentum K. Expressions (8.25), (8.26), can be combined to get a symbolic notation

$$\alpha = \alpha_*(p, K) . \tag{8.27}$$

To remind, in the optimal swing control problems (see chapter 5) the extremums of related derivative were found in analytic form, and thus the optimal control law was built explicitly as a feedback. In problems of optimal control of a double pendulum the feedback can be realized only by finding the corresponding extremums numerically.

Below the results of some numerical experiments are provided. These experiments were carried out with numeric values of pendulum parameters as in (8.21) and control parameter α following limitations (8.22).

Figure 8.7 illustrates functions $K(t)$, $p(t)$, $\alpha(t)$ and $\varphi(t)$ that correspond to the optimal pendulum damping process starting from initial state

$$p(0) = -3\pi/4 \ (135°), \quad K(0) = 0 .$$

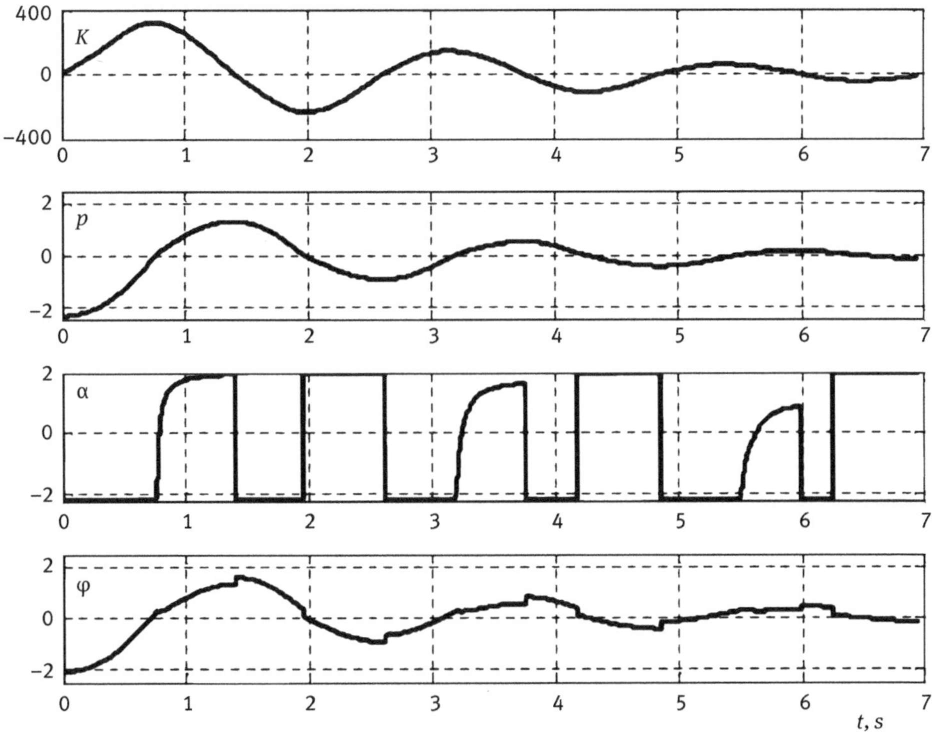

Fig. 8.7. Dynamics of variables K, p, α, φ during pendulum damping.

The optimal control law is built according to expressions (8.25), (8.26) or, equivalently, according to combined expression (8.27).

Inspecting Figure 8.7, one can see that the magnitude of oscillations of variables K, p and φ decreases as time flows. Angle α mainly takes its terminal values $\alpha_{\min} = -2\pi/3$ and $\alpha_{\max} = 0$. Function $\alpha(t)$ has discontinuities when angular momentum K changes its sign. It also has discontinuities in other points, many of them close to extremum points of the angular momentum. Yet on some time intervals function $\alpha(t)$ is not constant, but changes continuously. Functions $K(t)$ and $p(t)$ are continuous. Angle $\varphi(t)$ changes in phase with reduced angle $p(t)$. On those intervals where $\alpha(t) = 0$, equality $\varphi(t) = p(t)$ takes place. On other intervals angle φ is different from angle p, but the two angles are "close". Whenever control function $\alpha(t)$ has a discontinuity, angle φ takes a step change, because function $F(\alpha)$ (8.8) has a discontinuity.

Figure 8.8 shows a trajectory of optimal pendulum damping in phase plane (p, K). The same trajectory in terms of variables φ and K is shown in Figure 8.9.

In Figures 8.8 and 8.9 it can be also seen that the provided control law makes the oscillations of the double pendulum fade. In Figure 8.9, better than in Figure 8.7, the discontinuity points of function $\varphi(t)$ are visible. These points correspond in Figure 8.9

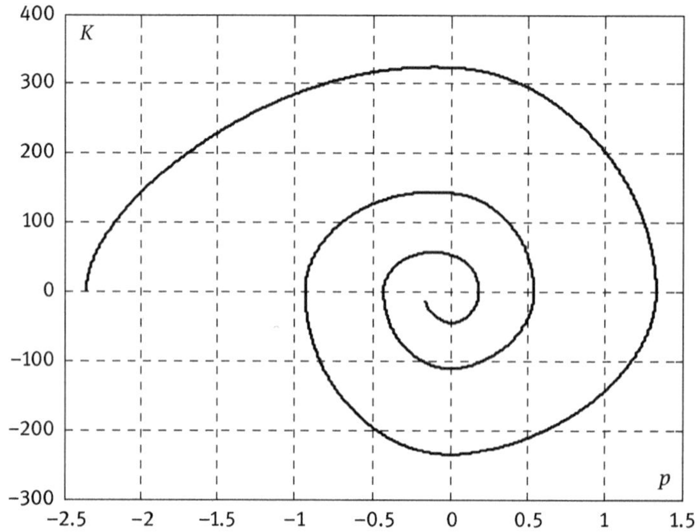

Fig. 8.8. Dynamics of phase variables p and K in the process of pendulum damping.

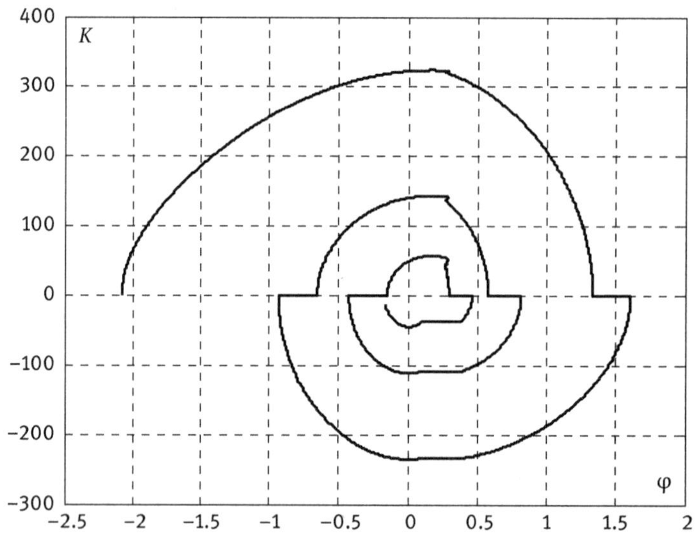

Fig. 8.9. Dynamics of variables φ and K in the process of pendulum damping.

to horizontal straight segments. Such segments are located on the abscissa, where angular momentum K changes its sign, and they also appear in other locations of phase plane (φ, K). Some of these locations are close to the extremum points of the angular momentum.

8.5 On translating the pendulum from its bottom equilibrium to the top one

Using the optimal control law that swings a double pendulum, it is possible to implement a law that translates it from the bottom, stable equilibrium

$$\varphi = 0, \quad \alpha = 0 \quad (p = 0), \quad K = 0\,, \tag{8.28}$$

when both links hang vertically down, to the top, unstable one

$$\varphi = \pm\pi, \quad \alpha = 0 \quad (p = \pm\pi), \quad K = 0\,, \tag{8.29}$$

when both links are stretched vertically up. Let angle α (the control parameter) be constrained as in (8.22).

It is possible to translate the pendulum from position (8.28) to position (8.29) in a finite time, for example, by using the following technique. Let Π_0 denote potential energy of the double pendulum when it is in position

$$p = -\pi, \quad \alpha = \alpha_{\min} \quad (\varphi = -\pi - F(\alpha_{\min}))\,. \tag{8.30}$$

Note that if $\alpha_{\min} \neq 0$, then in position (8.30) the pendulum is "bent". Energy Π_0 can be calculated according to the second expression in (8.1). The pendulum is then swung starting from position (8.28), like it is shown in Figures 8.2 and 8.5, by means of control law (8.20). During this process, at each time instant t the total energy of the pendulum is evaluated, that is, kinetic together with potential energy, supposing that the control angle α at time instant $t+0$ were instantaneously set equal $\alpha_{\min}$ and left in that state. The evaluation of this "virtual" energy is done as follows. The potential energy of the pendulum at time $t+0$ is calculated in accordance with the second expression in (8.1). The double pendulum "frozen" with $\alpha = \alpha_{\min}$ moves as if it were a single-link pendulum, influenced only by the gravity force. Its kinetic energy equals

$$\frac{1}{2} j_1(\alpha_{\min})\dot{\varphi}^2 = \frac{1}{2} j_1(\alpha_{\min})\left(\frac{K}{j_1(\alpha_{\min})}\right)^2 = \frac{1}{2}\frac{K^2}{j_1(\alpha_{\min})}\,, \tag{8.31}$$

where $\dot{\varphi}$ is the angular velocity of a single-link pendulum, and K is its angular momentum, that changes during its rotation.

At time instant t (before the step-change of angle α) the angular momentum of the double pendulum is known. At time $t+0$, after angle α changes stepwise, angular momentum K of the system remains the same, only its derivative changes (stepwise), in accordance with equation (8.13). So, knowing the angular momentum of the double pendulum before the step of control parameter α, and using expression (8.31), one can evaluate the "virtual" kinetic energy of the pendulum at time $t + 0$. Adding together the potential and the kinetic energy of the pendulum at time instant $t + 0$ yields the total energy of the single-link pendulum that remains constant during its motion with

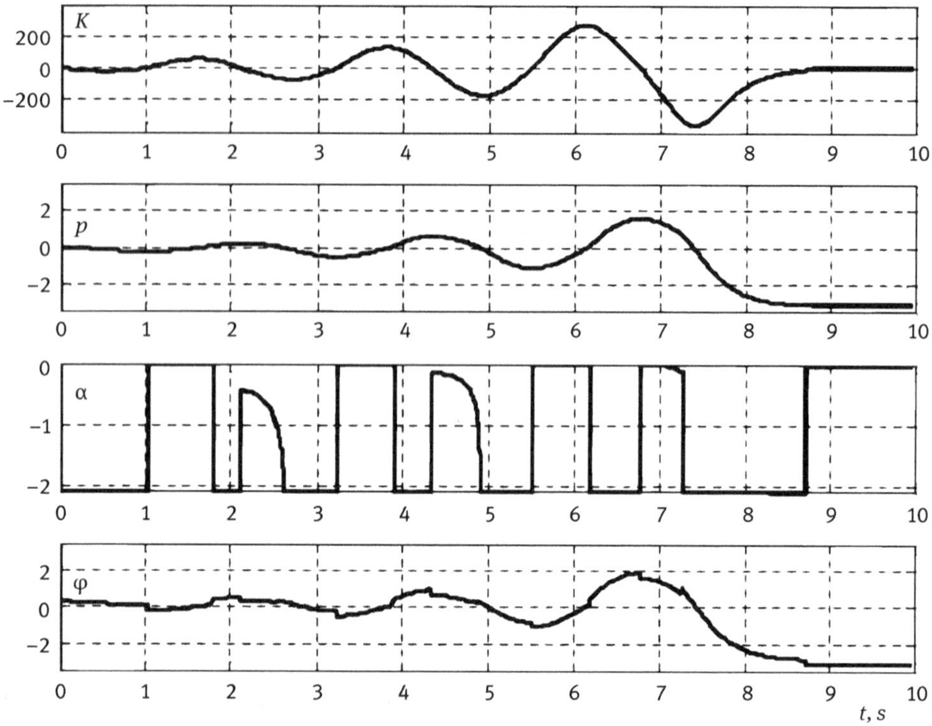

Fig. 8.10. Dynamics of angular momentum K, angles p, α and φ in process of translating a pendulum from its bottom equilibrium to the top one.

$\alpha \equiv \alpha_{\min}$. This "virtual" energy is evaluated at each time instant during the process of swinging the double pendulum, and it is compared to the value of Π_0.

Numerical experiments show that this "virtual" energy increases as time passes. When this energy becomes equal to Π_0, angle α is instantaneously set $\alpha = \alpha_{\min}$. When $\alpha \equiv \alpha_{\min}$, it is a single-link the pendulum. Let this single-link pendulum move freely until its angular momentum K becomes zero. If state (8.30) is not a state of equilibrium, this will happen in a finite time. At that, the single-link pendulum reaches state (8.30). At this time instant, angle α is changed instantaneously from value $\alpha = \alpha_{\min}$ to value $\alpha = 0$. The reduced angle p (see (8.10) for reference) does not change with this step (it is still equal to $-\pi$), angular momentum K remains zero, and angle φ, in accordance with expression (8.12), becomes equal $-\pi$. To explain, at the very last moment the pendulum straightens instantaneously, and it remains inverted. Thus the double pendulum can be driven to the desired top (unstable) position in a finite time.

The described control algorithm for translating the pendulum from its bottom equilibrium to the top one was realized in a software model. Figure 8.10 illustrates dynamics of functions $K(t)$, $p(t)$, $\alpha(t)$ and $\varphi(t)$ recorded during pendulum motion.

It can be seen from Figure 8.10 that the pendulum straightens at time $t \approx 8.63$ s.

The problem of stabilizing the pendulum in its top, unstable equilibrium is not discussed in the current chapter. The next chapter proposes a solution to the problem of translating the double pendulum from bottom to top in a more practical case, when not the inter-link angle α is taken as the control parameter, but a torque that is applied between the two links. A solution to the problem of stabilizing the top equilibrium by means of this torque is also given in chapter 9.

9 Global stabilization of an inverted pendulum controlled by torque in the inter-link joint

The current chapter considers a double pendulum with one control torque that is applied in the inter-link joint. This torque is limited in absolute value. The problem solved here is to design such control law that would translate the pendulum from the bottom, stable equilibrium to the top, unstable one. The control law should then stabilize the pendulum in the unstable equilibrium. The double pendulum with a torque applied in the inter-link joint is often called "acrobot" [124–126, 145, 166].

The control design proposed in the current chapter relies on results got in the previous chapters, 7 and 8. In chapter 8, the pendulum is controlled by the inter-link angle, by directly assigning its value. This angle is assumed to vary in a given range. The variation law for this angle was proposed as a feedback, so that the magnitude of the pendulum oscillations about its bottom equilibrium increases at each oscillation semiperiod. The control law suggested here has several stages and several levels. First, for the inter-link angle α such a control law is designed that the magnitude of the pendulum oscillations about its bottom equilibrium increases "quickly". For this purpose, the variation range for this angle is made "large", and the desired angle value is calculated at each time instant in accordance with the algorithm proposed in chapter 8. The control torque acts like a *servosystem* to track the desired angle value. The *tracking* is done by means of a linear feedback with saturation. The parameters used in the feedback are the difference between the desired angle and the real one, and the corresponding angular velocity. Such feedback is typically called a proportional-differential controller, or the PD-controller. The feedback may get saturated, due to the limitation imposed on the control torque. At this first stage, the pendulum bends significantly, because the range of allowable values for angle α is set "large". When the pendulum is "far enough" from the bottom equilibrium, the next control stage begins. At this stage, the range for the inter-link angle is made narrower. The pendulum continues to increase its swing magnitude, but it happens slower than in the first stage, in order for the pendulum not to bypass the desired top equilibrium. When in process of swinging the pendulum gets close enough to the top equilibrium, the range for the inter-link angle is set even smaller. There may be several stages of narrowing the angle range. The last stage should have this range narrow enough so that the pendulum would not miss the domain of attraction. If the allowed range of the inter-link angle is small, then the pendulum behaves almost like a single-link one, and its oscillation magnitude increases not much from cycle to cycle. If the magnitude increase is small, then at some time instant the pendulum reaches the domain of attraction of the top equilibrium. After that, the final control stage begins – the stage of local stabilization. At this stage, in accordance with the algorithm proposed in chapter 7, the inter-link torque is set as a linear feedback with saturation that involves the two unstable Jordan variables. These variables enter in motion equations linearized about the top equilibrium. This feedback is built so (see chapter 7) that to maximize, if possible, the domain of attraction of the desired top equilibrium, which the pendulum should reach during the swinging process.

Thus the pendulum swinging is done by a control law that has two levels. At the first level, the value of the inter-link angle is calculated. At the second level, the inter-link angle is tracked. As a result, the designed control law is a feedback that involves four phase variables of the system.

9.1 Mathematical model

Figure 9.1 shows the double pendulum that is studied here, and the generalized coordinates (the angles) used for its investigation. Like in Figure 7.1, the pendulum is shown in a position where both its links are "almost" upside down. Unlike chapter 7, as generalized coordinates, not the angles φ_1 and φ_2 are chosen that represent link deviation from the vertical, but the angles φ and α. The former, φ, designates deviation of the first link from the vertical, and the latter, α, designates deviation of the second link from the first one. Similar angles were introduced in chapter 8 (see Figure 8.1). However, in Figure 8.1, the angle φ represents the deviation of the first link of the pendulum from its bottom position. Angle α is used here as the second generalized coordinate because of the control law, where at first the law of controlling the inter-link angle is built that increases the oscillation magnitude (see expression (8.20)).

The expression for kinetic energy T of the pendulum is the same as written in (8.1), and as for expression for potential energy Π, it is obtained from (8.1) by replacing angle φ with $\varphi + \pi$

$$T = \frac{1}{2}\left[a_{11}\dot{\varphi}^2 + 2a_{12}\dot{\varphi}\,(\dot{\varphi} + \dot{\alpha})\cos\alpha + a_{22}\,(\dot{\varphi} + \dot{\alpha})^2\right], \quad \Pi = b_1\cos\varphi + b_2\cos(\varphi + \alpha)\,. \tag{9.1}$$

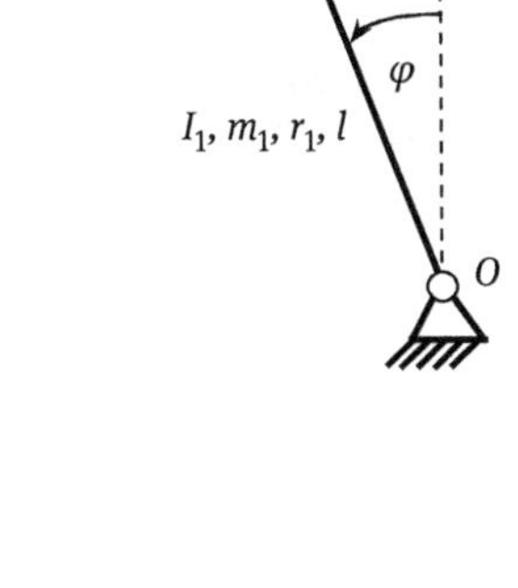

Fig. 9.1. The double pendulum.

Expressions for coefficients a_{11}, a_{22}, a_{12}, b_1, b_2 and their meanings are the same as in the chapters 7 and 8. The virtual work of torque L in the inter-link joint is described by the last one of relations (8.1).

Using relations (9.1), the motion equations for a double pendulum with control torque L applied in the inter-link joint can be written as follows:

$$j_1(\alpha)\ddot{\varphi} + j_2(\alpha)\ddot{\alpha} - 2a_{12}\dot{\varphi}\dot{\alpha}\sin\alpha - a_{12}\dot{\alpha}^2\sin\alpha = b_1\sin\varphi + b_2\sin(\varphi+\alpha)\,, \tag{9.2}$$

$$j_2(\alpha)\ddot{\varphi} + a_{22}\ddot{\alpha} + a_{12}\dot{\varphi}^2\sin\alpha = b_2\sin(\varphi+\alpha) + L\,. \tag{9.3}$$

Equations (9.2) and (9.3) can also derived from (8.2) and (8.3), respectively, by substituting $\varphi + \pi$ for angle φ. Expressions for coefficients $j_1(\alpha)$ and $j_1(\alpha)$ remain the same as in chapter 8.

Like in chapter 7, let W denote a set of piecewise-continuous functions $L(t)$ that are limited in absolute value (see inequality (7.4)) by constant L_0.

With $L \equiv 0$ system (9.2), (9.3) has a trivial solution

$$\varphi = \dot{\varphi} = \alpha = \dot{\alpha} = 0 \quad (\varphi = 2\pi,\, \dot{\varphi} = \alpha = \dot{\alpha} = 0)\,, \tag{9.4}$$

that corresponds to the unstable equilibrium of an uncontrolled pendulum with two inverted links. The purpose of the current chapter is to design a control law that would stabilize state (9.4).

9.2 Cascade form of dynamic equations

Equation (9.2), like in chapter 8, can be presented as

$$\dot{K} = b_1\sin\varphi + b_2\sin(\varphi+\alpha)\,, \tag{9.5}$$

where K is the angular momentum of the system about pivot point O, that is described by expression (8.5). Equation (9.5), compared to equation (8.4), has a different sign of the right-hand side, because in the current chapter angle φ is replaced everywhere with $\varphi+\pi$. Applying relations (8.6)–(8.10), a new variable p is introduced – the reduced angle. This new variable helps derive the following system of equations

$$\dot{p} = K/j_1(\alpha)\,, \tag{9.6}$$

$$\dot{K} = f(p,\alpha), \quad f(p,\alpha) = b_1\sin[p - F(\alpha)] + b_2\sin[p - F(\alpha) + \alpha]\,. \tag{9.7}$$

Equations (9.6) and (9.7) fully coincide with equations (8.11) and (8.13), respectively, with only one exception – the right-hand side of equation (9.7) has a different sign.

Relations (9.6), (9.7) may be considered as a second-order system of equations in terms of phase variables p and K. Angle α in equations (9.6), (9.7) is an input variable, that can be considered an intermediate control parameter.

In studies [124–126] it is proposed to exclude from equations (9.2), (9.3) the second derivative $\ddot{\varphi}$. After such transposition, the equations look like

$$d(\alpha)\ddot{\alpha} + h(\varphi, \dot{\varphi}, \alpha, \dot{\alpha}) = j_1(\alpha)L \, . \tag{9.8}$$

Function $d(\alpha) = j_1(\alpha)a_{22} - j_2^2(\alpha)$ is positive for all values of α, because it is actually a determinant of positive-definite matrix of kinetic energy (9.1), $h(\varphi, \dot{\varphi}, \alpha, \dot{\alpha})$ is a nonlinear function that takes four arguments, it can be evaluated by means of equations (9.2), (9.3). Using relationship

$$L = j_1^{-1}(\alpha)[d(\alpha)v + h(\varphi, \dot{\varphi}, \alpha, \dot{\alpha})] \tag{9.9}$$

it is convenient to introduce a new control parameter v, and to transpose equation (9.8) as

$$\ddot{\alpha} = v \, . \tag{9.10}$$

To remind, $j_1(\alpha) = a_{11} + a_{22} + 2a_{12} \cos\alpha$ is the moment of inertia of the pendulum with respect to pivot point O, and thus $j_1(\alpha) \neq 0$ for all values of angle α. Equations (9.6), (9.7), (9.9), (9.10) are the so-called “cascade” form of the original equation system (9.2), (9.3).

If control $\alpha(p, K)$ is built that solves one or another task concerning equations (9.6), (9.7), then by means of equation (9.10) it is possible to find a control function v that modifies angle α in an appropriate way, this control function being actually a feedback law. Then by means of relation (9.9) it is possible to build a control law that involves torque L. Thus torque L is sought as a feedback effectively involving phase variables $\varphi, \dot{\varphi}, \alpha, \dot{\alpha}$. However, such approach does not allow taking into account limitation like (7.4) that is imposed on control torque L. Current study utilizes a different approach, which regards such limitation. Control function L is built as a linear feedback (a PD-controller) that involves current deviation of angle α from its desired value, and derivative $\dot{\alpha}$. Limitation (7.4) is regarded by introducing saturation in linear feedback. Such feedback is used to track the desired dynamics of angle α.

9.3 Control law that swings the pendulum

Consider system of equations (9.6), (9.7), regarding angle α as a control parameter. Let this parameter be limited in absolute value

$$|\alpha| \leq \alpha_0, \quad \alpha_0 = \text{const} > 0 \, . \tag{9.11}$$

Similar to (8.18), one can get the expression to describe control law that swings the pendulum on time interval when angular momentum $K > 0$

$$\alpha(p) = \arg \max_{|\alpha| \leq \alpha_0} [f(p, \alpha) j_1(\alpha)] \, . \tag{9.12}$$

When $K < 0$, control law that swings the pendulum is described by expression

$$\alpha(p) = \arg\min_{|\alpha|\leq\alpha_0} [f(p,\alpha)j_1(\alpha)] . \tag{9.13}$$

Expression (9.13) is similar to (8.19). Interleaving control laws (9.12) and (9.13), it is possible to swing the double pendulum, providing it the maximum magnitude at the end of each semiperiod. The symbolic notation for a control law that swings the pendulum, like in (8.20) can be obtained by combining expressions (9.12), (9.13). This yields $\alpha = \alpha^*(p, K)$.

9.4 Tracking the desired inter-link angle dynamics

The inter-link angle α, as it was mentioned above, is considered as an intermediate control parameter. The actual control is performed by torque L applied in the inter-link joint. Using this torque, it is possible to track a desired dynamics of the inter-link angle $\alpha^*(p, K)$. Such tracking may be provided by means of a linear feedback (a PD-controller) with saturation:

$$L = \begin{cases} L_0 & \text{when} \quad -c_1[\alpha - \alpha^*(p,K)] - c_2\dot{\alpha} \geq L_0 \\ -c_1[\alpha - \alpha^*(p,K)] - c_2\dot{\alpha} & \text{when} \quad |-c_1[\alpha - \alpha^*(p,K)] - c_2\dot{\alpha}| \leq L_0 \\ -L_0 & \text{when} \quad -c_1[\alpha - \alpha^*(p,K)] - c_2\dot{\alpha} \leq -L_0 . \end{cases} \tag{9.14}$$

Here c_1 and c_2 are the constant feedback coefficients, respectively for deviation of angle α from its desired value $\alpha^*(p, K)$ and for angular velocity $\dot{\alpha}$. These coefficients are to be assigned. The presence of angular velocity $\dot{\alpha}$ in feedback (9.14) makes the tracking process stable. The feedback signal limitation emerges naturally due to restriction (7.4).

The results of numerical experiments are provided below. These results show that with appropriate choice of feedback coefficients c_1 and c_2 the tracking of the inter-link angle is satisfactory. Better quality of tracking can be achieved with smaller variation range of angle $\alpha^*(p, K)$, that is, when the value of α_0 in inequality (9.11) is smaller, and when the control resource is wider, that means, when the value of L_0 is larger.

So the feedback used to swing the pendulum includes two levels. At the first level, by means of expressions (9.12), (9.13) the desired (program) value $\alpha^*(p, K)$ for angle α is calculated. At the second level, this program value $\alpha^*(p, K)$ is tracked in accordance with law (9.14). The tracking is certainly done with an error. However, at each time instant the value of $\alpha^*(p, K)$ is calculated based on the current values of p and K. So the first level regards the tracking errors that occur at the second level, and these errors can be negated to a certain extent.

9.5 Local stabilization of an inverted pendulum

Suppose that during the pendulum swinging it reaches some sufficiently small neighborhood of the top, unstable equilibrium (9.4). Once this happens, the task is to stabilize equilibrium (9.4) locally. When designing a stabilization law, it is appropriate to maximize the domain of attraction of the desired equilibrium, because the larger the domain of attraction, the easier it is for the pendulum to get inside that domain during swinging process. In other words, the larger the domain of attraction, the more robust will the control algorithm be.

In chapter 7 a method of designing a control algorithm is proposed that provides a "large" domain of attraction. According to it, first, the motion equations (9.2), (9.3) need to be linearized about the desired equilibrium (9.4) (see system (7.6)). The resulting system has, without control, two positive and two negative eigenvalues. This system is then transposed to Jordan form, and two equations are separated that correspond to the positive eigenvalues (see, for example, equations (7.15)). To realize the maximum possible domain of attraction, the feedback should only involve Jordan variables that correspond to the positive eigenvalues (see feedback law (7.34)). The larger in absolute value feedback gain γ in control law (7.34), the closer is the border of the attraction domain of the linearized system to the border of its controllability domain (see Figures 7.3 and 7.4), which is described by equations (7.19), (7.20). Since the system can be transposed to Jordan form in different ways, the feedback can involve different variables, but, anyway, they must be the "unstable" (without control) variables.

9.6 Numerical experiments

Let at the initial time instant $t = 0$ both pendulum links hang down, and their angular velocities be equal to zero.

$$\varphi(0) = \pi, \alpha(0) = 0, \quad \dot{\varphi}(0) = \dot{\alpha}(0) = 0 . \tag{9.15}$$

Figure 9.2 illustrates the process of translating the pendulum from stable equilibrium (9.15) into unstable one (9.4) by means of control law (9.12), (9.13), (9.14), (7.34). The pendulum links are taken similar and homogeneous (see relations (7.21)) with parameters (7.26).

Figure 9.2 has four plots. The top (first) one shows time dynamics of angle φ, the next one (second) – the dynamics of inter-link angle α, the third one – torque L applied in the inter-link joint, and, finally, the bottom plot illustrates the dynamics of the total energy of the system $E = T + \Pi$ (see expressions (9.1) for kinetic energy T and potential energy Π). At the beginning of the process, when the pendulum is in equilibrium (9.15), the total energy $E = E(0) = \Pi(0) = -b_1 - b_2$. In state (9.4) $E = b_1 + b_2 = -E(0)$. The total energy of the pendulum during its swinging increases "almost" monotonically.

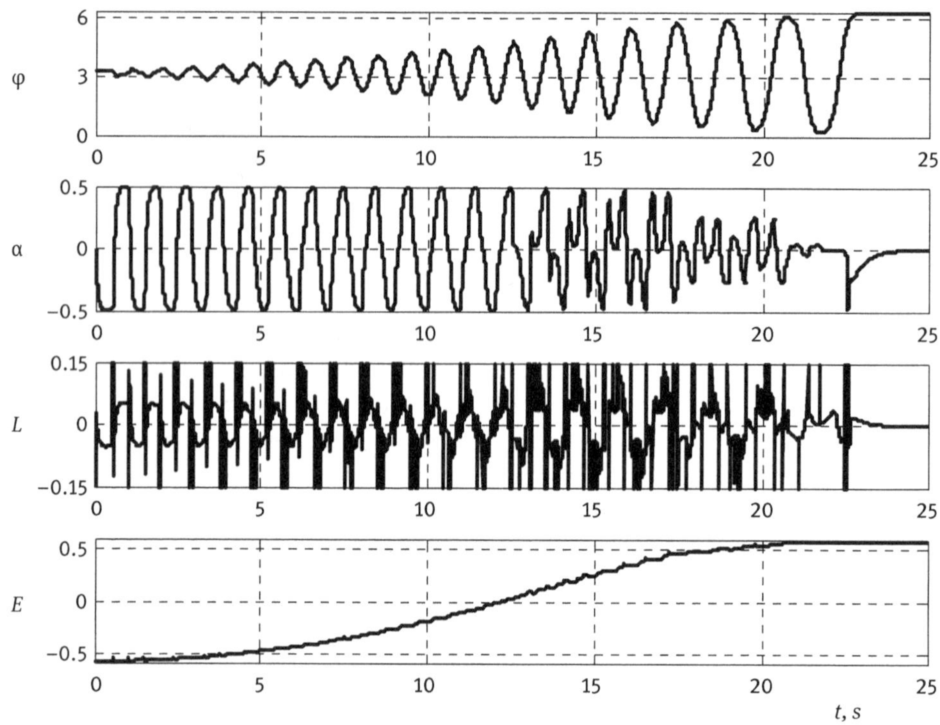

Fig. 9.2. Translating the pendulum from the bottom equilibrium to the top one.

In the end of the process illustrated in Figure 9.2 $\varphi \to 2\pi$ and $\alpha \to 0$, i.e. the pendulum asymptotically reaches its equilibrium (9.4). The program dynamics $\alpha^*(p, K)$ of inter-link angle α at the beginning stage of control process is built according to expression (9.12), so that the angular momentum K at the first oscillation semiperiod, when $t > 0$, would be positive, at the next semiperiod it would be negative, and so on. Instead of (9.12), law (9.13) can be used at the beginning, so that at the first semiperiod, when $t > 0$, angular momentum K would be negative, then positive, and so on. As limitation (9.11) imposed on angle α is symmetric with respect to zero, the pendulum oscillation process in the first case is symmetric to the process in the second case with respect to the vertical. In other words, the sequence of pendulum shapes in the one case can be got by mirroring with respect to the vertical axis the corresponding sequence in the other case.

The value of α_0 at the beginning of the process is chosen relatively large ($\alpha_0 = 0.5$) in order to make the pendulum swing faster. The program law $\alpha^*(p, K)$ is tracked by means of feedback (9.14) with coefficients $c_1 = 25\,\mathrm{N \cdot m}$, $c_2 = 1.5\,\mathrm{N \cdot m \cdot s}$, that were adjusted during numerical experiments. The program value $\alpha^*(p, K)$ of angle α is tracked accurately enough. Control torque L most of the time at the beginning of the process takes the terminal values $|L| = L_0$, the magnitude of pendulum oscillations (in terms of angle φ) grows rapidly. During the swinging process, the value of α_0 is reduces in a step-like manner as the full energy $E = T + \Pi$ increases, in order for the pendulum not to miss the desired equilibrium. When the potential energy of the pendulum becomes $3/4$ of the value of potential energy at the state of equilibrium (9.4), the value of α_0 in control law $\alpha^*(p, K)$ is reduced from 0.5 to 0.25. Then the value of α_0 is reduced once again. As α_0 decreases, the magnitude of pendulum oscillations rises slower and slower (see Figure 9.2).

When the value of angle α is small, the double pendulum behaves as if it were a single link. The magnitude increase rate may be made arbitrarily small by choosing the value of α_0 small enough. When this rate is small, the swing process becomes longer, but it is made certain that the system can get inside the domain of attraction. Indeed, at an end of each semiperiod the angular momentum of the pendulum becomes zero: $K(\tau) = 0$. Angle α can vary in a small range near zero; if angular velocity $\dot{\alpha}(\tau)$ is also close to zero, then, in accordance with relation (8.5), angular velocity $\dot{\varphi}(\tau)$ is also close to zero. When the magnitude of oscillations of angle φ (at ends of semiperiods) is small, then in the process of pendulum swinging there will be a time instant τ, when angle φ will be close enough to 0 or 2π. So it can be expected that there exists such small value of α_0, that at some time instant τ the values of $\varphi(\tau)$, $\dot{\varphi}(\tau)$, $\alpha(\tau)$, $\dot{\alpha}(\tau)$ become close enough to zero (or angle $\varphi(\tau)$ to 2π), so that system (9.2), (9.3) gets into the domain of attraction of the state of equilibrium (9.4). In the process illustrated in Figure 9.2 it happens exactly like that: when $t \approx 22.51\,s$ the system reaches the domain of attraction. After that, control law (7.34) is activated that stabilizes the double pendulum in equilibrium (9.4).

The algorithms of control for the pendulum at different stages of its motion are built analytically, for example, the algorithm of varying the inter-link angle, the algorithm of local stabilization. Yet the functionality of the overall control algorithm as a whole can be verified only by numerical investigation.

The numerical experiments involving the control algorithm that was stated above were conducted with various initial conditions of the pendulum – not only with both links hanging down and their angular velocities being zero, but with many other states, both close to the bottom equilibrium and far from it. The control algorithm proves its functionality for initial deviations of the pendulum from vertical even up to 45°. Concerning global stability, the following can be said. From any initial position, with any initial velocities of the links the pendulum can be translated to the bottom equilibrium, or to a state close to this equilibrium, by applying in the inter-link joint a torque that damps oscillations. If the control algorithm proposed above drives the pendulum from the bottom to the top equilibrium, this provides stability of that equilibrium under all initial conditions, that means, global stability of the inverted pendulum.

Without control, the double pendulum has four equilibriums. The state when both links hang vertically down is stable, the other three are unstable. When control law is applied as proposed in the current chapter, the pendulum has only one equilibrium (9.4), with both links stretched vertically up, and this equilibrium is asymptotically stable.

10 Global stabilization of an inverted pendulum controlled by torque in the pivot

In the current chapter, like in the previous one, chapter 9, the task of creating a control algorithm that provides global stability of an inverted pendulum is solved. The only control torque this time is applied in the pivot point, while in chapter 9 the similar task was considered with control torque applied in the inter-link joint. The system is underactuated, it has two degrees of freedom and only one control input. A double pendulum with control torque in its pivot is often called “pendubot” [1, 67, 124–126, 127, 146, 169]. Motion equations of a pendulum controlled from its pivot cannot be reduced to a relatively simple “cascade” form with an intermediate control parameter, as it was done in chapters 8 and 9. So here the task is solved a little differently. Still, like in chapter 9, the control algorithm is designed in such a way that while the pendulum is swung, it is kept “almost” straightened, and its stabilization is done according to the methods developed in chapter 7 (see also [15]).

10.1 Mathematical model

For deriving the equations of pendulum motion, angles φ_1 and φ_2 will be taken as generalized coordinates. These angles represent deviations of the first and the second link from the vertical. Similar variables were chosen in chapter 7 (see Figure. 7.1). Kinetic and potential energy of the system are like in (7.1). System (7.3) can be presented as two scalar equations

$$\begin{aligned} &a_{11}\ddot{\varphi}_1 + a_{12}\cos(\varphi_2 - \varphi_1)\ddot{\varphi}_2 - a_{12}\sin(\varphi_2 - \varphi_1)\dot{\varphi}_2^2 - b_1 \sin\varphi_1 = L\,, \\ &a_{12}\cos(\varphi_2 - \varphi_1)\ddot{\varphi}_1 + a_{22}\ddot{\varphi}_2 + a_{12}\sin(\varphi_2 - \varphi_1)\dot{\varphi}_1^2 - b_2 \sin\varphi_2 = 0\,. \end{aligned} \tag{10.1}$$

Parameters $a_{11}, a_{22}, a_{12}, b_1, b_2$ in equations (10.1) have the same meaning as in chapter 7.

The angular momentum of the pendulum K with respect to pivot O looks as follows:

$$K = \frac{\partial T}{\partial \dot{\varphi}_1} = a_{11}\dot{\varphi}_1 + a_{12}\cos(\varphi_2 - \varphi_1)\dot{\varphi}_2\,. \tag{10.2}$$

Like in chapter 9 (see Figure 9.1), α denotes the angle between the line continuing the first link and the second link: $\alpha = \varphi_2 - \varphi_1$. Like before, torque L that is applied in the pivot point is considered to be limited in its absolute value (like in inequality (7.4)).

System (10.1) with $L = 0$ has an unstable equilibrium with both links upside down,

$$\varphi_1 = 0, 2\pi, \quad \varphi_2 = 0, 2\pi\,. \tag{10.3}$$

Besides, it has one stable equilibrium where both links hang down,

$$\varphi_1 = \pi, \quad \varphi_2 = \pi\,. \tag{10.4}$$

The task considered here is to translate the pendulum from equilibrium (10.4) into unstable equilibrium (10.3) and to stabilize it in that state.

10.2 Swinging the pendulum

The algorithm of pendulum swinging is designed basing on the "energy" approach [17, 48, 49, 68, 141], and also on the common sense.

Translation of the pendulum from state (10.4) into state (10.3) includes several stages. First, the pendulum is swung, by increasing its total mechanical energy $T + \Pi$. In the desired state of equilibrium (10.3) the total energy is equal to only potential energy $\Pi = b_1 + b_2$ (see expressions (7.1)).

In region

$$\pi - \Delta\varphi_1 \leq \varphi_1 \leq \pi + \Delta\varphi_1 \tag{10.5}$$

the following control law is used:

$$L = \begin{cases} L_0/\sigma & \text{when} \quad \dot{\varphi}_1 > 0 \\ -L_0/\sigma & \text{when} \quad \dot{\varphi}_1 \leq 0. \end{cases} \tag{10.6}$$

Here $\Delta\varphi_1 = \text{const} > 0$, $\sigma > 1$ – the parameters that are defined at the simulation. With control law (10.6) (in region (10.5)) the total energy of the system increases monotonically, because its time derivative behaves according to the following relation

$$\frac{d(T + \Pi)}{dt} = L\dot{\varphi}_1 = |\dot{\varphi}_1|\, L_0/\sigma\,,$$

and it is non-negative. This derivative becomes zero only at times when the angular velocity of the first link $\dot{\varphi}_1$ becomes zero.

Numerical experiments show that instead of control law (10.6), within range (10.5) the following law can be used

$$L = \begin{cases} L_0/\sigma & \text{when} \quad K > 0 \\ -L_0/\sigma & \text{when} \quad K \leq 0 \end{cases}, \tag{10.7}$$

where K is the angular momentum (10.2).

At the beginning of control process, when both links of the pendulum are hanging down, inequality (10.5) takes place, therefore control law (10.6) or (10.7) is used.

10.3 Straightening the pendulum

Outside region (10.5) the inter-link angle α will be attempted to keep "close" to zero. Note that in the desired equilibrium (10.3), angle $\alpha = 0$. In other words, outside range

(10.5) the effort will be made to "straighten" the double pendulum. If angle α is close to zero, then the double pendulum is similar to a single-link one, and a single-link pendulum is appropriate to swing using a relay control law (10.6) or (10.7).

Angle α outside region (10.5) can be demanded to behave according to equation

$$\ddot{\alpha} = -c_1\dot{\alpha} - c_2\alpha\,, \tag{10.8}$$

where c_1 and c_2 are constant positive ($c_1, c_2 > 0$) coefficients. Linear differential equation of the second order (10.8) has a trivial solution

$$\alpha(t) = 0\,. \tag{10.9}$$

Under any initial conditions $\alpha(0)$, $\dot{\alpha}(0)$ the solution to equation (10.8) asymptotically converges to the trivial one (10.9).

Equation (10.8) can be treated like a constraint imposed on the solution of system (10.1). In equations (10.1), angle φ_2 (and its derivatives) are replaced with angle α (and its derivatives) using relation $\varphi_2 = \varphi_1 + \alpha$. In resulting equations, the second derivative $\ddot{\alpha}$ is replaced with expression $-c_1\dot{\alpha} - c_2\alpha$, following equation (10.8). Then, after excluding the second derivative $\ddot{\varphi}$ in the equations, torque L can be found

$$L = L_d = -a_{12}\,(\dot{\varphi}_1 + \dot{\alpha})^2 \sin\alpha - b_1 \sin\varphi_1 + \\ + \frac{(c_1\alpha + c_2\dot{\alpha})(a_{11}a_{22} - a_{12}^2\cos^2\alpha) + (a_{11} + a_{12}\cos\alpha)[b_2\sin(\varphi_1 + \alpha) - a_{12}\dot{\varphi}^2\sin\alpha]}{a_{12}\cos\alpha + a_{22}}\,. \tag{10.10}$$

The denominator of expression (10.10) is non-zero if angle α is close to zero. With control torque (10.10) applied, angle α changes in accordance with equation (10.8), and therefore, system (10.1) (regardless of behavior of angle φ_1) has an asymptotically stable solution (10.9) where $\varphi_2(t) = \varphi_1(t)$. So feedback (10.10) helps "segregate" one equation (10.8) from the original nonlinear system (10.1). This "segregated" equation is linear.

Designing a feedback (generally, a nonlinear one) that renders the original system or its part linear is referred to as "feedback linearization" [96, 142].

Relation (10.10) describes a nonlinear feedback that can be realized if the parameters of the double pendulum are known.

With limitation (7.4) taking place, instead of control law (10.10), a saturated control law should be considered

$$L = \begin{cases} L_0 & \text{when} \quad L_d \geq L_0 \\ L_d & \text{when} \quad |L_d| \leq L_0 \\ -L_0 & \text{when} \quad L_d \leq -L_0\,. \end{cases} \tag{10.11}$$

If in expression (10.10) angle α together with its derivative $\dot{\alpha}$ is assigned zero, $\alpha = \dot{\alpha} = 0$, it becomes as follows:

$$L = L_d = \left(\frac{a_{11} + a_{12}}{a_{12} + a_{22}} b_2 - b_1\right)\sin\varphi_1\,. \tag{10.12}$$

System (10.1) with control law (10.12), like with law (10.10), has trivial solution (10.9). Numerical experiments show that the pendulum can be straightened using a less bulky control law than (10.10):

$$L = \left(\frac{a_{11} + a_{12}}{a_{12} + a_{22}} b_2 - b_1\right) \sin\varphi_1 + c_1\alpha + c_2\dot{\alpha}\,, \tag{10.13}$$

where c_1 and c_2 are constant coefficients. Of course, if there is restriction (7.4), the control signal (10.13) needs to be limited.

When the original system is simplified by introducing a feedback, usually an intermediate control parameter is introduced. Consider instead of (10.8) the following second-order equation:

$$\ddot{\alpha} = v\,, \tag{10.14}$$

where v is a new control parameter. According to equation (10.8), this control parameter follows relation $v = -c_1\alpha - c_2\dot{\alpha}$. However, its behavior can be assigned in a different way. Let the control parameter v be limited in absolute value: $|v| \le v_0$, where $v_0 = \text{const}$ is some terminal value. It is known [27, 130] that control law $v(\alpha, \dot{\alpha})$ that drives system (10.14) into the origin $\alpha = \dot{\alpha} = 0$ in the least possible time is as follows:

$$v = -v_0 \operatorname{sign}(2v_0\alpha + \dot{\alpha}|\dot{\alpha}|)\,. \tag{10.15}$$

From practical point of view, in order to avoid sliding mode, it is better not to use relay control law (10.15), but a continuous law instead

$$v = -v_0 th[\tau(2v_0\alpha + \dot{\alpha}|\dot{\alpha}|)]\,, \tag{10.16}$$

where τ is a parameter. Substituting in (10.10), expression (10.16) instead of $-c_1\alpha - c_2\dot{\alpha}$ yields a different control law

$$L = L_d = -a_{12}(\dot{\varphi}_1 + \dot{\alpha})^2 \sin\alpha - b_1 \sin\varphi_1 + \\ + \frac{v_0 th[\tau(2v_0\alpha + \dot{\alpha}|\dot{\alpha}|)](a_{11}a_{22} - a_{12}^2\cos^2\alpha) + (a_{11} + a_{12}\cos\alpha)[b_2\sin(\varphi_1 + \alpha) - a_{12}\dot{\varphi}^2\sin\alpha]}{a_{12}\cos\alpha + a_{22}}\,. \tag{10.17}$$

Numerical experiments show that control law (10.17) (with saturation), with appropriately chosen parameters v_0 and τ, successfully works at stages when the double pendulum requires to be straightened.

During pendulum swinging, control laws (10.6) (or (10.7)) and the one that straightens the pendulum interleave. At the last stage, when the system gets into domain of attraction, the control switches to the law that provides local stabilization of equilibrium (10.3). The next question discussed further will be this local stabilization law.

10.4 Linear model, local stabilization

Linearization of motion equations (10.1) about equilibrium (10.3) yields a system

$$a_{11}\ddot{\varphi}_1 + a_{12}\ddot{\varphi}_2 - b_1\varphi_1 = L, \quad a_{12}\ddot{\varphi}_1 + a_{22}\ddot{\varphi}_2 - b_2\varphi_2 = 0\,. \tag{10.18}$$

Equations (10.18) may be rewritten as a fourth-order system

$$\dot{z} = Gz + pL\,, \tag{10.19}$$

where

$$z = \begin{Vmatrix} z_1 \\ z_2 \\ z_3 \\ z_4 \end{Vmatrix} = \begin{Vmatrix} \varphi_1 \\ \varphi_2 \\ \dot{\varphi}_1 \\ \dot{\varphi}_2 \end{Vmatrix}, \quad G = \begin{Vmatrix} 0 & 0 & 1 & 0 \\ 0 & 0 & 0 & 1 \\ -A_0^{-1} & B & & \mathbf{0}_{2\times 2} \end{Vmatrix} = \begin{Vmatrix} 0 & 0 & 1 & 0 \\ 0 & 0 & 0 & 1 \\ a_{22}b_1/\Delta & -a_{12}b_2/\Delta & 0 & 0 \\ -a_{12}b_1/\Delta & a_{11}b_2/\Delta & 0 & 0 \end{Vmatrix},$$

$$p = \begin{Vmatrix} 0 \\ 0 \\ A_0^{-1}\begin{Vmatrix} 1 \\ 0 \end{Vmatrix} \end{Vmatrix} = \frac{1}{\Delta}\begin{Vmatrix} 0 \\ 0 \\ a_{22}b_1 \\ -a_{12}b_1 \end{Vmatrix}. \tag{10.20}$$

In expressions (10.20), in turn,

$$A_0 = \begin{Vmatrix} a_{11} & a_{12} \\ a_{12} & a_{22} \end{Vmatrix}, \quad B = \begin{Vmatrix} -b_1 & 0 \\ 0 & -b_2 \end{Vmatrix}, \quad \mathbf{0}_{2\times 2} = \begin{Vmatrix} 0 & 0 \\ 0 & 0 \end{Vmatrix}, \quad \Delta = \det A_0 = a_{11}a_{22} - a_{12}^2\,. \tag{10.21}$$

Matrices A_0 and B appear also in chapter 7. The value of $\Delta \neq 0$, because it is a determinant of matrix A_0 of system's kinetic energy (see (7.1)) with $\varphi_1 = \varphi_2$.

The expression for determinant of Kalman's controllability matrix is as follows. The derivation is omitted:

$$\det \|p \;\; Gp \;\; G^2p \;\; G^3p\| = -\frac{a_{12}^2 b_1^4 b_2^2}{\left(a_{11}a_{22} - a_{12}^2\right)^4} = -\frac{m_2^4 r_2^4 l^2 (m_1 r_1 + m_2 l)^4 g^6}{\left(a_{11}a_{22} - a_{12}^2\right)^4}\,.$$

If $r_2 \neq 0$, then this determinant is non-zero (it is natural to consider $m_2 \neq 0$ and $l \neq 0$), and thus system (10.18) or (10.19), (10.20) is completely controllable.

Note that chapter 7 also discusses the matter of controllability of a double pendulum with control torque applied in the pivot point. However, the problem in chapter 7 is different from the problem investigated here. The matter of controllability in chapter 7 is considered only after reducing the system to normal coordinates.

The characteristic equation of system (10.18) or (10.19), (10.20) is biquadratic (μ is the spectral parameter)

$$(a_{11}a_{22} - a_{12}^2)\mu^4 - (a_{11}b_2 + a_{22}b_1)\mu^2 + b_1 b_2 = 0\,, \tag{10.22}$$

because system (10.1), and hence system (10.18), with $L = 0$ is conservative. This equation, after substituting $\lambda = \mu^2$, coincides with (7.7). Equation (10.22) has two real positive roots $\mu_1 > 0$, $\mu_2 > 0$ and two roots that have equal absolute values and opposite signs compared to the first two, $-\mu_1 < 0$, $-\mu_2 < 0$, because $a_{11}b_2 + a_{22}b_1 > 0$, $b_1 b_2 > 0$.

Following the method developed in chapter 7, to build an algorithm to stabilize an unstable equilibrium that has a "large" domain of attraction, first of all, it is required to extract from system (10.18) two equations that correspond to the unstable modes. In chapter 7, as a first step in order to do so, the system is reduced to normal coordinates [40, 41]. Then, from equations written in terms of normal coordinates, the two desired equations are selected. However, it is possible to take such two equations bypassing system reduction to normal coordinates. By applying a nondegenerate linear transformation of the form $z = Sy$ where S is a constant matrix, system (10.19), (10.20) can be "immediately" reduced to Jordan form

$$\dot{y} = \Lambda y + dL \,, \tag{10.23}$$

where Λ is a diagonal matrix with eigenvalues in its main diagonal and zeros in the other positions

$$\Lambda = S^{-1}GS = \begin{Vmatrix} \mu_1 & 0 & 0 & 0 \\ 0 & \mu_2 & 0 & 0 \\ 0 & 0 & -\mu_1 & 0 \\ 0 & 0 & 0 & -\mu_2 \end{Vmatrix}, \quad d = S^{-1}p = \begin{Vmatrix} d_1 \\ d_2 \\ d_3 \\ d_4 \end{Vmatrix}. \tag{10.24}$$

(Note that mathematical software that can automatically find a Jordan form of a system. This software includes "Matlab" and many other programs). The first two scalar differential equations of system (10.23), (10.24) are the ones that correspond to the positive eigenvalues μ_1 and μ_2

$$\dot{y}_1 = \mu_1 y_1 + d_1 L, \quad \dot{y}_2 = \mu_2 y_2 + d_2 L \,. \tag{10.25}$$

Equations (10.25) differ from equations (7.15) only in the form of coefficients at control parameter L. Multiplying the first equation of system (7.15) by μ_1, the second one by μ_2, and taking instead of y_1, y_2, as a new Jordan variables products $\mu_1 y_1$, $\mu_2 y_2$, yields system (10.25), that is hence equivalent to (7.15).

System (10.18), and, consequently, (10.23), (10.24) is completely controllable in Kalman's sense. This implies that subsystem (10.25) is also fully controllable [89–92], and that $d_1 \neq 0$, $d_2 \neq 0$.

The boundary of controllability domain S of system (10.25) (in chapter 7 the domain of controllability in case when torque L is applied in the pivot is designated as $S^{(1)}$) consists of two integral trajectories of this system that correspond to $L(t) \equiv L_0$ and $L(t) \equiv -L_0$ [27, 53, 158]. These trajectories are symmetric to each other with respect to the origin. One of these trajectories that is built for $L(t) \equiv L_0$ begins with $t = 0$

at point

$$y_1 = \frac{d_1}{\mu_1} L_0, \quad y_2 = \frac{d_2}{\mu_2} L_0\,, \tag{10.26}$$

and ends when $t \to -\infty$ at point

$$y_1 = -\frac{d_1}{\mu_1} L_0, \quad y_2 = -\frac{d_2}{\mu_2} L_0\,. \tag{10.27}$$

The curve is described by the following parametric equations:

$$y_1(t) = \frac{d_1 L_0}{\mu_1}(2e^{\mu_1 t} - 1), \quad y_2(t) = \frac{d_2 L_0}{\mu_2}(2e^{\mu_2 t} - 1) \quad (-\infty < t \le 0)\,. \tag{10.28}$$

Equivalently, it can be said that trajectory (10.28) begins at $t = -\infty$ at point (10.27) and ends at $t = 0$ at point (10.26).

The other trajectory, built for $L(t) \equiv -L_0$, begins at $t = -\infty$ at point (10.26) and at $t = 0$ it ends at point (10.27). Its equations look like

$$y_1(t) = -\frac{d_1 L_0}{\mu_1}(2e^{\mu_1 t} - 1), \quad y_2(t) = -\frac{d_2 L_0}{\mu_2}(2e^{\mu_2 t} - 1) \quad (-\infty < t \le 0)\,. \tag{10.29}$$

Similarly to (7.34), the feedback that is applicable to system of equations (10.25) is described by expression

$$L = \begin{cases} -L_0 & \text{when} \quad \gamma\left(\frac{d_2}{\mu_2} y_1 - \frac{d_1}{\mu_1} y_2\right) \le -L_0 \\ \gamma\left(\frac{d_2}{\mu_2} y_1 - \frac{d_1}{\mu_1} y_2\right) & \text{when} \quad \left|\gamma\left(\frac{d_2}{\mu_2} y_1 - \frac{d_1}{\mu_1} y_2\right)\right| \le L_0 \\ L_0 & \text{when} \quad \gamma\left(\frac{d_2}{\mu_2} y_1 - \frac{d_1}{\mu_1} y_2\right) \ge L_0\,. \end{cases} \tag{10.30}$$

Expressions (10.26)–(10.30) can be obtained from expressions (7.17), (7.16), (7.19), (7.20), (7.34), by ignoring in the latter ones superscript (i) and substituting d_1/μ_1 and d_2/μ_2, with d_1 and d_2, respectively.

The characteristic equation for system (10.25) with linear control function

$$L = \gamma\left(\frac{d_2}{\mu_2} y_1 - \frac{d_1}{\mu_1} y_2\right)$$

(see the middle line in expression (10.30)) looks like (λ is the spectral parameter)

$$\lambda^2 + \lambda\left[\frac{\gamma d_1 d_2}{\mu_1 \mu_2}(\mu_2 - \mu_1) - (\mu_1 + \mu_2)\right] + \mu_1 \mu_2 = 0\,. \tag{10.31}$$

As $\mu_1 > \mu_2 > 0$, the roots of equation (10.31) are located in the left complex semiplane if and only if the coefficient at the first degree of the spectral parameter in this equation is positive, that is, when

$$\gamma d_1 d_2 < -\frac{\mu_1 \mu_2 (\mu_1 + \mu_2)}{\mu_1 - \mu_2}\,. \tag{10.32}$$

Equation (10.31) and inequality (10.32) can be derived from equation (7.35) and inequality (7.36), if in the latter ones the superscript (i) is ignored, values d_1 and d_2 are substituted for d_1/μ_1 and d_2/μ_2, respectively.

It follows from the results of chapter 7 that when the value of $|\gamma|$ is large enough, the domain of attraction B of system (10.25) controlled by feedback (10.30) is close enough to domain of controllability S. If the feedback gain γ satisfies inequality

$$|\gamma| > \frac{\mu_1\mu_2(\mu_1+\mu_2)}{|d_1d_2|(\mu_1-\mu_2)}, \tag{10.33}$$

i.e. its absolute value is large enough, then, according to inequality (10.32), it must satisfy condition

$$\operatorname{sign}\gamma = -\operatorname{sign}(d_1d_2). \tag{10.34}$$

Requirement (10.33) for system (10.25), (10.30) is similar to requirement (7.38), that was derived in chapter 7 for system (7.15), (7.34), and condition (10.34) is similar to (7.37).

10.5 Numerical experiments

Like in the previous chapter, the experiments were conducted on a pendulum with similar homogeneous links (see relations (7.21)), its parameters described by expressions (7.26).

Pendulum control law includes several stages. Inside region (10.5) control law (10.6) or (10.7) was used. The width of region (10.5) is determined by the value of $\Delta\varphi_1$, for numerical experiments it was taken equal to $\pi/12$. Experiments show, however, that the control law remains functional also for other values of $\Delta\varphi_1$. The values of $\Delta\varphi_1$ within range $(\pi/24, \pi/6)$ are acceptable. If the value of $\Delta\varphi_1$ is reduced, then interval (10.5), and also the time of energy pumping, become smaller. This makes the time to translate the pendulum to the desired equilibrium longer.

Outside region (10.5), for pendulum "straightening" control law (10.11) was used. To define torque L, both expressions (10.10), and expressions (10.13) or (10.17) were used. When using control law (10.11), (10.13) coefficients were assigned $c_1 = 2$, $c_2 = 0.4\,\mathrm{s}$. The control law remains efficient when these coefficients vary in a wide range.

Stabilization of the pendulum in its top equilibrium is done by means of control law (10.30). As it was mentioned many times above, different ways are possible to take from linear system (10.18) or from system (10.19) equations (10.25) that correspond to the positive eigenvalues μ_1 and μ_2. Different ways of transposing the system to Jordan form result in different values of coefficients d_1 and d_2 in control function L. Of course, eigenvalues μ_1 and μ_2, do not depend of transposition, they are determined solely by characteristic equation (7.7) and relations (7.14): $\mu_1 = 18.5611$, $\mu_2 = 6.92$. In numerical experiments, equations (10.25) were derived by means of an appropriate program in "Matlab" package. The values of coefficients d_1 and d_2 were determined

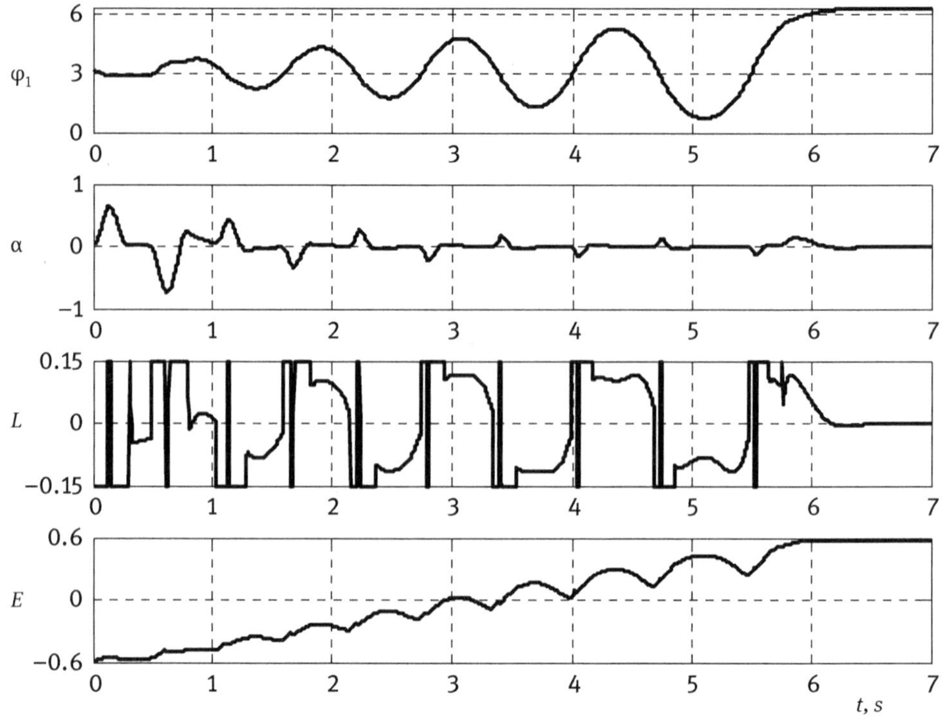

Fig. 10.1. Pendulum translation from the bottom equilibrium to the top one.

as $d_1 = -368.1672$, $d_2 = -56.8625$. In stabilization law (10.30) the gain was taken $\gamma = -0.02$. This value, as it can be easily checked, satisfies criterion (10.32).

Figure 10.1 illustrates the process of pendulum translation from stable equilibrium (10.4) to unstable (10.3) using control law (10.7), (10.11), (10.13), (10.30).

The first (topmost) plot shows the dynamics of the angle that represents deviation from vertical of the first link, φ_1. Function $\varphi_1(t)$ has a number of oscillations about value $\varphi_1 = \pi$, then it asymptotically converges to value $\varphi_1 = 2\pi$. The pendulum is swinging from side to side, the magnitude of oscillations increasing, approaching the top equilibrium. After several oscillations, the pendulum gets stabilized in that position. The second plot from the top shows dynamics of inter-link angle α. At the beginning of the process, this angle reaches values of up to $0.6 \div 0.75$. Later on, its peak values become smaller and smaller, and the plot for $\alpha(t)$ "clings" to axis $\alpha = 0$. At the end of the process, angle $\alpha \to 0$. Most of the time, the pendulum is straightened, and it behaves itself like a single-link one. The third plot shows how control torque L changes in time. All resources are used during the control process. The last (bottom) plot shows total energy $E = T + \Pi$ of the double pendulum. Its value "in average" increases monotonically. At the beginning of the pendulum swing process, when it is located in bottom equilibrium (10.4), total energy $E = E(0) = \Pi(0) = -b_1 - b_2$. At the end of this process, when it arrives in top equilibrium (10.3), the total energy becomes equal to $E = b_1 + b_2 = -E(0)$.

Like it was mentioned earlier, Figure 10.1 illustrates the process where the pendulum is swung by control law (10.7). The value of σ in that law is to be assigned manually. At first, the value of σ is taken as small as possible, that is, $\sigma = 1$. As the pendulum is swung up and its energy increases, the value of σ is increased several times, in a step-like manner, in order for the pendulum not to miss the desired equilibrium. When the value of the total energy of the system reaches the level of $3/4$ of the potential energy corresponding to equilibrium (10.3), the value of σ becomes equal to 5. The process of pendulum swinging becomes less intensive, its oscillation magnitude rises slower. When the speed of magnitude rise is slow, the swing process takes longer time, but there is a certain level of confidence that the system will not "miss" desired equilibrium (10.3), and that it will get into the domain of attraction. When the total energy becomes equal to $7/8$ of the potential energy in equilibrium (10.3), the value of σ increases once again, it becomes equal to 10. If the value of $\sigma > 1$, then, whenever the system is within range (10.5), the control resources are not utilized fully. When the system gets outside this region, the pendulum starts its straightening, and the control resources are used completely. At time instant $t \approx 5.75$ s (see Figure 10.1) the system crosses the boundary of the attraction domain, and it gets inside this domain. Then control law (10.30) comes into action. It stabilizes the pendulum in unstable equilibrium $\varphi_1 \equiv 2\pi$, $\alpha \equiv 0$.

Suggested control algorithm is capable to drive the pendulum to the top equilibrium not only from the bottom equilibrium, from (10.4). Numerical experiments show that this control law is efficient when initially the pendulum is significantly deviated from equilibrium (10.4). For example, the task of stabilization is successfully fulfilled when $\varphi_1(0) = \pi$, $\alpha(0) = \pi/3$, $\dot{\varphi}_1(0) = \dot{\alpha}(0) = 0$.

Note that the pendulum can be translated to the bottom equilibrium, or to a state that is close to that equilibrium, from any initial state. This can be done by applying in the pivot point a torque that damps pendulum oscillations. If the control algorithm described above can translate the pendulum from the bottom to the top equilibrium, it is then capable to provide global stability of that equilibrium under all initial conditions. Thus the suggested control algorithm provides global stability of the inverted pendulum.

In chapter 7 it is pointed out that, from the numerical experiments, it may seem that the torque applied in the pivot point is more efficient than the torque applied in the inter-link joint. Comparing results of the previous chapter 9 and results of the current chapter, one can get the same impression. The point is that the torque applied in the inter-link joint makes the double pendulum move from the bottom to the top in approximately 22.5 seconds (see Figure 9.2). The same pendulum, driven by a torque with the same limitations, but applied in the pivot point, moves to its top equilibrium in approximately 6.5 seconds (see Figure 10.1).

It is shown in chapter 7 that as $|\gamma|$ grows, the domain of attraction of the desired equilibrium of the linearized system increases, infinitely approaching the domain of controllability. However, it is easy to see that if $|\gamma| \to \infty$, then, following stability criteria (10.32) or (10.34), one of the roots of characteristic equation (10.31) approaches $-\infty$, while the other one, remaining negative, approaches 0. When an eigenvalue is close to zero, the transients in the closed-loop system become longer. Numerical experiments demonstrate such transient prolongation as the absolute value of gain γ grows. Therefore, it is better to avoid "extreme" values of $|\gamma|$.

11 Multi-link pendulum on a moving base

The main source of interest for investigation of pendulum-based systems is the possible application of the results to construction of monocycle machines [87, 109–113], development of tower cranes [10, 37], vibration robots [26, 171]. One of the newly developed personal transportation devices is the "Segway" [165]. It is effectively a platform based on two coaxial wheels [20], and it can be in some sense called a monocycle (see discussion in Part 1). A person riding such platform can be considered an inverted pendulum – single- or multi-link, with joints that include viscous friction elements and (or) some control torques.

The current chapter considers a plane motion of a *multi-link* pendulum that is mounted via a joint on a moving base – a wheel or a cart [114, 115]. The control torque applied between the base and the first link of the pendulum does not depend on base position or velocity, and it is limited in absolute value. The system coordinate that corresponds to the base position is cyclic. At that, the mathematical model allows separation of equations extracting those that describe *only* the pendulum motion. The resulting equations are different from known motion equations of a pendulum with a fixed base. The difference is in equation structure as well as in the parameters involved. The chapter also demonstrates the phase picture of free (without control) single-link pendulum motion on a wheel or on a cart. A feedback control is built that provides global stabilization of its top, unstable equilibrium. The time-optimal control law is designed.

11.1 Multi-link pendulum on a wheel

Consider a system consisting of a wheel and an n-link pendulum. The wheel can roll without slipping over a horizontal plane. The pendulum links can move in the same vertical plane with the wheel. Like in chapter 2, M will denote the wheel mass, R its radius, ρ its radius of gyration with respect to wheel center O, that will also be assumed the center of mass of the wheel. The first link of the pendulum is connected to the wheel by a joint in point O (Figure 11.1). The joint at point O will be counted *the first* one. At the other end of the first link another joint is located, where the second link connects to the first one, and so on. The links will be numbered in sequence, as they are connected to each other further away from point O, and the joint will be numbered likewise.

Angle β_0 denotes counter-clockwise rotation of some distinguished (in the wheel) radius that was initially (at the beginning of motion) aligned together with horizontal axis X (in chapter 2 this wheel rotation angle is denoted as φ). The movement of wheel

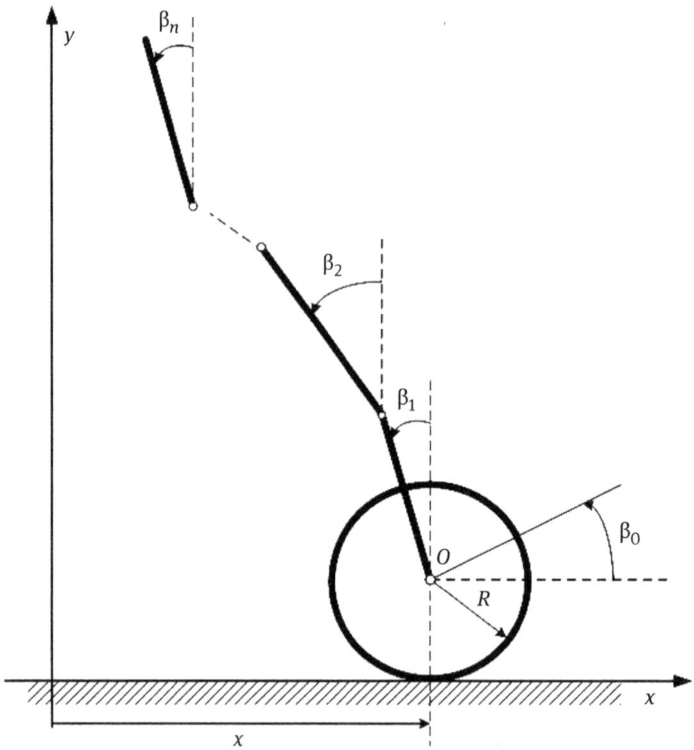

Fig. 11.1. Multi-link pendulum on a wheel.

center O along the horizontal axis will be denoted as x, so that $\dot{x} = -R\dot{\beta}_0$. The angle between the vertical line and the k–th link will be denoted as β_k $(k = 1, 2, \dots, n)$. An angle $\beta_k > 0$ if the link is turned counter-clockwise. The mass of the k–th link is denoted as m_k, the length of this link (the distance between the k–th joint and the joint number $k + 1$) – as l_k, the radius of gyration with respect to the k–th joint – as r_k. Let the center of mass of each link be located on the segment that connects its ends (the joints), and the distance to this center of mass from the k–th joint be b_k.

At joint O a control torque L is applied to the first link. If $L > 0$, this torque tends to turn this link counter-clockwise. The same (in absolute value) torque, but directed clockwise, is applied to the wheel. Such torque is generated, for example, by an electric DC motor with its stator attached to the wheel and its armature – to the first link of the pendulum. All joints are considered ideal (frictionless).

The kinetic energy of the wheel is

$$T_0 = \frac{1}{2}M(R^2 + \rho^2)\dot{\beta}_0^2 \,. \tag{11.1}$$

The kinetic energy of the k–th link can be presented as

$$T_k = \frac{1}{2}m_k\left[\left(R\dot\beta_0 + \sum_{i=1}^{k-1} l_i\dot\beta_i\cos\beta_i\right)^2 + \left(\sum_{i=1}^{k-1} l_i\dot\beta_i\sin\beta_i\right)^2 + \right.$$
$$\left. +2b_kR\dot\beta_0\dot\beta_k\cos\beta_k + 2\sum_{i=1}^{k-1} l_ib_k\dot\beta_i\dot\beta_k\cos(\beta_k-\beta_i) + r_k^2\dot\beta_k^2\right]. \qquad (11.2)$$

With $k = 1$ sums like $\sum_{i=1}^{k-1}\dots$ that enter in expression (11.2) and in further expressions "disappear". The potential energy Π and the virtual work δW of torque L are (g is the gravity acceleration)

$$\Pi = \sum_{k=1}^{n} m_kg\left(\sum_{i=1}^{k-1} l_i\cos\beta_i + b_k\cos\beta_k\right), \qquad \delta W = L(\delta\beta_1 - \delta\beta_0). \qquad (11.3)$$

Length l_n of the last link does not enter in expressions (11.2), (11.3), but these expressions involve distance b_n between the n–th joint and the mass center of the n–th link. Using relations (11.1)–(11.3) to build up the Lagrange's function $\mathscr{L} = T - \Pi$, the $n + 1$ equations of motion of the system can be derived.

Looking closer at expression (11.2), one can see that terms $Rl_k\dot\beta_0\dot\beta_k\sin\beta_k$ that appear in Lagrange's equations from calculating derivatives $d(\partial\mathscr{L}/\partial\dot\beta_k)/dt$ are reduced together with similar terms that result from derivatives $\partial\mathscr{L}/\partial\beta_k$ Thus angle β_0 enters in the equations of motion only as its second derivative $\ddot\beta_0$.

Let $\boldsymbol{q} = \|\beta_0, \boldsymbol{\beta}^{\mathrm{T}}\|^{\mathrm{T}}$ be the vector of generalized coordinates of the whole system, $\boldsymbol{\beta} = \|\beta_1, \dots, \beta_n\|^{\mathrm{T}}$ – the vector of generalized coordinates of the pendulum. The superscript ^{T} represents transposition. Then the motion equations can be written as follows

$$\boldsymbol{A}(\boldsymbol{\beta})\ddot{\boldsymbol{q}} + \boldsymbol{B}(\boldsymbol{\beta})\|\dot{\boldsymbol{q}}^2\| - \boldsymbol{G}\|\sin\boldsymbol{\beta}\| = \boldsymbol{Q}. \qquad (11.4)$$

Here $\boldsymbol{A}(\boldsymbol{\beta})$ is the symmetric matrix of kinetic energy, its dimensions are $(n + 1)\times(n + 1)$:

$$\boldsymbol{A}(\boldsymbol{\beta}) = \begin{Vmatrix} a_{00} & a_{01}\cos\beta_1 & a_{02}\cos\beta_2 & \cdot & a_{0n}\cos\beta_n \\ * & a_{11} & a_{12}\cos(\beta_1-\beta_2) & \cdot & a_{1n}\cos(\beta_1-\beta_n) \\ * & * & a_{22} & \cdot & a_{2n}\cos(\beta_2-\beta_n) \\ \cdot & \cdot & \cdot & \cdot & \cdot \\ * & * & * & \cdot & a_{nn} \end{Vmatrix}$$

(the asterisks in representation of matrix $\boldsymbol{A}(\boldsymbol{\beta})$ are placed wherever the omitted expressions can be restored assuming that the matrix is symmetric),

$$a_{00} = M(R^2+\rho^2) + R^2\sum_{i=1}^{n} m_i, \quad a_{0s} = R\left(b_sm_s + l_s\sum_{i=s+1}^{n} m_i\right) \quad (s = 1, \dots, n),$$
$$a_{jj} = r_j^2m_j + l_j^2\sum_{i=j+1}^{n} m_i, \qquad a_{js} = l_j\left(b_sm_s + l_s\sum_{i=s+1}^{n} m_i\right) \quad (j, s = 1, \dots, n)$$

(when $j, s = n$ sums $\sum_{i=j+1}^{n} m_i$, $\sum_{i=s+1}^{n} m_i$ "disappear"),

$$B(\beta) = \begin{Vmatrix} 0 & -a_{01}\sin\beta_1 & -a_{02}\sin\beta_2 & \cdot & -a_{0n}\sin\beta_n \\ 0 & 0 & a_{12}\sin(\beta_1-\beta_2) & \cdot & a_{1n}\sin(\beta_1-\beta_n) \\ 0 & -a_{12}\sin(\beta_1-\beta_2) & 0 & \cdot & a_{2n}\sin(\beta_2-\beta_n) \\ \cdot & \cdot & \cdot & \cdot & \cdot \\ 0 & -a_{1n}\sin(\beta_1-\beta_n) & -a_{2n}\sin(\beta_2-\beta_n) & \cdot & 0 \end{Vmatrix},$$

$$G = \frac{g}{R}\,\mathrm{diag}(0, a_{01}, \dots, a_{0n}), \quad \|\dot{q}^2\| = \|\dot\beta_0^2, \quad \dot\beta_1^2, \quad \dots, \quad \dot\beta_n^2\|^{\mathrm{T}},$$

$$\|\sin\beta\| = \|\sin\beta_0, \quad \sin\beta_1, \quad \dots, \quad \sin\beta_n\|^{\mathrm{T}}, \quad Q = \|-L, \quad L, \quad 0, \quad \dots, \quad 0\|^{\mathrm{T}}.$$

The first column in matrix $B(\beta)$, as well as in G, is zero. This is natural, since, like it was mentioned above, the motion equations do not involve angle β_0 or its velocity $\dot\beta_0$. Note that equations (11.4) include the second powers of generalized velocities $\dot\beta_k$ and do not include their products; and the submatrix of the n-th order that is in the right-bottom corner of matrix $B(\beta)$ is skew-symmetric. Relatively simple structure of equations (11.4) is a result of the choice of generalized coordinates: angles β_k of link deviation from the vertical line were chosen, and not the inter-link angles. The mathematical model of a human-like mechanism studied in [55, 65] has a similar structure.

Solving equation (11.4) for the highest order derivatives yields

$$\ddot{q} = A(\beta)^{-1}[Q - B(\beta)\|\dot{q}^2\| + G\|\sin\beta\|]\,. \tag{11.5}$$

The last n equations in system (11.5) involve only angles β_k ($k = 1, 2, \dots, n$) with their derivatives. These equations can be separately written as a system

$$\ddot\beta = F(\beta, \dot\beta, L)\,. \tag{11.6}$$

If torque L is generated by an electric DC motor with its stator attached to the wheel and its armature attached to the first pendulum link, then this torque depends on the relative angular velocity of the first link with respect to the wheel $\dot\beta_1 - \dot\beta_0$, because the counter-EMF that is proportional to this velocity is generated in the motor coil. Thus the expression that describes this torque involve the wheel angular velocity $\dot\beta_0$. However, if this counter-EMF is negligible and can be omitted in the mathematical model of the motor, then velocity $\dot\beta_0$ does not enter in the expression for torque L. Now, if the voltage supplied to the motor does not depend on wheel turn angle β_0 or its angular velocity $\dot\beta_0$, system (11.6) describes *exclusively* the motion of pendulum. Such decomposition is possible in more general situation, when the torques applied in *all* joints do not depend on angle β_0 and velocity $\dot\beta_0$. For example, these can be torques developed by viscoelastic springs. System (11.6) *is different* from the known system of equations used to describe motion of an n-link pendulum with a stationary base [95, 99, 151]. In particular, it includes equivalent moment of inertia of the wheel $M(R^2 + \rho^2)$, and with $n = 1$ the squared angular velocity of the pendulum $\dot\beta_1^2$ (see chapter 2 and the next section).

11.2 Single-link pendulum on a wheel

With $n = 1$ the original system consists of a wheel and a single-link pendulum mounted on it (see Figure 2.1). It has two degrees of freedom, and, according to (11.4), it is described by equations

$$\left[M(R^2+\rho^2)+m_1R^2\right]\ddot{\beta}_0 + m_1Rb_1\ddot{\beta}_1\cos\beta_1 - m_1Rb_1\dot{\beta}_1^2\sin\beta_1 = -L\,, \tag{11.7}$$

$$m_1Rb_1\ddot{\beta}_0\cos\beta_1 + m_1r_1^2\ddot{\beta}_1 - m_1b_1g\sin\beta_1 = L\,. \tag{11.8}$$

System (11.7), (11.8) is equivalent to system (2.4) or to (2.5), (2.6). In the current chapter, the results that were obtained in chapter 2 are updated and developed further.

The angular acceleration $\ddot{\beta}_0$ can be extracted from equation (11.7) and substituted into equation (11.8). This makes equation (11.6) a scalar one. The nondimensional time τ and nondimensional torque μ are introduced similarly to (1.3) as

$$\tau = t\frac{\sqrt{gb_1}}{r_1}, \quad \mu = \frac{L}{m_1gb_1}\,, \tag{11.9}$$

This transposes the scalar equation that involves only angle β_1 (in case when $n = 1$ the subscript index at variable β_1 is omitted) to system

$$\beta' = \omega, \quad (1-d^2\cos^2\beta)\omega' + d^2\omega^2\sin\beta\cos\beta - \sin\beta = (1+e^2\cos\beta)\mu\,. \tag{11.10}$$

Prime mark $'$ represents differentiating with respect to nondimensional time τ,

$$d^2 = \frac{m_1b_1^2}{M(R^2+\rho^2)+m_1R^2}\left(\frac{R}{r_1}\right)^2 < 1, \quad e^2 = \frac{m_1Rb_1}{M\,(R^2+\rho^2)+m_1R^2}\,. \tag{11.11}$$

System (11.10) is equivalent to second-order equation (2.7).

If wheel mass $M \to \infty$, then $d \to 0$, $e^2 \to 0$ (see expressions (11.11)), and equations (11.10) *convert* into motion equations of a "regular" pendulum with stationary pivot, that is natural. System (11.10) is *different* from these equations. It involves factor $1-d^2\cos^2\beta$ at derivative ω', which does not appear in equations that describe motion of a regular pendulum (refer to (1.4)). This factor effectively reduces (with $\beta \neq \pi/2$) the inertia moment of the pendulum as compared to the same pendulum with a stationary pivot. It is like the wheel-based pendulum has less inertia.

System (11.7), (11.8) with $L = 0$ can be linearized about the top position of the pendulum ($\beta = 0$, $\omega = 0$) or the bottom position ($\beta = \pi$, $\omega = 0$), and then transposed to normal coordinates [40, 41]. One of the characteristic measures (the so-called Poincaree's stability coefficients [40, 41], or the roots of the *secular equation* that the frequencies of small oscillations must satisfy) of the system will then be zero, and the other one – negative, if the system is linearized about state ($\beta = 0$, $\omega = 0$). The latter measure will be positive if the system is linearized about state ($\beta = \pi$, $\omega = 0$). Let the pivot of the pendulum be locked, or, in other words, let deviation $\Delta\beta_0$ about the equilibrium be zero, i.e. $\Delta\beta_0 = 0$ or $\Delta x = 0$. Locking the pendulum pivot is actually adding an additional constraint to the system. It is easy to verify that, in terms of normal coordinates, constraint $\Delta\beta_0 = 0$ involves both of these coordinates. Thus the *only* characteristic measure of the linearized model of a pendulum with this constraint is located *exactly* between the zero and the other characteristic measure of the linearized pendulum model without the constraint [40, 41]. Therefore, the oscillation frequency about the bottom equilibrium is lower for the pendulum with a stationary pivot than for the same pendulum that hangs from a wheel center. This statement can also be verified directly. It follows from equations (11.10) that the oscillation frequency (in terms of nondimensional time τ) about the bottom equilibrium of the pendulum that hangs on a wheel is equal to $1/\sqrt{1-d^2}$. This frequency is *higher* than that of the stationary-based pendulum, which is equal to one. Assume that the mass of the first link is concentrated entirely in its center of mass, that is, $b_1 = r_1$. Then, if the wheel mass $M \to 0$, then $d \to 1$ (see (11.11)), and the frequency of the pendulum oscillations about its bottom equilibrium rises to ∞. The described relation between the frequencies is true also for a load that hangs from a crane [10]. If the crane in its free motion (in absence of any controlling force or resistance) can move on the surface, then the frequency of oscillations of the load is higher than the oscillation frequency of the same load that hangs from a "locked" crane.

The second term in the second equation in (11.10) appears as a consequence of the centrifugal force that results from the pendulum rotation. This force is directed along the pendulum from the pivot point O outwards. The projection of this force onto axis X results in acceleration of the base (the wheel), and the moment of the inertia force applied to the center of mass of the pendulum is equal to this term. Term $e^2 \cos\beta$ enters in the right-hand side of the second equation in (11.10) because torque L is applied not only to the pendulum, but also to the wheel (see equation (11.7)), thus accelerating point O, that, in turn, influences the pendulum motion. In range $-\pi/2 < \beta < \pi/2$ this acceleration is directed one way, and in range $\pi/2 < \beta < 3\pi/2$ with the same value of control parameter μ it is directed the other way. Therefore, term $e^2 \cos\beta$ amplifies the control input when $-\pi/2 < \beta < \pi/2$, and it attenuates this input when $\pi/2 < \beta < 3\pi/2$.

With $\mu = \text{const}$ system (11.10) has a first integral

$$\frac{1}{2}(1 - d^2\cos^2\beta)\omega^2 + \cos\beta - (\beta + e^2 \sin\beta)\mu = \text{const}\,, \tag{11.12}$$

that is the energy integral when $\mu = 0$

$$E = \frac{1}{2}(1 - d^2\cos^2\beta)\omega^2 + \cos\beta = \text{const}\,. \tag{11.13}$$

Figure 11.2 illustrates the phase picture of solutions for system (11.10) in stripe $-\pi \le \beta \le \pi$ with $\mu = 0$ and $d = 0.8$. This phase picture, like the phase picture of a "regular" pendulum with a stationary pivot, has a saddle point ($\beta = 0$, $\omega = 0$) and the separatrices that pass through this point. On that picture there is also center ($\beta = \pm\pi$, $\omega = 0$) that is surrounded by looped trajectories; and there are also trajectories that

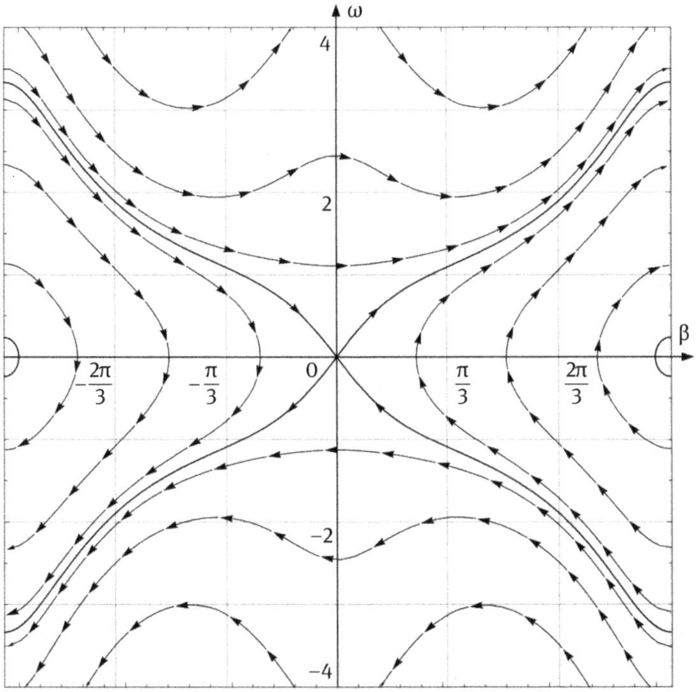

Fig. 11.2. Phase picture of an uncontrolled single-link pendulum on a wheel.

correspond to rotational motion modes of the pendulum. However, curves $\omega(\beta)$ that lie in the upper semiplane ($\omega > 0$) and have a "large" amount of energy have a local maximum at point $\beta = 0$, and two local minimums to the left and to the right of this maximum. The reason for this fact is the factor $1 - d^2 \cos^2 \beta$ at term ω^2 in energy integral (11.13). Such local maximums and minimums are present also in the phase picture of a free jointed double-link mechanism that is described in [143].

11.3 Global stabilization of the inverted pendulum

Let control input μ be limited in absolute value

$$|\mu| \leq \mu_0, \quad \mu_0 = \text{const}\,. \tag{11.14}$$

System (11.10) with $\mu = 0$ has an unstable equilibrium

$$\beta = 0, 2\pi \quad \omega = 0\,, \tag{11.15}$$

that corresponds to the inverted pendulum. States $\beta = 2\pi m$ ($m = 0, \pm 1, \pm 2, \ldots$) correspond to one and the same pendulum position, so it is reasonable to investigate the pendulum motion in a cylindrical phase space. In state (11.15) the total energy of the pendulum (see expression (11.13)) is $E = 1$.

Consider the problem of design of a control function $\mu = \mu(\beta, \omega)$ that drives the pendulum to the unstable equilibrium (11.15) and stabilizes it there. This control function must satisfy condition (11.14). Let in initial state $\beta(0)$, $\omega(0)$ the total energy $E(0) \leq 1$.

Full time derivative of the energy E with respect to system (11.10) is

$$\frac{dE}{d\tau} = (1 + e^2 \cos\beta)\omega\mu \, . \tag{11.16}$$

Derivative (11.16) is maximal when the control function is

$$\mu = \mu_0 \operatorname{sign}[(1 + e^2 \cos\beta)\omega] \, . \tag{11.17}$$

If $e^2 < 1$ then the sign of control input μ is determined entirely by the sign of velocity ω, that is, (11.17) is reduced to

$$\mu = \mu_0 \operatorname{sign}\omega \, . \tag{11.18}$$

If $e^2 > 1$ then control input μ changes its sign also in the "vicinity" of position $\beta = \pi$, regardless of the sign of velocity ω.

With relay control (11.17) (or (11.18)) energy E monotonically increases, and at some time instant it becomes equal to one. At this time, the representative point arrives at the separatrix (see Figure 11.2). Let after this instant

$$\mu = 0 \, . \tag{11.19}$$

Then the point on the phase plane will move along the separatrix towards the saddle point (11.15), and as $t \to \infty$, it will approach this point.

Equations (11.10), being linearized about the saddle point (11.15), look as follows:

$$\beta' = \omega, \quad (1 - d^2)\omega' - \beta = (1 + e^2)\mu \, . \tag{11.20}$$

System (11.20) can be transposed to a diagonal Jordan form (see equations (2.10), (2.11)). This transposition separates an equation that corresponds to the unstable motion mode

$$y' = \frac{1}{\sqrt{1 - d^2}} y + \frac{1 + e^2}{\sqrt{1 - d^2}} \mu \quad \left(y = \beta + \sqrt{1 - d^2}\,\omega\right) . \tag{11.21}$$

Domain of controllability P of system (11.20) is limited only in the "unstable" variable y (see inequality (2.12) and Figure 2.2)

$$|y| < (1 + e^2)\mu_0 \quad \left(\left|\beta + \sqrt{1 - d^2}\,\omega\right| < (1 + e^2)\mu_0\right) . \tag{11.22}$$

A linear (saturated) feedback that involves the unstable variable y (see expression (1.17))

$$\mu = \begin{cases} -\mu_0 & \text{when } \gamma\left[\beta + \left(\sqrt{1-d^2}\omega\right)\right] \le -\mu_0 \\ \gamma\left[\beta + \left(\sqrt{1-d^2}\omega\right)\right] & \text{when } \left|\gamma\left[\beta + \left(\sqrt{1-d^2}\omega\right)\right]\right| \le \mu_0 \\ \mu_0 & \text{when } \gamma\left[\beta + \left(\sqrt{1-d^2}\omega\right)\right] \ge \mu_0 \end{cases} \quad \left(\gamma < -\frac{1}{1+e^2}\right). \tag{11.23}$$

provides asymptotic stability of equilibrium (11.15) of linear system (11.20) with all initial states within area (11.22). Thus control law (11.23) provides maximal domain of attraction of equilibrium (11.15) of linear system (11.20). In accordance with Lyapunov's theorem, control (11.23) also provides asymptotic stability of solution (11.15) to nonlinear system (11.10).

If the time instant when the representative point of system (11.10), (11.19) crosses the separatrix is determined with an error, this point can then "miss" the saddle point. So it is appropriate to use control law (11.23) at the last stage of stabilization. If the error in determining the separatrix crossing instant is small, then control law (11.17), (11.19) drives system (11.10) within a finite time to domain of attraction of state (11.15), and then this system asymptotically reaches this state.

Thus control (11.17), (11.19), (11.23) moves system (11.10) from a given initial state to equilibrium (11.15).

If in initial state total energy $E(0) > 1$ then the pendulum requires to be damped, for example, by means of a relay control law that is opposite to law (11.17)

$$\mu = -\mu_0 \operatorname{sign}[(1 + e^2 \cos\beta)\omega], \tag{11.24}$$

so that the energy E will become equal to one, and then control (11.19), (11.23) can be applied. If $e^2 < 1$ then control law (11.24) is equivalent to law

$$\mu = -\mu_0 \operatorname{sign}\omega. \tag{11.25}$$

Thus control (11.17), (11.19), (11.23) or (11.24), (11.19), (11.23) provides the global stability of the inverted pendulum.

If the time when the separatrix is crossed is determined with a "large" error, then system (11.10) controlled by law (11.17), (11.19) or (11.24), (11.19) can move "far" from state (11.15), so that it will fail to get to its domain of attraction. So, a different control law is more "reliable" at the last stage of stabilization.

Consider the trajectories of system (11.10) in phase plane (β, ω) that pass through point $\beta = 0$, $\omega = 0$ when $\mu = \mp\mu_0$. Applying integral (11.12) yields the equations describing these curves

$$\omega = \sqrt{2[1 - \cos\beta - (\beta + e^2 \sin\beta)\,\mu_0]/(1 - d^2\cos^2\beta)}, \tag{11.26}$$

$$\omega = -\sqrt{2[1 - \cos\beta + (\beta + e^2 \sin\beta)\,\mu_0]/(1 - d^2\cos^2\beta)}. \tag{11.27}$$

When a phase point moves along curve (11.26) or (11.27) (driven by a corresponding control input $\mu = -\mu_0$ or $\mu = \mu_0$), it reaches state (11.15) in a finite time. Consider a part of trajectory (11.26) with $\beta \leq 0$ and a part of trajectory (11.27) with $\beta \geq 0$. These two parts are combined into a *switch curve K*.

It follows from investigation of integral (11.12) that $E = 1$ at point (11.15) and $E > 1$ at all other points of curve K. As the point moves closer to point (11.15) along curve K, energy E of the system strictly monotonically decreases. Inspection of the phase picture with $\mu = \mp\mu_0$ and $e^2 < 1$ shows that if $E(0) \leq 1$, then under control law (11.17) (or (11.18)) the trajectory of system (11.10) crosses the separatrix, and then it arrives at curve K. At that, of the phase point gets onto section (11.26) of this curve, which lies in semiplane $\omega > 0$, then control input μ must be switched from μ_0 to $-\mu_0$. If the phase point gets onto section (11.27), which lies in semiplane $\omega < 0$, then control input μ must be switched from $-\mu_0$ to μ_0. After the switching, system (11.10) reaches state (11.15) in a finite time.

In Figure 11.3, a blue line at $0 \leq \beta \leq 2\pi$ represents a trajectory along which nonlinear system (11.10) moves from its stable equilibrium $\beta = \pi$, $\omega = 0$ to the unstable one $\beta = 0$, $\omega = 0$. This trajectory is built for control (11.17), but after the representative point arrives at switch curve (11.27), the control input is set $\mu = \mu_0$. Switch curve K (see

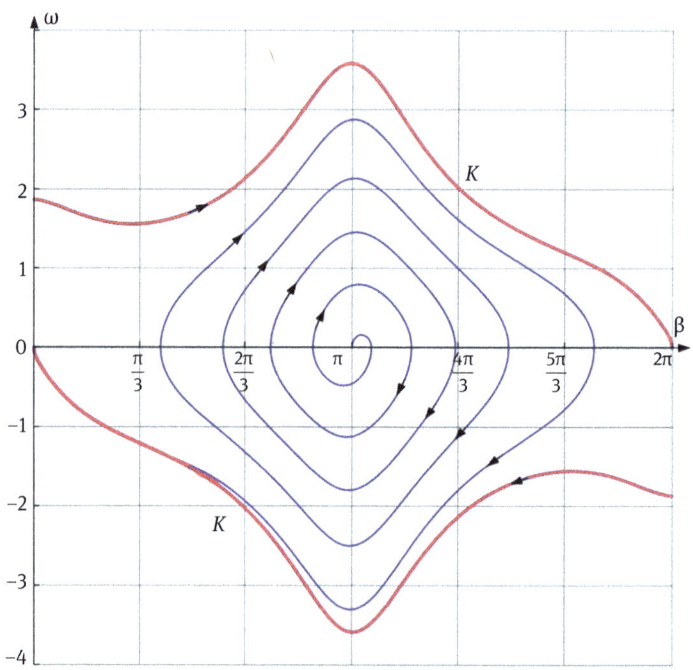

Fig. 11.3. Trajectory of a controlled pendulum motion from its bottom equilibrium to the top equilibrium.

pink lines) consists of two sections (11.26) and (11.27). System parameters are assigned so that $d = 0.8$, $e = 0.2$, $\mu_0 = 0.1$.

If $E(0) > 1$, two cases may occur. If at $t = 0$ the phase point is located in the area between the separatrix and the switch curve, then control law (11.17) will drive it at some time instant to the switch curve, and then control input $\mu = \mu_0$ or $\mu = -\mu_0$ will drive it in a finite time to point $\beta = 0$, $\omega = 0$ or to $\beta = 2\pi$, $\omega = 0$. If at $t = 0$ the phase point is located outside that area, then it is required to "slow" the pendulum applying control law (11.24). Still system (11.10) at some time instant will get onto the switch curve, and move to one of points (11.15).

11.4 Domain of controllability

Expression (11.22) describes the set of initial states from which the *linearized* system (11.20) can be translated to the origin $\beta = 0$, $\omega = 0$, – the so-called domain of controllability. Now the task is to build in stripe $-\pi \le \beta \le \pi$ of the phase plane (β, ω) a set D of states from which *nonlinear* system (11.10) can be translated to equilibrium $\beta = 0$, $\omega = 0$ *without oscillations* about position $\beta = \pi$. The task of finding such a domain is important for investigation of motion of a device like "Segway", because when a passenger is stabilized on such a device, any large angular deviations from vertical are unacceptable. And, of course, angular deviations greater than $\pi/2$ are impossible.

The sought domain of controllability D is apparently symmetric with respect to point $\beta = 0$, $\omega = 0$, because the set of possible values of control input (11.14) is symmetric with respect to value $\mu = 0$. As for velocities of system (11.10), when the position changes from point (β, ω) to point $(-\beta, -\omega)$, and the control input changes from μ to $-\mu$, the phase velocity vector (β', ω') becomes $(-\beta', -\omega')$. Thus it is enough to find only one border of domain D in phase plane (β, ω), for example, the right one. To build the right border of domain D, consider system (11.10) with $\mu = -\mu_0$. The phase picture of this system has a *saddle* point lying on axis β, the value of coordinate β being the least positive root of equation

$$\sin\beta = (1 + e^2 \cos\beta)\mu_0 . \tag{11.28}$$

Two separatrices come into this saddle point with $t \to +\infty$, and two more with $t \to -\infty$. Applying integral (11.12) and expression (11.28) helps finding these four separatrices. Figure 11.4 shows these separatrices built numerically with parameters $d = 0.8$, $e = 0.5$, $\mu_0 = 0.5$. The arrows indicate directions of motion of a phase point as time increases. Green lines represent the separatrices that come to the saddle point with $t \to +\infty$, blue lines – the separatrices that come to that point with $t \to -\infty$. Also by blue lines in Figure 11.4 the trajectories of system (11.10) with $\mu = -\mu_0$ that pass "close" to the saddle point (11.28) are shown. These trajectories look like hyperbolas. The separatrices that come to the saddle point with $t \to +\infty$ form the *right* border of domain D [104]. One more green line in Figure 11.4 shows the *left* border of domain D,

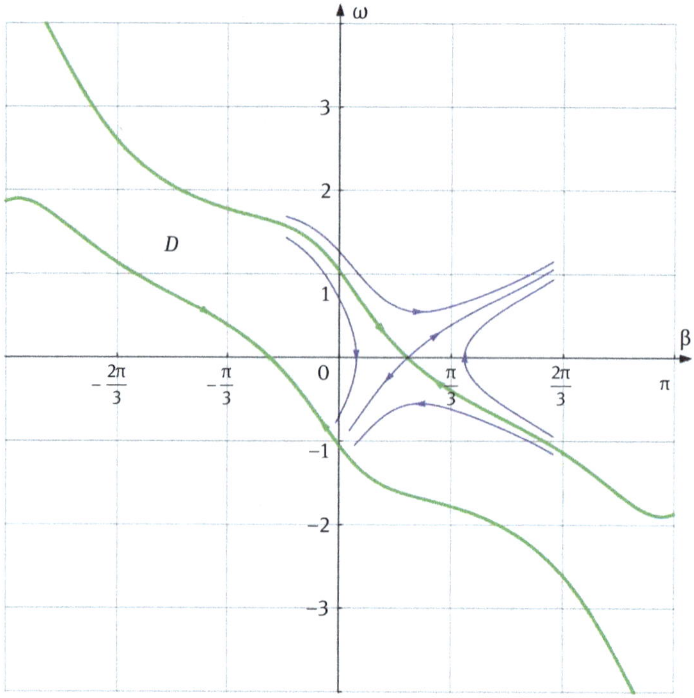

Fig. 11.4. Separatrices of motion equations with constant control input.

that can be found by rotating the right one 180° about point $\beta = 0$, $\omega = 0$. Thus the domain of controllability is complete, and its borders are shown by the green lines in Figure 11.4.

Inspection of Figure 11.4 shows that blue trajectories built for $\mu = -\mu_0$ that begin to the left of the right boundary of domain D remain in this area. Trajectories that begin to the right of this boundary go away from it. It will be shown that the boundary itself does not belong to domain D, i.e. that domain D is open. Consider an arbitrary point with $\omega > 0$ lying on the separatrix that comes into the saddle point with $t \to +\infty$. The phase velocity vector at this point is a geometric sum of two vectors

$$(\omega, 0) \quad \text{and} \quad (0, (1 + e^2 \cos\beta)\mu - d^2\omega^2 \sin\beta\cos\beta + \sin\beta)\,. \tag{11.29}$$

The first of vectors in (11.29) is parallel to the abscissa and directed rightwards. The second one is parallel to the ordinate axis. When $\mu = -\mu_0$, the second vector is directed downwards, because when $\mu = -\mu_0$, the vector sum is tangent to the separatrix. With $-\mu_0 < \mu \le \mu_0$, the ordinate of the second vector is greater than with $\mu = -\mu_0$, in which case the vector sum is directed rightwards from the right boundary of domain D. Therefore, none of the phase velocity vectors starting on this separatrix with $-\mu_0 \le \mu \le \mu_0$ is directed inside domain D. Now consider the separatrix that comes to the saddle point from semiplane $\omega < 0$ with $\mu = -\mu_0$. In the same manner as above, it can

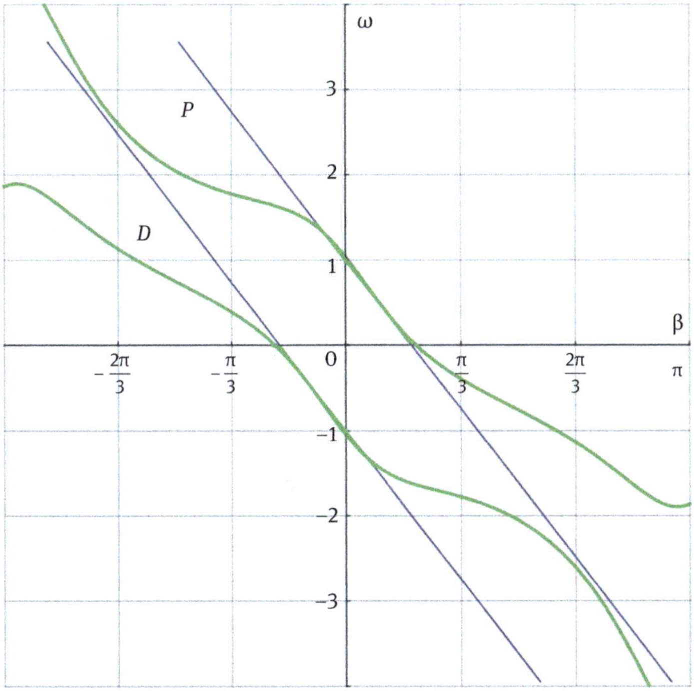

Fig. 11.5. Controllability domains *D* of the nonlinear model and *P* of the linear one.

be shown that none of the phase velocity vectors on this separatrix with $-\mu_0 \le \mu \le \mu_0$ is directed inside domain D. This means that it is impossible to get inside domain D from its boundary point, i.e. that domain D is open.

Figure 11.5 illustrates domain of controllability D for nonlinear system (11.10). Its borders are shown by green lines. Straight blue lines are the borders of controllability domain P (11.22) of linearized system (11.20). At Figure 11.5 one can see that domain (11.22) in the neighborhood of the desired equilibrium (11.15) encloses a "considerable" part of the exact domain D.

11.5 Designing time-optimal trajectories

In this section, a time-optimal control design will be considered. The optimal trajectories will be built inside domain of controllability D. It is assumed that time-optimal control functions exist for all points within this domain. To remind, domain D is built inside stripe $-\pi \le \beta \le \pi$. To find the optimal control law, the maximum principle will be used [27, 130].

The Hamiltonian function H looks like

$$H = \omega\psi_1 + \frac{\psi_2}{1 - d^2\cos^2\beta}\left[-d^2\omega^2 \sin\beta\cos\beta + \sin\beta + (1 + e^2\cos\beta)\mu\right], \tag{11.30}$$

where ψ_1 and ψ_2 are the costate variables. Hence the optimal control law (that maximizes function H) is determined by expression (remember that $d^2 < 1$)

$$\mu = \mu_0 \operatorname{sign}[(1 + e^2\cos\beta)\psi_2]. \tag{11.31}$$

If $e^2 < 1$ then instead of (11.31) the following expression can be used

$$\mu = \mu_0 \operatorname{sign}\psi_2. \tag{11.32}$$

Differential equations for costate variables are as follows

$$\begin{aligned}
\psi_1' &= \psi_2 \frac{d^2\omega^2\cos 2\beta - \cos\beta - d^4\omega^2\cos^2\beta + d^2\cos\beta(1+\sin^2\beta) + \sin\beta[d^2\cos\beta(2+e^2\cos\beta)+e^2]\mu}{(1-d^2\cos^2\beta)^2} \\
\psi_2' &= -\psi_1 + \psi_2\frac{d^2\omega\sin 2\beta}{1 - d^2\cos^2\beta}.
\end{aligned} \tag{11.33}$$

An optimal control law cannot include sections of singular mode, namely, where $\psi_2(t) \equiv 0$ [104]. Indeed, if on some time interval this identity takes place, then, as it follows from the second equation of system (11.33), on this interval $\psi_1(t) \equiv 0$. However, according to the maximum principle [27, 130], the vector function $\psi(t) = (\psi_1(t), \psi_2(t))$ cannot become zero. Thus only the terminal values are allowed for the optimal control function, $\mu = \pm\mu_0$.

The optimal trajectories will be built using the reverse motion. In other words, equations (11.10), (11.31), (11.33) will be numerically integrated in reverse time, starting from state $\beta = 0$, $\omega = 0$, $\psi_1 = \cos\delta$, $\psi_2 = \sin\delta$. The parameter δ here should take all values from within interval $[0, 2\pi)$. In practice, a discrete step can be used. The reverse time integration is stopped when the trajectory of system (11.10) leaves stripe $-\pi \le \beta \le \pi$.

Figure 11.6 illustrates the controlled time-optimal motion inside domain D of the system with numerical values of parameters $d = 0.8$, $e = 0.5$, $\mu_0 = 0.5$. Domain of controllability D is shown by its green borders. Switch curve K is described by equations (11.26), (11.27), and it is a pink line in Figure 11.6. Blue lines on the same figure are the time-optimal trajectories. The phase point at first moves along a blue line until it reaches switch curve K. Then it moves along this curve K all the way to the origin.

As $M \to \infty$, the picture shown in Figure 11.6 transforms into the time-optimal control design picture for a regular pendulum with a fixed pivot [133, 134].

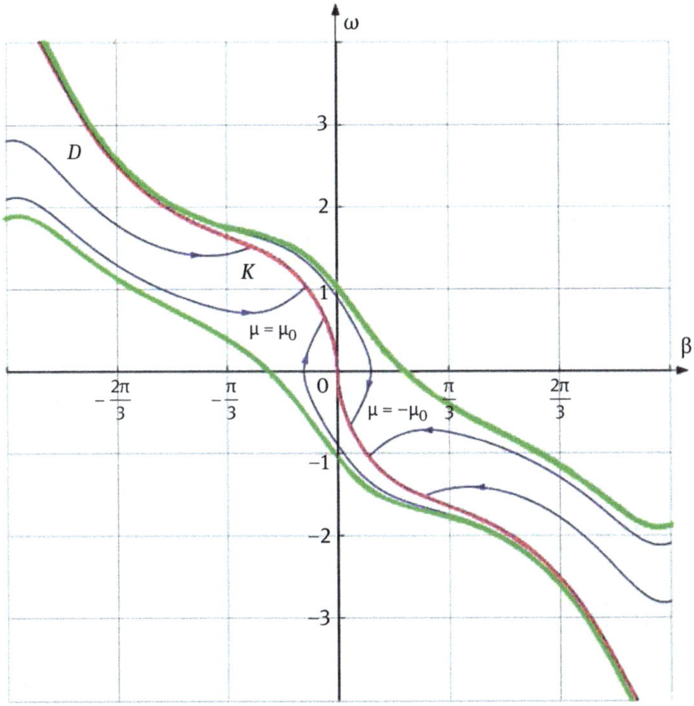

Fig. 11.6. Time-optimal motion trajectories of a wheel-based pendulum.

11.6 A pendulum on a cart

This section discusses a system that consists of a cart and an n-link pendulum. The pendulum is attached to the cart by a joint (Figure 11.7). The cart may move along a horizontal straight line X without any friction or other resistance. Its position on axis X is represented by coordinate x. A cart with an n-link pendulum that is hanging down can be used as a model of a crane system with a load that hangs from the crane arm on a flexible wire rope. Unlike the system considered here, the crane system is driven by a control input that is applied to the cart and moves it in order to position the load to a desired location.

Similar to the pendulum on a wheel, it will be assumed that a torque L is applied to the first link. However, unlike the wheel-based pendulum, this torque does not *directly* influence the cart motion, and thus it does not enter into the equation of motion that correspond to the generalized coordinate of the cart x. To derive equations of motion of the multi-link pendulum on a cart, the angular acceleration $\ddot{\beta}_0$ in equations (11.4) must be replaced with linear acceleration $\ddot{x}$ using relation $\ddot{\beta}_0 = -\ddot{x}/R$, and the first element $-L$ in column-matrix $\boldsymbol{Q}$ must be set equal to zero. To make use of the expressions for coefficients of matrices $\boldsymbol{A}(\boldsymbol{\beta})$ and $\boldsymbol{B}(\boldsymbol{\beta})$, the value of R must be set equal

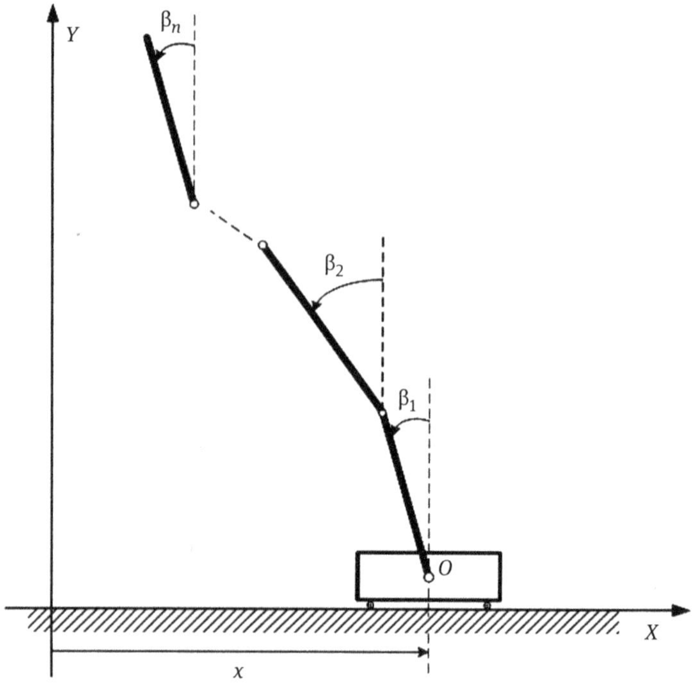

Fig. 11.7. Multi-link pendulum on a cart.

to one, and ρ – to zero. If torque L does not depend on coordinate x of the cart nor on its velocity $\dot{x}$ then in motion equations displacement value x will be represented only by its second derivative $\ddot{x}$. Excluding acceleration $\ddot{x}$ from these equations yields a system of n equations like (11.6), that involves only angles β_k $(k = 1, 2, \dots, n)$ and their derivatives. The resulting equations are different from equations of motion of a pendulum with a stationary base [95, 99, 151].

As for equations of motion of a cart with attached single-link ($n = 1$) pendulum, they can be derived from equations (11.7), (11.8) if the angular acceleration $\ddot{\beta}_0$ is replaced with linear one $\ddot{x}$ by means of relation $\ddot{\beta}_0 = -\ddot{x}/R$, and also if in equation (11.7) the right-hand side is set equal to zero. Excluding acceleration $\ddot{x}$ from these equations, introducing nondimensional time (11.9) and nondimensional parameters (11.11) again yields system (11.10), (11.11), yet here it is already assumed that $\rho = 0$, $e = 0$. In virtue of equality $e = 0$, this system is simpler than the system that describes motion of the wheel-based pendulum. Control laws (11.17), (11.24) and (11.31) with $e = 0$ turn into (11.18), (11.25) and (11.32), respectively; equation (11.28) can be solved for the pendulum deviation angle: $\beta = \arcsin \mu_0$. The results formulated above for the wheel-based pendulum and illustrated in figures 11.3–11.6 are qualitatively the same for the cart-based pendulum.

11.7 Frequency lowering as a result of constraining

Let the torque applied to the n-link pendulum ($n > 1$) at its pivot point be zero, $L = 0$. The considered mechanical system is then conservative. The original complete equations of system (11.4) can be linearized about equilibrium $\beta_k \equiv \pi$ ($k = 1, 2, \ldots, n$), that is, when all the links hang down. Coordinate β_0 (or x) is cyclic, thus the linearized model has a zero characteristic measure. Besides, the linear model has n positive measures. Each characteristic measure has a corresponding natural frequency. Among these frequencies, there is a zero frequency that will *not* be regarded further. Now suppose that the cart (or, equivalently, the wheel) is fixed, i.e. a constraint is imposed on the system, $x = 0$. Then, in accordance with the theorem of constraint influence [40, 41, 43, 71], each characteristic measure of the linearized constrained system is located between the measures of the unconstrained system, or coincides with one of them. The zero characteristic measure "disappears" when a constraint is added. At that, a positive measure appears, that can be less than, or equal to, the positive measure of the original unconstrained system that is closest to zero. This means, the frequencies of the constrained system are *less than or equal to* the frequencies of the unconstrained system. If deviation Δx is a linear combination of *all* principal coordinates (the constraint involves all principal coordinates), then introducing constraint $\Delta x = 0$ makes characteristic measures of the constrained system interleave with measures of the unconstrained one [40, 41]. Besides, adding constraint $\Delta x = 0$ moves all frequencies to the *left*. In this case, the frequencies of the pendulum oscillations about the bottom, stable equilibrium become *lower*.

At first glance, the provided conclusions concerning the frequency change in the considered system is in contradiction with Rayleigh's theorem that claims that whenever a constraint is added to a system, its frequencies *cannot become lower* [43, 71]. However, there is no contradiction. Actually, a zero characteristic measure is present, along with a corresponding zero natural frequency, that usually is *not considered* as a frequency. Since the zero measure "disappears" as the constraint is added, and all the other measures do not become larger, the frequencies also do not rise. Examples of such systems with a zero characteristic measure can be found, for instance, in study [128].

Part III: **Ball on a beam**

This part discusses the problem of stabilizing a ball that can roll without slipping on a rod, or a beam [12–14, 61, 139, 154, 155]. The beam may turn about its pivot point that is located below it. Thus the beam is similar to an inverted pendulum. The considered system has two degrees of freedom. A torque developed by an electric DC motor is applied in the pivot. Voltage supplied to the motor is assumed limited in absolute value. The system has an unstable (when no control torque is applied) equilibrium that is to be stabilized by means of the motor. Two cases are investigated. In one case, the beam is considered straight. Then the linearized system has one unstable motion mode, i.e. system's degree of instability is equal to one. In the other case, the beam is curvilinear. More exactly, it is a part of a circle. When the radius of this circle is small enough, that is, the curvature is large, the linearized model has two unstable modes, and the degree of instability is equal to two. In this latter case, the system is more difficult to stabilize than in the first case.

12 Stabilization of a ball on a straight beam

The current chapter considers the problem of stabilization of a ball that can roll without slipping on a *straight* beam. In equilibrium that is to be stabilized the beam is located horizontally, and the ball is in the middle of the beam, just above its pivot point. Without control this position is unstable. The system has two degrees of freedom, and it is controlled by only a single torque, thus it is underactuated. When solving the stabilization problem, it is required to build a feedback in such a way that the domain of attraction of the equilibrium would be maximum possible.

The system with a straight beam is typically referred to as the "beam-and-ball system" [49, 82, 125, 139, 154, 155]

12.1 Mathematical model of the system

Consider a mechanical system illustrated in Figure 12.1. It includes a homogeneous straight beam and a ball on it. Point A is the middle of the beam, and at the same time it is its center of mass. The ball can roll on the beam without slipping. Point C_1 is the center of mass of the beam together with its support OA. Point C_2 is the geometric center of the ball, and at the same time it is its center of mass; r is the radius of the ball. The mass of the beam together with support OA and the mass of the ball are denoted respectively as m_1 and m_2. Let ρ_1 and ρ_2 designate the inertia radiuses, so $I_1 = m_1\rho_1^2$ and $I_2 = m_2\rho_2^2$ are moments of inertia of the beam together with the support OA with respect to pivot point O and of the ball with respect to its center C_2. Let $OA = l$ and $OC_1 = a$ $(a < l)$.

The position of the system is determined by two generalized coordinates – angles θ and φ (see Figure 12.1). Angle θ represents deviation of support OA from the vertical, or, equivalently, deviation of the rod from the horizontal. Angle φ describes the rotation of the ball; and $\varphi = 0$ when the ball is located in the center A of the beam. The

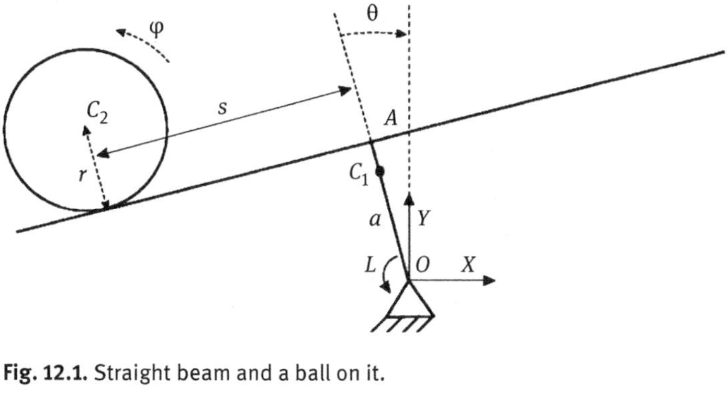

Fig. 12.1. Straight beam and a ball on it.

position of the ball on the beam can also be described by distance $s = r\varphi$. The motion of the system takes place in a vertical plane, i.e. the problem discussed here is plane. So it may be as well considered as the problem of stabilizing a *wheel* on a beam.

An electric DC motor is installed in pivot point O. This motor is capable of producing torque L. The torque is proportional to the electric current in the armature coil. Neglecting inductivity of the coil (that determines an electromagnetic time constant of the armature circuit) the torque may be written as [42, 76, 108]

$$L = c_u u - c_v \dot{\theta} \,. \tag{12.1}$$

Here u is the voltage supplied to the motor. The positive constants c_u and c_v for a particular motor can be calculated knowing its technical specification: the stall torque, nominal voltage, no-load torque and no-load speed of the armature [76]. Product $c_v\dot{\theta}$ is the counter-electromotive torque. Viscous friction torque in pivot O, if at all present, is also proportional to angular velocity $\dot{\theta}$.

The voltage supplied to the motor will be assumed limited in absolute value

$$|u| \leq u_0, \quad u_0 = \text{const} \,. \tag{12.2}$$

If the counter-EMF $c_v\dot{\theta}$ in relation (12.1) is neglected, then (12.2) becomes a limitation imposed on torque L that is applied in the pivot. This is precisely the type of limitation considered in the chapters above.

The next step is to build up differential equations that describe motion of this mechanical system. Expressions for kinetic energy T and potential energy Π are

$$\begin{aligned} T &= \frac{1}{2}\left\{m_1\rho_1^2\dot{\theta}^2 + m_2\left[r^2\varphi^2 + (r+l)^2\right]\dot{\theta}^2 + 2m_2 r(r+l)\,\dot{\varphi}\dot{\theta} + m_2(r^2+\rho_2^2)\dot{\varphi}^2\right\}, \\ \Pi &= m_1 g a\cos\theta + m_2 g[(r+l)\cos\theta - r\varphi\sin\theta] \,. \end{aligned} \tag{12.3}$$

Using relations (12.3), the system motion equations can be derived by means of Lagrangian approach of the second kind [16, 41, 73]. Expression (12.1) for torque L is then substituted in these equations:

$$\begin{aligned} &\left[m_1\rho_1^2 + m_2(r+l)^2 + m_2 r^2\varphi^2\right]\ddot{\theta} + m_2 r(r+l)\ddot{\varphi} + 2m_2 r^2\varphi\dot{\varphi}\dot{\theta} - \\ &\qquad - g[m_1 a + m_2(r+l)]\sin\theta - m_2 g r\varphi\cos\theta = c_u u - c_v\dot{\theta} \,, \end{aligned} \tag{12.4}$$

$$r(r+l)\ddot{\theta} + (r^2+\rho_2^2)\ddot{\varphi} - r^2\varphi\dot{\theta}^2 - gr\sin\theta = 0 \,. \tag{12.5}$$

If the viscous friction torque in pivot O is significant and it needs to be regarded, then in the right-hand side of equation (12.4) term $-f\dot{\theta}$ enters, constant f representing the viscous friction coefficient.

If $u = 0$, then system (12.4), (12.5) has an unstable equilibrium

$$\theta = 0, \quad \varphi = 0 \quad (s = 0), \quad \dot{\theta} = 0, \quad \dot{\varphi} = 0 \quad (\dot{s} = 0) \,. \tag{12.6}$$

In state (12.6) the beam is positioned horizontally, and the ball is in its middle, just above pivot O.

The task is to design a feedback that renders state (12.6) asymptotically stable. This feedback should make the domain of attraction of this state as large as possible.

12.2 Linearized model

Linearizing equations of motion (12.4), (12.5) about unstable equilibrium (12.6) yields

$$\left[m_1\rho_1^2 + m_2(r+l)^2\right]\ddot{\theta} + m_2 r(r+l)\ddot{\varphi} - g[m_1 a + m_2(r+l)]\theta - m_2 g r\varphi = c_u u - c_v\dot{\theta}\,, \tag{12.7}$$

$$r(r+l)\ddot{\theta} + \left(r^2 + \rho_2^2\right)\ddot{\varphi} - gr\theta = 0\,. \tag{12.8}$$

Differential equations (12.7), (12.8) may be presented in Cauchy's form, as a system of four first-order equations. This system can be then written in matrix form

$$\dot{x} = Ax + bu = \begin{Vmatrix} 0_{2\times 2} & I_{2\times 2} \\ D^{-1}E & D^{-1}\begin{Vmatrix} -c_v & 0 \\ 0 & 0 \end{Vmatrix} \end{Vmatrix} x + \begin{Vmatrix} 0 \\ 0 \\ D^{-1}\begin{Vmatrix} c_u \\ 0 \end{Vmatrix} \end{Vmatrix} u\,. \tag{12.9}$$

Here $x = \|\theta, \varphi, \dot{\theta}, \dot{\varphi}\|^{\mathrm{T}}$ is the vector of system state variables, $0_{2\times 2}$ is a zero matrix of size 2×2, $I_{2\times 2}$ is an identity matrix of the same size. The superscript ^{T} denotes transposition. Matrices D and E are as follows:

$$D = \begin{Vmatrix} m_1\rho_1^2 + m_2(r+l)^2 & m_2 r(r+l) \\ r(r+l) & r^2 + \rho_2^2 \end{Vmatrix}, \qquad E = g\begin{Vmatrix} m_1 a + m_2(r+l) & m_2 r \\ r & 0 \end{Vmatrix}. \tag{12.10}$$

Matrix D is positive-definite, because it is effectively a matrix of kinetic energy of the system with $\varphi = 0$.

Looking at the determinant of controllability matrix with $c_v = 0$, it is clear that in this case system (12.9) (or system (12.7), (12.8)) is completely controllable in Kalman's sense [89–92]. Thus it is completely controllable when $c_v \neq 0$ as well. Indeed, it follows from the fact of controllability with $c_v = 0$ that for any given initial state $x^*(0)$ and any final state $x^*(T)$ there can be found a corresponding control function $u^*(t)$ (or control torque $L^*(t) = c_u u^*(t)$). This control function makes trajectory of the system $x^*(t)$ that starts from that given point $x^*(0)$ reach the final point $x^*(T)$ in a given time T. During this motion, dynamics of angle θ is described by some pattern $\theta^*(t)$. Now, let $c_v \neq 0$. Using expression (12.1), a new control function can be assigned

$$u(t) = \frac{1}{c_u}[L^*(t) + c_v\dot{\theta}^*(t)]\,. \tag{12.11}$$

If $x(0) = x^*(0)$, then system (12.9) (or (12.7), (12.8)) with control function (12.11) has the same solution like the system with $c_v = 0$ and control function $u = u^*(t)$. This actually means that with control function (12.11) the system reaches a given state $x^*(T)$ in a finite time T, that is, it is completely controllable.

Using a nondegenerate linear transformation $x = Sy$ with a constant matrix S system (12.9) can be transposed to make a Jordan form

$$\dot{y} = \Lambda y + du\,, \tag{12.12}$$

where Λ is a diagonal matrix with eigenvalues of matrix A filling its main diagonal, and zeros outside that diagonal

$$\Lambda = S^{-1}AS = \begin{Vmatrix} \lambda_1 & 0 & 0 & 0 \\ 0 & \lambda_2 & 0 & 0 \\ 0 & 0 & \lambda_3 & 0 \\ 0 & 0 & 0 & \lambda_4 \end{Vmatrix}, \quad d = S^{-1}b = \begin{Vmatrix} d_1 \\ d_2 \\ d_3 \\ d_4 \end{Vmatrix}. \tag{12.13}$$

Here values $\lambda_1, \lambda_2, \lambda_3, \lambda_4$ are roots of the characteristic equation

$$a_0\lambda^4 + a_1\lambda^3 + a_2\lambda^2 + a_3\lambda + a_4 = 0\,, \tag{12.14}$$

where

$$\begin{aligned} &a_0 = \det D > 0, \quad a_1 = c_v(r^2 + \rho_2^2) > 0\,, \\ &a_2 = m_2 g(r + l)(r^2 - \rho_2^2) - m_1 ga(r^2 + \rho_2^2), \quad a_3 = 0, \quad a_4 = \det E = -m_2 g^2 r^2 < 0\,. \end{aligned} \tag{12.15}$$

With $c_v = 0$, equation (12.14) is biquadratic, and its roots are located on the complex plane symmetrically to one another with respect to the origin. To be more exact, the equation has two roots that are equal in absolute value and have different signs, one positive and one negative, and two roots that are purely imaginary. This follows from the fact that with $c_v = 0$ and $u = 0$ nonlinear system (12.4), (12.5), and hence linear system (12.7), (12.8), both are conservative. In further discussion, it will be considered that $c_v > 0$.

Equation (12.14) always has at least one positive root, because $a_0 > 0$, and $a_4 < 0$, and also it has one negative root. The sequence of coefficients a_0, a_1, a_2, a_3, a_4 has exactly *one* sign change (see relations (12.15)), independently of the sign of coefficient a_2 (it can also be equal to zero). Therefore, in accordance with Descartes rule of signs [97], this equation has *exactly one positive* root. Equation (12.14) does not have a zero root, because $a_4 \neq 0$. It also does not have purely imaginary roots of type $\pm i\omega$. Indeed, substituting in equation (12.14) value $i\omega$ for the spectral parameter and setting the imaginary part of this equation equal to zero yields equation $a_1\omega^3 = 0$. However, $a_1 \neq 0$, thus $\omega = 0$. To completely resolve the question of root location on the complex plane, Routh–Hurwitz criterion can be applied [40, 41, 97]. According to this criterion, the number of roots of equation (12.14) that have their real part positive is equal to the number of sign changes in sequence

$$T_0, \quad T_1, \quad T_1T_2, \quad T_2T_3, \quad T_3T_4\,, \tag{12.16}$$

where values T_0, T_1, T_2, T_3, T_4 are

$$T_0 = a_0, \quad T_1 = a_1, \quad T_2 = \begin{vmatrix} a_1 & a_0 \\ 0 & a_2 \end{vmatrix} = a_1 a_2,$$

$$T_3 = \begin{vmatrix} a_1 & a_0 & 0 \\ 0 & a_2 & a_1 \\ 0 & a_4 & 0 \end{vmatrix} = -a_1^2 a_4, \quad T_4 = \begin{vmatrix} a_1 & a_0 & 0 & 0 \\ 0 & a_2 & a_1 & a_0 \\ 0 & a_4 & 0 & a_2 \\ 0 & 0 & 0 & a_4 \end{vmatrix} = a_4 T_3 .$$

Values T_2, T_3, T_4 are effectively the Routh–Hurwitz determinants written for case $a_3 = 0$. The sequence of values (12.16) is presented below. The signs of these values are written wherever they can be found from signs of coefficients (12.15)

$$T_0 = a_0 > 0, \quad T_1 = a_1 > 0, \quad T_1 T_2 = a_1^2 a_2, \quad T_2 T_3 = -a_1^3 a_2 a_4, \quad T_3 T_4 = a_4 T_3^2 < 0. \tag{12.17}$$

To remind, $a_4 < 0$, and the sign of coefficient a_2 is unknown. Suppose that $a_2 > 0$, then $T_1 T_2 > 0$, $T_2 T_3 > 0$ and hence sequence (12.17) has exactly one sign change. If $a_2 < 0$, then $T_1 T_2 < 0$, $T_2 T_3 < 0$ and therefore sequence (12.17) again has exactly one sign change. Case $a_2 = 0$ is not considered here; it is very specific. So there is only one root of equation (12.14) located in the right semiplane of the complex plane.

To summarize, equation (12.14) has *one real positive root* and three roots with negative real part.

12.3 Feedback design

Let λ_1 be the real positive root of equation (12.14). All the other roots of this equation, as it was discussed above, are located in the left half of the complex plane, i.e. $\mathrm{Re}\lambda_i < 0$ $(i = 2, 3, 4)$. The first equation of system (12.12), that corresponds to eigenvalue λ_1 looks like

$$\dot{y}_1 = \lambda_1 y_1 + d_1 u . \tag{12.18}$$

If system (12.9) is completely controllable in Kalman's sense, then subsystem (12.18) (being just a scalar equation) of this system is also completely controllable. Then scalar value $d_1 \neq 0$, as well as all the other values d_i $(i = 2, 3, 4)$. Let W designate a set of piecewise-continuous functions $u(t)$ that satisfy inequality (12.2). Then subsystem (12.18) can be driven to point $y_1 = 0$ by means of an admissible control function $u(t) \in W$ if and only if the initial condition is compliant with inequality [53]

$$|y_1| < |d_1| \frac{u_0}{\lambda_1} . \tag{12.19}$$

Inequality (12.19) is easily derived directly from equation (12.18) and limitation (12.2). The set of initial states from which the system can be translated to the origin by applying an admissible control function $u(t) \in W$ is the domain of controllability. It is inequality (12.19) that describes the controllability domain of subsystem (12.18). Like before, Q will denote the domain of controllability of the fourth-order system (12.9) or (12.12). As it is shown in study [53], inequality (12.19) describes not only the controllability domain of subsystem (12.18), but controllability domain Q of the original system (12.12) or (12.9) of the fourth order, since $\mathrm{Re}\lambda_i < 0$ $(i = 2, 3, 4)$.

Instability of coordinate y_1 can be "suppressed" by means of linear feedback

$$u = \gamma y_1 \tag{12.20}$$

if restriction is met

$$\lambda_1 + \gamma d_1 < 0\,. \tag{12.21}$$

When system (12.9) (or (12.12)) is looped by feedback (12.20), (12.21), positive eigenvalue λ_1 "turns" to negative $\lambda_1 + \gamma d_1$. Other eigenvalues remain equal to λ_2, λ_3 and λ_4.

Coming back to limitation (12.2), instead of linear feedback (12.20) the following saturated feedback should be used:

$$u = \begin{cases} -u_0 & \text{when} \quad \gamma y_1 \le -u_0 \\ \gamma y_1 & \text{when} \quad |\gamma y_1| \le u_0\,, \qquad \lambda_1 + \gamma d_1 < 0\,. \\ u_0 & \text{when} \quad \gamma y_1 \ge u_0 \end{cases} \tag{12.22}$$

In region

$$y_1 \ge |d_1| \frac{u_0}{\lambda_1} \tag{12.23}$$

the right-hand side of equation (12.18) looped by control (12.22) is non-negative, thus derivative $\dot{y}_1 \ge 0$. Therefore, control law (12.22) is unable to drive system (12.18) from region (12.23) to state $y_1 = 0$. In region

$$y_1 \le -|d_1| \frac{u_0}{\lambda_1}\,, \tag{12.24}$$

as it is easy to see, derivative $\dot{y}_1$ with respect to system (12.18), (12.22) is non-positive. So from points located within region (12.24) control law (12.22) cannot translate system (12.18) into state $y_1 = 0$. That means that from points located outside area (12.19) system (12.18), (12.22) cannot move to state $y_1 = 0$.

As for area (12.19), inside it, when $y_1 > 0$ derivative $\dot{y}_1$ with respect to system (12.18), (12.22) is negative, and when $y_1 < 0$ it is positive. So as $t \to \infty$, any solution to system that begins inside area (12.19) converges to 0. Thus domain of controllability (12.19) is at the same time domain of attraction when the system is controlled by law described as (12.22).

Similar reasoning was used in chapter 1, where the discussion is illustrated (see Figures 1.2–1.6).

If $y_1(t) \to 0$ when $t \to \infty$, then, according to expression (12.22), also $u(t) \to 0$ as $t \to \infty$. The second, third and fourth equations of system (12.12) are now considered as equations with nonhomogeneous terms $d_i u(t)$ $(i = 2, 3, 4)$. Functions $d_i u(t)$ converge to zero as $t \to \infty$. Therefore, solutions $y_i(t)$ $(i = 2, 3, 4)$ of the second, third and fourth equations with *any* initial conditions $y_i(0)$ converge to zero when $t \to \infty$. This assertion results from the following statement [45]. Suppose that in a linear system of differential equations with constant coefficients all nonhomogeneous terms converge to zero as $t \to \infty$, and the real parts of all eigenvalues of the homogeneous system are negative. Then, all solutions to the nonhomogeneous system converge to zero as $t \to \infty$. Remember that in the system considered here $\mathrm{Re}\lambda_i < 0$ for $i = 2, 3, 4$. So no restrictions are imposed on initial conditions $y_i(0)$ $(i = 2, 3, 4)$.

It follows from this discussion that domain of attraction B of system (12.12) with feedback (12.22) is described by inequality (12.19), and thus it coincides with domain of controllability Q. This effectively means that control law (12.22) provides the largest possible attraction domain for system (12.12), and therefore, for system (12.9).

In view of transformation $x = Sy$ or $y = S^{-1}x$, variable y_1 depends, generally, on all of the original phase variables – angles θ, φ and their angular velocities $\dot{\theta}$, $\dot{\varphi}$. So expression (12.22) describes a feedback that depends on original variables θ, φ, $\dot{\theta}$, $\dot{\varphi}$. When putting this feedback into practice, all these variables must be measured.

So the trivial solution of linearized system (12.7), (12.8) or (12.9) looped by control law (12.22) is asymptotically stable. Basing on Lyapunov's theorem [40, 96], it can be concluded that equilibrium (12.6) of nonlinear system (12.4), (12.5) with feedback (12.22) is also asymptotically stable.

As the absolute value of feedback gain γ grows, negative eigenvalue $\lambda_1 + \gamma d_1$ in the closed-loop system decreases. However, the following fact should be kept in mind (see chapter 1). With any delay in system that may result, for example, in measuring the phase variables, the increase in absolute gain value $|\gamma|$ leads to reduction of delay time that can be allowed from stability point of view.

The next chapter contains some results of numerical experiments. In particular, it gives estimates to initial deviations of some variables that still allow driving the system to the desired equilibrium.

12.4 Numerical experiments

For numerical experiments, the following values were assigned to the system parameters:

$$\begin{aligned} &m_1 = 1.0\,\mathrm{kg}\,, \quad m_2 = 0.2\,\mathrm{kg}\,, \quad g = 9.81\,\mathrm{m/s^2}\,, \quad r = 0.05\,\mathrm{m}\,, \quad l = 0.2\,\mathrm{m}\,, \\ &a = 0.15\,\mathrm{m}\,, \quad \rho_1 = 0.2179\,\mathrm{m}\,, \quad \rho_2 = 0.1414\,\mathrm{m}\,, \quad c_u = 0.007\,\mathrm{N\cdot m/V}\,, \\ &c_v = 0.0001\,\mathrm{N\cdot m/s}\,, \quad u_0 = 19\,\mathrm{V}\,. \end{aligned} \tag{12.25}$$

The roots of characteristic equation (12.14) with values of parameters as in (12.25) are the following:

$$\lambda_1 = 5.7202, \quad \lambda_2 = -5.7218, \quad \lambda_{3,4} = -2.8\cdot 10^{-7} \pm 1.0558i\,. \tag{12.26}$$

Absolute values of roots λ_1 and λ_2 are “close” to one another, and the real part of complex conjugate roots λ_3 and λ_4 is “close” to zero. The reason is that counter-EMF coefficient c_v has a small nonzero value. If $c_v \to 0$, the real roots become equal in absolute value, and the complex roots become purely imaginary. When the system is looped by feedback (12.22), roots λ_2, λ_3 and λ_4, keep their values, as it was discussed above. As λ_3 and λ_4 are close to the imaginary axis, the transients in the system with feedback (12.22) are lengthy. Suppose that viscous friction is present in joint O. The friction torque is determined as $f\dot{\theta}$ (f = const), applied in this joint. Introduction of torque $f\dot{\theta}$ results in change of term $c_v\dot{\theta}$ in expression (12.1) to $(c_v + f)\dot{\theta}$. With friction coefficient $f = 0.4\,\mathrm{N\cdot m\cdot s}$, for example, the eigenvalues of the open-loop system (with no feedback applied) become as follows instead of (12.26):

$$\lambda_1 = 3.4001, \quad \lambda_2 = -10.0181, \quad \lambda_{3,4} = -0.1041 \pm 1.0297i\,. \tag{12.27}$$

With $u = u_0$ nonlinear equations (12.4), (12.5) have a steady-state solution

$$\theta = \dot{\theta} = \dot{\varphi} = 0, \quad \varphi = -\frac{c_u u_0}{r m_2 g} \quad \left(s = -\frac{c_u u_0}{m_2 g}\right), \tag{12.28}$$

and with $u = -u_0$ – steady-state solution

$$\theta = \dot{\theta} = \dot{\varphi} = 0, \quad \varphi = \frac{c_u u_0}{r m_2 g} \quad \left(s = \frac{c_u u_0}{m_2 g}\right). \tag{12.29}$$

Points (12.28), (12.29) are symmetric to each other with respect to origin (12.6). Linear equations (12.7), (12.8) with $u = \pm u_0$ also have steady states (12.28), (12.29). On the other hand, these steady-state solutions can be found by assuming $u = \pm u_0$ in linear equations (12.12). In terms of variables y_i (i = 1 through 4), these states are presented as follows:

$$y_i = \mp\frac{d_i u_0}{\lambda_i} \quad (i = 1 \div 4).$$

Thus it is clear that the steady state points (12.28), (12.29) are located on the borders $y_1 = \mp d_1 u_0/\lambda_1$ of attraction domain B, and interval

$$\theta = \dot{\theta} = \dot{\varphi} = 0, \quad -\frac{c_u u_0}{r m_2 g} < \varphi < \frac{c_u u_0}{r m_2 g} \quad \left(-\frac{c_u u_0}{m_2 g} < s < \frac{c_u u_0}{m_2 g}\right) \tag{12.30}$$

belongs completely to this domain. Each of the states in (12.30) is an equilibrium when the control voltage has an appropriate value, $u = -m_2 g s/c_u$, and, with no feedback applied, these equilibriums are unstable.

Existence of steady states (12.28), (12.29) can be easily explained judging from physical matters. Product $c_u u_0$ is the maximum torque that the motor can produce being stalled. This torque acts counter-clockwise. If $\theta = 0$, then product $s m_2 g$ is the moment of the gravity force with respect to pivot O, produced by the mass of the ball that is located at distance s to the left of point A (see Figure 12.1). This moment also acts counter-clockwise. Equality $-m_2 g s = c_u u_0$ also represents the balance of these two torques, but for the case when the ball is located to the right of point A at distance $|s|$.

With numerical values of parameters of the system as in (12.25), relations (12.30) look like

$$\theta = \dot{\theta} = \dot{\varphi} = 0\,, \quad -77.679^\circ < \varphi < 77.679^\circ \quad (-0.0678\,\mathrm{m} < s < 0.0678\,\mathrm{m})\,. \quad (12.31)$$

If the initial state is within interval (12.30), then as $t \to \infty$ the solution to linear system (12.9) (or (12.12)) with feedback (12.22) apparently converges asymptotically to equilibrium (12.6). Numerical experiments show that with initial conditions within interval (12.30) the solutions of *nonlinear* system (12.4), (12.5) with feedback (12.22) also converge to equilibrium (12.6) when $t \to \infty$. If the initial state lies outside interval (12.30), that is, it is in region

$$\theta = \dot{\theta} = \dot{\varphi} = 0\,, \quad |\varphi| > \frac{c_u u_0}{r m_2 g} \quad \left(|s| > \frac{c_u u_0}{m_2 g}\right)\,, \quad (12.32)$$

then the solution of linear system (12.9) (or (12.12)) with feedback (12.22) certainly does not converge to (12.6). The numerical experiments show that with initial conditions belonging to set (12.32), solutions to *nonlinear system* (12.4), (12.5), (12.22) also do not converge to equilibrium (12.6). Thus if $\theta(0) = \dot{\theta}(0) = \dot{\varphi}(0) = 0$, nonlinear system (12.4), (12.5) with feedback (12.22) comes to equilibrium (12.6) only when initial values of angle φ lie within interval (12.30). Numerical experiments provide evidence to presume that there is no admissible control function $|u(t)| \leq u_0$ capable of translating the nonlinear system to equilibrium (12.6) from points of region (12.32). Such presumption can be justified by both numerical computations and physical reasoning. Actually, if $\theta(0) = \dot{\theta}(0) = \dot{\varphi}(0) = 0$ and $|\varphi(0)| > c_u u_0/(r m_2 g)$, the gravity moment of the ball with respect to pivot point O cannot be compensated by the moment produced by the electric motor.

Figure 12.2 illustrates transients in terms of deviation angle θ of the beam from the horizontal and angle φ of the ball rotation. These transients refer to nonlinear system (12.4), (12.5) with feedback (12.22). Feedback gain in the control law is $\gamma = -122$. Initial conditions are taken from within interval (12.31) close to its boundary – $\varphi(0) = 77.65^\circ$ (1.36). Viscous friction coefficient in the pivot is $f = 0.4\,\mathrm{N \cdot m \cdot s}$.

From inspection of Figure 12.2 it can be concluded that functions $\theta(t)$ and $\varphi(t)$ converge to zero as $t \to \infty$. Yet it can be seen that the transient includes significant oscillation, and it converges slowly. This can be explained by the fact that the initial state is close to the border of the attraction domain.

Figure 12.3 presents the corresponding time dynamics of voltage u (in volts) supplied to the electric motor.

At the initial time instant the voltage takes its minimal value (–19 V). Then it goes to zero "quickly", without any oscillations. At the same time, as it can be seen in Figure 12.2, variables θ and φ, and hence their derivatives as well, oscillate during the transient stage of motion. The inconsistence between phase variable behavior and

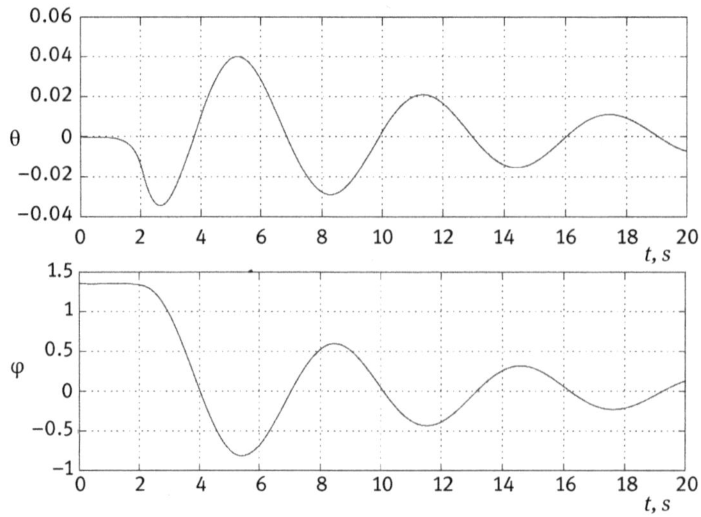

Fig. 12.2. Dynamics of angles θ and φ (in radians).

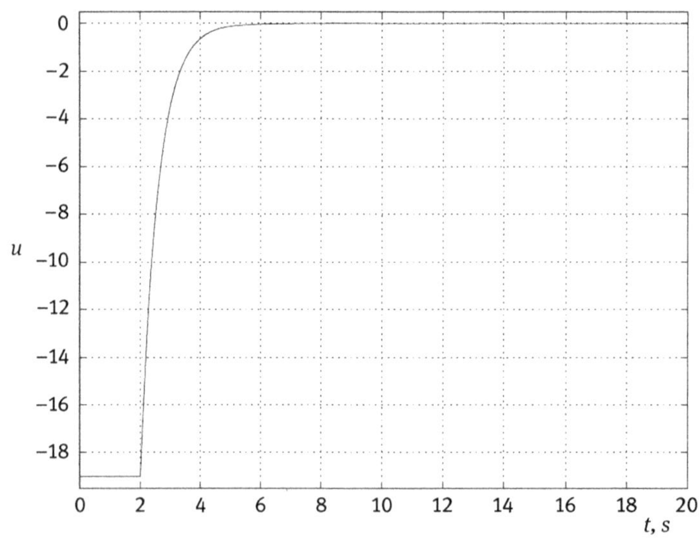

Fig. 12.3. Time dynamics of voltage u (in volts).

the voltage can be explained as follows. The feedback involves a single variable y_1. Its dynamics is determined by equations (12.18), (12.22). Inspecting these two equations, one can see that solution $y_1(t)$ to equation (12.18) with any initial conditions that lie within interval (12.24) is a strictly monotonic time function, because derivative (12.18) maintains its sign at all times. And if variable y_1 changes monotonically, so does voltage u, in virtue of equation (12.22).

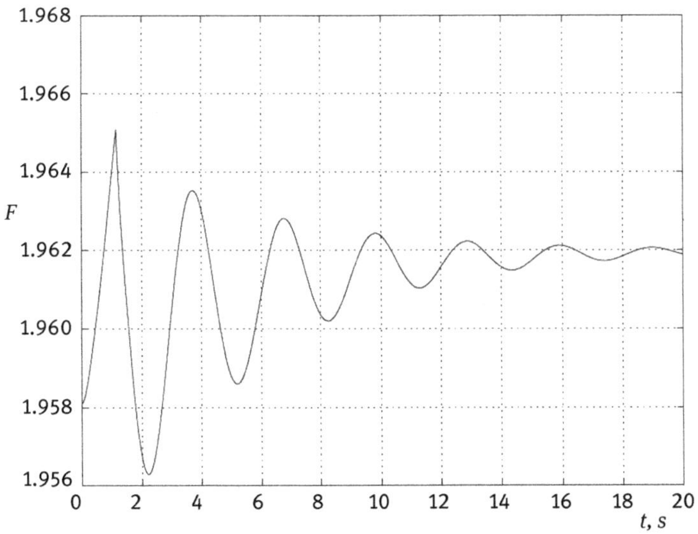

Fig. 12.4. Time dynamics of beam reaction force F.

The reaction force that is applied to the ball from the beam during the transient is also of interest. In real life, if this force becomes zero or even negative, the ball, not being attached to the beam, will lose contact with it. For the reaction force component F that is directed perpendicular to the beam, the following expression can be derived:

$$F = m_2[g\cos\theta - (l + r)\dot{\theta}^2 - 2r\dot{\varphi}\dot{\theta} - r\varphi\ddot{\theta}] . \tag{12.33}$$

In the numerical experiment described above, reaction force (12.33) is positive at all times. Its dynamics is shown in Figure 12.4. The value of the force is expressed in Newtons.

From Figure 12.4 it is clear that during the transient, force F deviates from the ball weight, that is equal to 1.962 N, only a small fraction.

For example, if $\varphi(0) = \dot{\varphi}(0) = \dot{\theta}(0) = 0$, then, by applying inequality (12.19), there can be found the upper boundary of initial values $\theta(0)$ from which control law (12.22) drives linear system (12.7), (12.8) to equilibrium (12.6). When numerical parameters of the system are like in (12.25), this boundary is $\theta(0) = 3.61°$. Numerical experiments show that for nonlinear system (12.4), (12.5) with control law (12.22) the value of this boundary is approximately $\theta(0) \approx 3.64°$. This value is close to the corresponding value for the linear system.

Numerical experiments show that with parameters of the system as in (12.25), inequality (12.19) can be used to evaluate the domain of attraction of the nonlinear system with a reasonable accuracy, at least, in terms of some of the variables. It might probably be admitted that such situation takes place with other parameter values as well.

13 Stabilization of a ball on a curvilinear beam

In the previous chapter, the problem of stabilizing a ball on a straight beam was discussed. This problem was investigated in a series of works [49, 82, 125, 139, 154, 155]. The current chapter considers a system of a *new kind*, where a curvilinear (bent) beam is taken instead of a straight one. While the system with a straight beam has only one unstable motion mode (in the linearized model), the system with a curvilinear beam, if its curvature is large enough, has two unstable modes. So the task of stabilization becomes more complicated than a similar task for the former system [12–14, 61].

13.1 Mathematical model of the system

Consider a problem of stabilizing a ball on a curvilinear beam. The beam is actually a segment of a circle that has radius R, and the center of that circle is located at point C. Such system that is comprised of a ball and a circular beam is illustrated in Figure 13.1.

The beam may turn around pivot point O, where torque L developed by an electric motor is applied. The value of torque L is determined by expression (12.1). Let the voltage supplied to the motor be denoted as u, and assume that it complies with inequality (12.2). Point A is the middle of the curvilinear beam, C_1 is the center of mass of the beam together with its support OA. Similarly to the case of a straight beam, m_1 and m_2 are respectively the mass of the beam with its support and the mass of the ball; ρ_1 and ρ_2 are the inertia radiuses of the beam (together with support OA) with respect to pivot O and of the ball with respect to its center C_2, $OA = l$ and $OC_1 = a$.

The generalized coordinates that describe the system state are angles θ and φ. The ball placement of the beam can also be determined by distance $s = r\varphi$. Angles φ and ψ (see Figure 13.1) are related by the following expression:

$$r\varphi = R\psi \,. \tag{13.1}$$

Kinetic energy T and potential energy Π are determined by expressions

$$\begin{aligned} 2T &= \left\{m_1\rho_1^2 + m_2\left[(R+r)^2 + (R-l)^2 - 2(R+r)(R-l)\cos\frac{r\varphi}{R}\right]\right\}\dot{\theta}^2 + \\ &\quad + m_2\left[(R+r)^2 + \left(\frac{\rho_2 R}{r}\right)^2\right]\left(\frac{r\dot{\varphi}}{R}\right)^2 + 2m_2\left[(R+r)^2 - (R+r)(R-l)\cos\frac{r\varphi}{R}\right]\dot{\theta}\frac{r\dot{\varphi}}{R}\,, \\ \Pi &= [m_1 a + m_2(l-R)]g\cos\theta + m_2 g(R+r)\cos\left(\frac{r}{R}\varphi + \theta\right). \end{aligned} \tag{13.2}$$

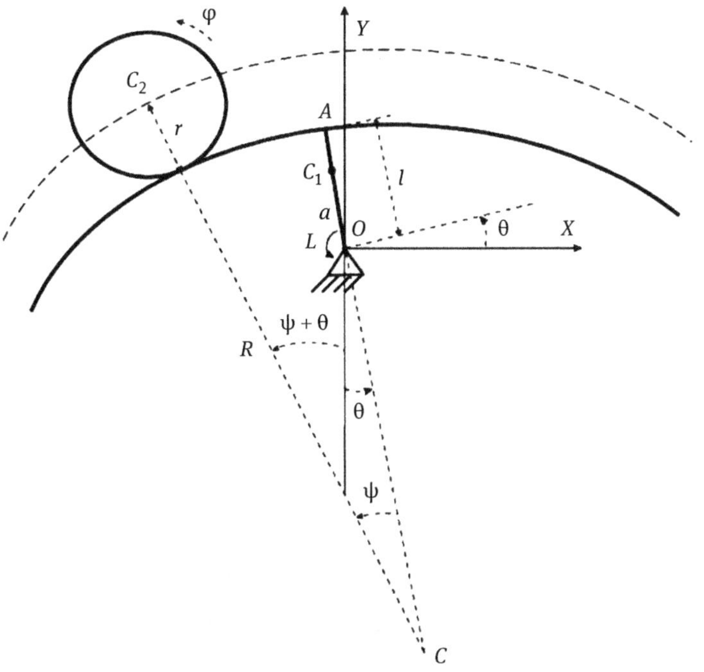

Fig. 13.1. A ball on a curvilinear beam.

Applying Lagrangian approach of the second kind, with expressions (13.2), the motion equations for the system can be derived

$$\left[m_1\rho_1^2 + m_2\left(r^2 + l^2 + 2rl\cos\frac{r\varphi}{R}\right) + 2m_2R(R+r-l)\left(1-\cos\frac{r\varphi}{R}\right)\right]\ddot{\theta} + \\ + m_2r\left(1+\frac{r}{R}\right)\left[R+r+(l-R)\cos\frac{r\varphi}{R}\right]\ddot{\varphi} + m_2r\left(1+\frac{r}{R}\right)(R-l)\left(2\dot{\theta}+\frac{r\dot{\varphi}}{R}\right)\dot{\varphi}\sin\frac{r\varphi}{R} - \\ - g[m_1a + m_2(l-R)]\sin\theta - m_2g(r+R)\sin\left(\theta+\frac{r\varphi}{R}\right) = c_u u - c_v\dot{\theta}\,, \quad (13.3)$$

$$r\left(1+\frac{r}{R}\right)\left[R+r+(l-R)\cos\frac{r\varphi}{R}\right]\ddot{\theta} + \left[\rho_2^2 + r^2\left(1+\frac{r}{R}\right)^2\right]\ddot{\varphi} + \\ + \left(1+\frac{r}{R}\right)(l-R)\dot{\theta}^2\sin\frac{r\varphi}{R} - gr\left(1+\frac{r}{R}\right)\sin\left(\theta+\frac{r\varphi}{R}\right) = 0\,. \quad (13.4)$$

Expressions (13.2) for kinetic and potential energy, as well as equations of motion (13.3), (13.4) involve term $r\varphi/R$, that is, according to relation (13.1), equal to ψ.

Note 1. *Consider a particular case. Let $R = l$, i.e. let the center of the circle C coincide with pivot point of the beam O. Besides, let all the mass of the ball be concentrated in its center C_2, i.e. $\rho_2 = 0$ (although $m_2 \neq 0$). With such assumptions, system (13.3), (13.4) becomes considerably more simple, and it splits into two independent equations*

$$m_1\rho_1^2\ddot{\theta} - m_1 ga \sin\theta = c_u u - c_v\dot{\theta}, \quad (R + r)\ddot{\alpha} - g\sin\alpha = 0\,,$$

where $\alpha = \theta + r\varphi/R$ ($\alpha = \theta + \psi$). The second equation of this system is effectively a motion equation of a single-link pendulum. The control signal does not influence dynamics of angle α. Thus when $R = l$ and $\rho_2 = 0$ nonlinear system (13.3), (13.4) is not controllable.

Note 2. *In case when the beam is stationary ($\theta \equiv 0$), motion equation of the ball on it is similar to the pendulum equation. This equation for ball dynamics can be derived by assuming in (13.4) $\theta = 0$,*

$$\left[\rho_2^2 + r^2\left(1 + \frac{r}{R}\right)^2\right]\ddot{\varphi} - gr\left(1 + \frac{r}{R}\right)\sin\frac{r\varphi}{R} = 0\,.$$

When $R \to \infty$, this equation transforms to $\ddot{\varphi} = 0$. The latter equation can also be derived from (12.5).

When $u = 0$, system (13.3), (13.4) has an unstable equilibrium (see relations (12.6))

$$\theta = 0, \quad \varphi = 0 \quad (s = 0), \quad \dot{\theta} = 0, \quad \dot{\varphi} = 0 \quad (\dot{s} = 0)\,. \tag{13.5}$$

In state (13.5), support OA is positioned vertically, and the ball is located in the middle of the beam, touching it at point A (angle $\psi = 0$). Like in the previous chapter, the feedback will be sought that stabilizes state (13.5) and provides the largest domain of attraction.

13.2 Linearized model

Linearizing equations (13.3), (13.4) about equilibrium (13.5) yields

$$\begin{aligned}
&\left[m_1\rho_1^2 + m_2(r+l)^2\right]\ddot{\theta} + m_2 r\left(1 + \frac{r}{R}\right)(r+l)\ddot{\varphi} - \\
&\qquad -g[m_1 a + m_2(r+l)]\theta - m_2 g(r+R)\frac{r\varphi}{R} = c_u u - c_v\dot{\theta}\,, \\
&r\left(1 + \frac{r}{R}\right)(r+l)\ddot{\theta} + \left[\rho_2^2 + r^2\left(1 + \frac{r}{R}\right)^2\right]\ddot{\varphi} - gr\left(1 + \frac{r}{R}\right)\left(\theta + \frac{r\varphi}{R}\right) = 0\,.
\end{aligned} \tag{13.6}$$

Equations (13.6) can be presented as a system of four differential equations of the first order (12.9), or, in other words, in normal form. However, matrices D and E, unlike those in (12.10), for curvilinear beam are

$$D = \begin{Vmatrix} m_1\rho_1^2 + m_2(r+l)^2 & m_2 r(r+l)(1 + r/R) \\ r(r+l)(1+r/R) & r^2 + \rho_2^2(1 + r/R)^2 \end{Vmatrix}, \quad E = g\begin{Vmatrix} m_1 a + m_2(r+l) & m_2 r(1 + r/R) \\ r(1 + r/R) & r(1+r/R)r/R \end{Vmatrix}. \tag{13.7}$$

The characteristic equation of system (13.6) with $u = 0$ (see equation (12.14)) is

$$a_0\lambda^4 + a_1\lambda^3 + a_2\lambda^2 + a_3\lambda + a_4 = 0\,. \tag{13.8}$$

The coefficients in this equation are:

$$\begin{aligned} &a_0 = \det D > 0, \quad a_1 = c_v\left[\rho_2^2 + \frac{r^2}{R^2}(r+R)^2\right] > 0\,, \\ &a_3 = -c_v g\frac{r^2}{R^2}(r+R) < 0, \quad a_4 = \det E = g\frac{r^2}{R^2}(R+r)[m_1 a + m_2(l-R)]\,. \end{aligned} \tag{13.9}$$

Expression for a_2 is omitted here. If $c_v = 0$, equation (13.8) becomes biquadratic, and its spectrum is symmetric with respect to the origin of the complex plane. We assume further that $c_v > 0$. If radius R is large enough, so that

$$m_1 a + m_2(l-R) < 0 \quad \left(R > l + \frac{m_1}{m_2}a\right)\,, \tag{13.10}$$

then, according to expressions (13.9), $a_0 > 0$, $a_4 < 0$, and equation (13.8) has at least one real positive root, and also one negative root. In a sequence comprised of coefficients a_0, a_1, a_2, a_3, a_4 of equation (13.8), as it can be seen from expressions (13.9), there is exactly *one* sign change, regardless of the sign of coefficient a_2 (it may as well be equal to zero). So, by virtue of Descartes' sign rule [97], this equation has *exactly one positive root*. Equation (13.8) does not have zero roots, because $a_4 \neq 0$. It also does not have purely imaginary roots of type $\pm i\omega$. To verify this, substitute in (13.8) value $i\omega$, then assume the imaginary part equal to zero. This yields equation $-a_1\omega^3 + a_3\omega = 0$. The signs of coefficients a_1 and a_3 are different, thus this equation has only one real root $\omega = 0$. In order to fully clarify the question of root locations for the characteristic equation, the Routh–Hurwitz criterion can be applied [40, 41, 97]. Consider a sequence

$$T_0, \quad T_1, \quad T_1T_2, \quad T_2T_3, \quad T_3T_4\,, \tag{13.11}$$

where values T_0, T_1, T_2, T_3, T_4 are the Routh–Hurwitz determinants

$$T_0 = a_0, \quad T_1 = a_1, \quad T_2 = \begin{vmatrix} a_1 & a_0 \\ a_3 & a_2 \end{vmatrix} = a_1a_2 - a_0a_3,$$

$$T_3 = \begin{vmatrix} a_1 & a_0 & 0 \\ a_3 & a_2 & a_1 \\ 0 & a_4 & a_3 \end{vmatrix} = a_3T_2 - a_1^2a_4, \quad T_4 = a_4T_3\,.$$

According to the Routh–Hurwitz criterion [40, 41, 97], the number of roots of equation (13.8) that have positive real parts is equal to the number of sign changes in sequence (13.11).

The sequence of values (13.11) is written below. The signs of the values are put wherever they can be determined from relations (13.9), (13.10)

$$\begin{aligned} &T_0 = a_0 > 0, \quad T_1 = a_1 > 0, \quad T_1T_2 = a_1T_2, \\ &T_2T_3 = T_2(a_3T_2 - a_1^2a_4), \quad T_3T_4 = a_4T_3^2 < 0\,. \end{aligned} \tag{13.12}$$

But the value of $a_4 < 0$. Assume that $T_2 > 0$, then $T_1T_2 > 0$; then sequence (13.12) has exactly one sign change, regardless of the sign of product T_2T_3 (when $T_3 = 0$, the investigation becomes more complicated, and it goes beyond the scope of the current research). Now assume that $T_2 < 0$. Then $T_1T_2 < 0$, $T_2T_3 < 0$ (because $T_3 > 0$) and therefore, sequence (13.12) also has exactly one sign change. Thus there is exactly one root of equation (13.8) located in the right half of the complex plane.

So with condition (13.10) taking place, equation (13.8) has one real positive root, and three roots with negative real part. The roots of the characteristic equation in the terminal case ($R = \infty$), when the beam is straight, are located similarly, that seems natural because when $R \to \infty$ condition (13.10) is still satisfied.

Now assume that

$$m_1 a + m_2(l - R) > 0 \quad \left(R < l + \frac{m_1}{m_2} a\right), \tag{13.13}$$

then $a_4 > 0$. Regardless of the sign of coefficient a_2, sequence of coefficients (13.9) has exactly two sign changes. Then, according to Descartes' sign rule [97], equation (13.8) has two real positive roots, or it does not have such roots at all.

To refine the number of roots of equation (13.8) that have positive and negative real parts, the Routh–Hurwitz criterion should be used.

If $a_4 > 0$, the last relation in sequence (13.12) would have a different sign. Then, instead of (13.12), the following relations take place

$$\begin{aligned} &T_0 = a_0 > 0, \quad T_1 = a_1 > 0, \quad T_1T_2 = a_1T_2, \\ &T_2T_3 = T_2(a_3T_2 - a_1^2a_4), \quad T_3T_4 = a_4T_3^2 > 0\,. \end{aligned} \tag{13.14}$$

If $T_2 > 0$, then $T_1T_2 > 0$, $T_2T_3 = T_2(a_3T_2 - a_1^2a_4) < 0$. In this case, sequence (13.14) has exactly two sign changes. If $T_2 < 0$, then $T_1T_2 < 0$, and sequence (13.14) also has exactly two sign changes, regardless of the sign of product T_2T_3.

Thus, when condition (13.13) takes place, characteristic equation (13.8) has two roots in the right complex semiplane and two roots in the left semiplane.

A nondegenerate transformation of kind $x = Sy$ with constant matrix S can transform system (13.6) to Jordan form (12.13), where $\lambda_1, \lambda_2, \lambda_3, \lambda_4$ are the roots of characteristic equation (13.8).

13.3 Feedback design

Suppose that inequality (13.13) takes place, and let λ_1 and λ_2 be the real positive roots of equation (13.8). The other two roots of this equation, as it was mentioned above, are located in the left half of the complex plane, i.e. $\mathrm{Re}\lambda_i < 0$ ($i = 3, 4$). To be definite, let $\lambda_1 > \lambda_2$.

The first two equations of system (12.12) that correspond to eigenvalues λ_1 and λ_2 are as follows:

$$\dot{y}_1 = \lambda_1 y_1 + d_1 u, \quad \dot{y}_2 = \lambda_2 y_2 + d_2 u\,. \tag{13.15}$$

If system (12.9) is completely controllable in Kalman's sense, then subsystem (13.15) of this system is also completely controllable. At that, scalar coefficients $d_1, d_2 \neq 0$, as well as the two other values d_3, d_4.

The border of controllability domain S of system (13.15) is comprised of two integral trajectories of this system. These trajectories result from assigning $u(t) \equiv u_0$ and $u(t) \equiv -u_0$ (see chapters 7, 10), and they are symmetric with respect to the origin. One of these trajectories is described by equations

$$y_1(t) = \frac{d_1 u_0}{\lambda_1}(2e^{\lambda_1 t} - 1), \quad y_2(t) = \frac{d_2 u_0}{\lambda_2}(2e^{\lambda_2 t} - 1) \quad (-\infty < t \leq 0)\,. \tag{13.16}$$

Equations of the other trajectory have different signs of the right-hand sides, compared to equations (13.16)

$$y_1(t) = -\frac{d_1 u_0}{\lambda_1}(2e^{\lambda_1 t} - 1), \quad y_2(t) = -\frac{d_2 u_0}{\lambda_2}(2e^{\lambda_2 t} - 1) \quad (-\infty < t \leq 0)\,. \tag{13.17}$$

Similar to (10.30), the expression for the feedback law for equation system (13.15) can be written as follows:

$$u = \begin{cases} -u_0 & \text{when} \quad \gamma\left(\frac{d_2}{\lambda_2}y_1 - \frac{d_1}{\lambda_1}y_2\right) \leq -u_0 \\ \gamma\left(\frac{d_2}{\lambda_2}y_1 - \frac{d_1}{\lambda_1}y_2\right) & \text{when} \quad \left|\gamma\left(\frac{d_2}{\lambda_2}y_1 - \frac{d_1}{\lambda_1}y_2\right)\right| \leq u_0 \\ u_0 & \text{when} \quad \gamma\left(\frac{d_2}{\lambda_2}y_1 - \frac{d_1}{\lambda_1}y_2\right) \geq u_0. \end{cases} \tag{13.18}$$

Following the same reasoning as in the chapters 7 and 10, it can be determined that if feedback gain γ is chosen large enough (in absolute value), then, assuming

$$\operatorname{sign}\gamma = -\operatorname{sign}(d_1 d_2)\,,$$

domain of attraction B of the origin of system (13.15) looped by control law (13.18) can be arbitrarily close to the domain of controllability that is determined by its borders (13.16), (13.17).

In the next section, a system is investigated with some of its parameters assigned numerically. Estimates are given for initial deviation of the system from its desired equilibrium (in terms of some of the variables) that still allows this system to be driven to that equilibrium.

13.4 Numerical experiments

Consider a system with its parameters having the following values:

$$\begin{gathered} m_1 = 1.0\,\text{kg}\,, \quad m_2 = 0.2\,\text{kg}\,, \quad g = 9.81\,\text{m/s}^2\,, \quad r = 0.05\,\text{m}\,, \\ R = 0.8\,\text{m}\,, \quad l = 0.2\,\text{m}\,, \quad a = 0.15\,\text{m}\,, \quad \rho_1 = 0.2646\,\text{m}\,, \quad \rho_2 = 0.1414\,\text{m}\,, \\ c_u = 0.007\,\text{N}\cdot\text{m/V}\,, \quad c_v = 0.0001\,\text{N}\cdot\text{m/s}\,, \quad u_0 = 19\text{V}\,. \end{gathered} \tag{13.19}$$

Most of the values are the same as in (12.25). If the parameter values are like in (13.19), equality (13.13) is satisfied. The roots of equation (13.8) include two real positive and two real negative values

$$\lambda_1 = 4.8959, \quad \lambda_2 = 0.4652, \quad \lambda_3 = -4.8971, \quad \lambda_4 = -0.4652\,. \tag{13.20}$$

It is worth investigating traces of the roots of characteristic equation (13.8) that corresponds to the open-loop system as the value of radius R changes. First, assume that the system has no friction. Then equation (13.8) is biquadratic. When the value of radius R is "large", it has one positive root, one negative root (that is equal to the first one in absolute value), and a pair of purely imaginary roots. As the value of R decreases, the positive and the negative roots keep their status, while the imaginary roots approach the real axis and when $R = l + (m_1/m_2)a$ they "transform" to a double zero root. As radius R continues to decrease, one of the zero roots becomes positive (it "drifts" to the right complex semiplane), and the other one becomes negative. If the system has viscous friction, then instead of purely imaginary roots there are complex conjugate eigenvalues with negative real part. As the value of R decreases, these roots transform to a double negative value. Then one of them "drifts" to the right complex semiplane.

Figure 13.2 shows in dashed line domain of controllability S of system (13.15), built by means of expressions (13.16), (13.17). Its border has two corner points that are symmetric with respect to the origin. These points correspond to steady-state solutions to equations (13.15) with $u = \pm u_0$

$$y_1 = \mp\frac{d_1 u_0}{\lambda_1}, \quad y_2 = \mp\frac{d_2 u_0}{\lambda_2}, \tag{13.21}$$

The border also includes two nearly vertical sections.

When $u = \pm u_0$, the same steady states can be found in terms of the original variables

$$\theta = \mp\frac{c_u u_0}{g[m_1 a + m_2(l-R)]}, \quad \dot\theta = 0, \quad \varphi = -\frac{R}{r}\theta \quad (s = -R\theta), \quad \dot\varphi = 0 \quad (\dot s = 0)\,. \tag{13.22}$$

by means of linearized motion equations (13.6). From nonlinear model (13.3), (13.4), the following expressions for the steady-state solutions can be found instead of (13.22):

$$\theta = \mp\arcsin\frac{c_u u_0}{g[m_1 a + m_2(l-R)]}, \quad \dot\theta = 0, \quad \varphi = -\frac{R}{r}\theta \quad (s = -R\theta), \quad \dot\varphi = 0 \quad (\dot s = 0)\,. \tag{13.23}$$

In steady state (13.22) or (13.23), support OA of the beam is rotated angle θ away from the vertical, while the ball coincides with the beam in its highest point, where a tangent line to the beam is horizontal. This statement follows from equality $r\varphi = -R\theta$ (see (13.23)). Note that with $u = 0$, in stationary state (13.5) the ball is located in the middle of the beam. Point A is the contact point of the ball and the beam, and the line that passes through point A tangent to the beam is horizontal.

Domain of attraction B of system (13.15) looped by control law (13.18) is shown in Figure 13.2. Its border is shown in solid line. The border of domain B is a periodic

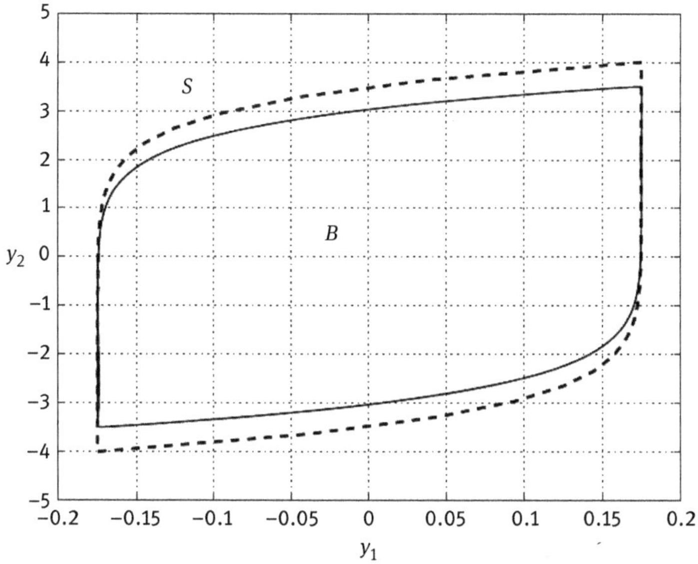

Fig. 13.2. Domain of controllability S and domain of attraction B.

solution (a cycle) to system (13.15), (13.18). This cycle can be found by solving system (13.15), (13.18) in reverse time. The initial state for reverse motion should be taken close to the origin $y_1 = y_2 = 0$ or outside domain of controllability S. Domain of attraction B depends on feedback gain γ. Figure 13.2 illustrates domain B built for $\gamma = 4000$.

Figure 13.3 shows dynamics of angles θ and φ when initial conditions are $\varphi(0) = 70.39°$, $\theta(0) = \dot{\theta}(0) = \dot{\varphi}(0) = 0$. Value $\varphi(0) = 70.39°$ is close to the upper boundary of initial values of angle $\varphi(0)$ where it is still possible to stabilize equilibrium (13.5) of nonlinear system (13.3), (13.4) by means of control law (13.18). The corresponding initial distance $s(0) = 0.061$ m.

Transients in variables θ and φ are aperiodic. This is explained by the fact that the eigenvaliues of the open-loop system are real.

Figure 13.4 shows dynamics of voltage $u(t)$ during the transient. This voltage is calculated according to expression (13.18).

In the beginning of the process, voltage u takes its least possible value of −19 V, then it switches to the largest possible value of +19 V, and finally, it asymptotically converges to zero, having no oscillations during this process.

The component of reaction force F that is applied in the contact point from the beam to the ball perpendicularly to the beam can be evaluated by the following expression:

$$F = m_2 \left[g \cos\left(\theta + \frac{r\varphi}{R}\right) - (R + r)\left(\dot{\theta} + \frac{r\dot{\varphi}}{R}\right)^2 + (R - l)\dot{\theta}^2 \cos\frac{r\varphi}{R} - (R - l)\ddot{\theta} \sin\frac{r\varphi}{R}\right] .$$

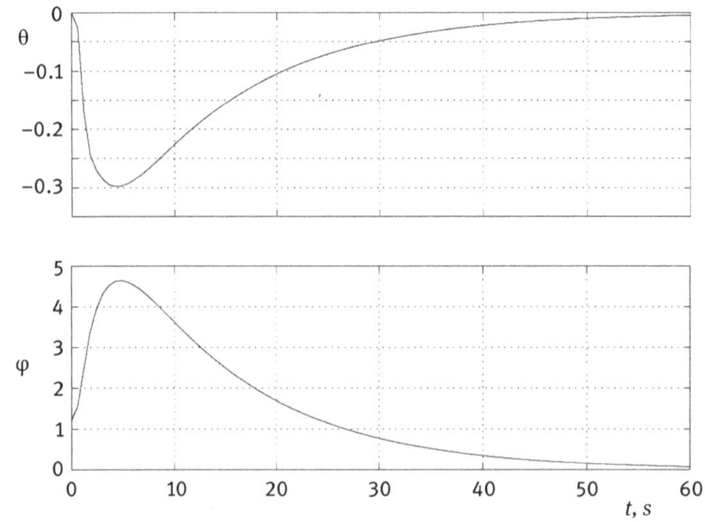

Fig. 13.3. Time dynamics of angles θ and φ.

In numerical experiment illustrated in Figure 13.3, 13.4, the value of F is positive at all times. Its value is equal to the weight of the ball with only minor deviations of less than one percent.

The values of angle θ and distance s calculated according to expressions (13.23) are equal to

$$\theta = 0.469, \quad s = -0.375\,\text{m}\,. \tag{13.24}$$

Let initial velocities be $\dot{\theta}(0) = 0$, $\dot{s}(0) = 0$ and let, according to the third equality in (13.22) or in (13.23), $s(0) = -R\theta(0)$. Then solving nonlinear equations (13.3), (13.4), (13.18) (with $\gamma = 4000$), yields the upper boundary of values of angle $\theta(0)$ that makes it is still possible to stabilize equilibrium (13.5), and the respective distance $s(0) = -R\theta(0)$ approximately

$$\theta(0) = 0.397, \quad s(0) = -0.318\,\text{m}\,. \tag{13.25}$$

Values (13.25) are 1.18 times less than values (13.24). If the value of gain in control law (13.18) is equal to $\gamma = 8000$, then the following values appear instead of (13.25):

$$\theta(0) = 0.430, \quad s(0) = -0.344\,\text{m}\,. \tag{13.26}$$

Values in (13.26) are closer to (13.24), than values in (13.25). Values in (13.26) are 1.09 times smaller than the respective values in (13.24). Numerical experiments show that with further growth of gain γ boundary initial values $\theta(0)$ and $s(0)$, that still make it possible to stabilize the equilibrium approach values (13.24). Thus expressions (13.22) or (13.23) may be used to evaluate the domain of attraction of equilibrium (13.5) of the original nonlinear system (13.3), (13.4).

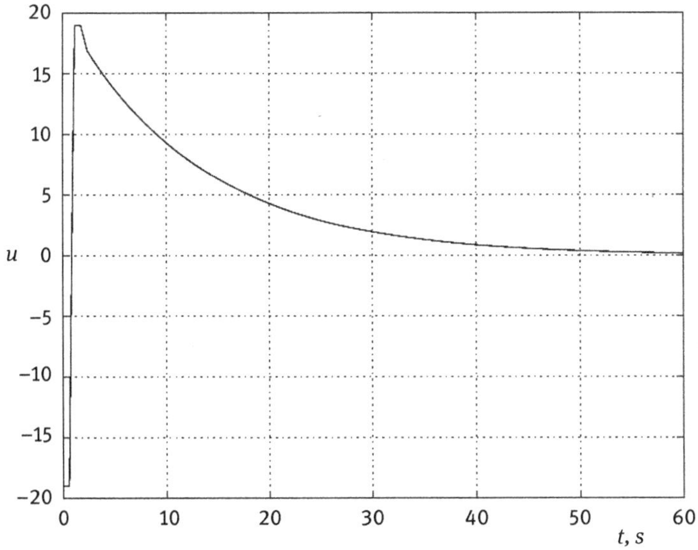

Fig. 13.4. Time dynamics of voltage u (in volts).

Numerical experiments also show that when the initial conditions are

$$\dot{\theta}(0) = \dot{\varphi}(0) = 0, \quad |\theta(0)| > \arcsin \frac{c_u u_0}{g[m_1 a + m_2(l - R)]}, \quad \varphi(0) = -\frac{R}{r}\theta(0) , \tag{13.27}$$

The solution to *nonlinear* system (13.3), (13.4), (13.18) does not converge to equilibrium (13.5). Further numerical investigation also lead to a presumption that there is no admissible control law $|u(t)| \leq u_0$ that would be capable of translating the nonlinear system to equilibrium (13.5) from points within region (13.27). In chapter 11, after investigation of the ball on a straight beam, similar thoughts were suggested.

Part IV: **Gyroscopic stabilization of a two-wheel bicycle**

It is known that the vertical position of a two-wheel bicycle is unstable. This makes it akin to an inverted pendulum. However, a person riding a bicycle may move not losing his balance – maintaining the vertical position. A cyclist can keep his balance during motion even not necessarily holding the handlebar. The bicycle will "automatically" stabilize itself in vertical position, because of specific design of its steering fork. The main purpose of the steerer is to direct the bicycle movement. Nevertheless, it also plays an important role in maintaining the bicycle upright position. The steering axis of the headset is inclined a certain angle with respect to the vertical, and the center of the front wheel is located ahead of that axis. The angular momentum of the spinning front wheel and such design of the steerer help stabilize the vertical position of the bicycle [77].

This part describes the two-wheel bicycles designed in the Institute of Mechanics of Lomonosov Moscow State University by A.V. Lenskii, the head of Mechatronics Lab [64, 102, 103]. The steering axis of the headset in each of the bicycles is vertical (when the bicycle body is vertical), and, besides, it passes through the center of the front wheel. Stability of such device is maintained by a controllable gyroscopic stabilizer, and the steerer is used only to direct its motion. It is worth emphasizing that the gyrostabilizer is *controllable*, not passive. A law is suggested for driving the motor that can control the gyrostabilizer. This motor generates torque that is applied to the gyro precession axis. So the bicycle stabilization can be realized by means of that gyrostabilizer.

14 Bicycle design

The current chapter describes composition of the two bicycles designed in the Institute of Mechanics. Both vehicles are equipped with gyroscopic systems of stabilization. One of the bicycles has its front wheel *steering* and *driving* at the same time. The construction is similar to a traditional bicycle, yet the rear wheel is *passive*. The other bicycle has *both* of the wheels (front and rear one) steering and driving at the same time.

14.1 Bicycle with one controlled wheel

Figure 14.1 shows the two-wheel bicycle that has its front wheel steering and driving at the same time.

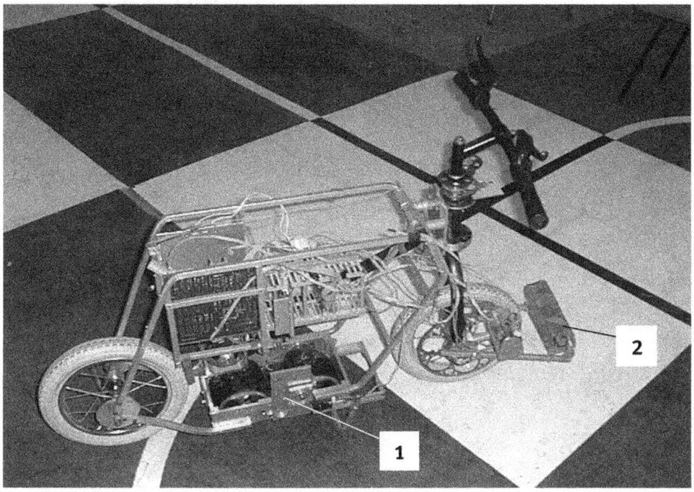

Fig. 14.1. Bicycle with one controlled wheel: 1 – gyrostabilizer, 2 – light detector array.

Motion of the bicycle that is shown in Figure 14.1 is realized by means of two electric DC motors. One motor spins the front wheel and thus provides the longitudinal motion. The other one turns the plane of the front wheel (the steerer) about the vertical axis, providing maneuvers of the vehicle. Such design makes the front wheel steering and driving at the same time. Each of the motors has an optical encoder. The encoder readings are used to calculate the path that the wheel has moved, and its current velocity. The other encoder is used to measure the steering angle. The steerer is turned automatically, by means of a servosystem. The rear wheel is passive, its plane of rotation is coincident with the plane of the frame (the chassis), similar to a traditional bicycle. The two motors and their respective servosystems make the bicycle capable of moving along a desired path with a desired speed. The path can be programmed in advance. It can also be provided as a stripe drawn on the supporting plane with an inverse color (see example at Figure 15.2).

Number 1 in Figure 14.1 marks the gyroscopic stabilizer that is attached to the bicycle in its lower part, approximately in the middle between the wheels. The stabilizer is discussed in detail below, in section 3.

Number 2 in Figure 14.1 marks the array of light detecting diodes that is installed in front of the front wheel. The array is firmly attached to the front fork. The light detectors are used to "see" the stripe that is drawn on the background to mark a program path. The servosystem that is used to control bicycle steering keeps the center of the array close to the middle of the stripe, and thus the vehicle, following its front wheel, moves along a desired track. When the front wheel moves along a curvilinear path, the rear wheel can move off that path. The rear wheel will only follow the front one at straight segments of the track.

The bicycle has three control circuits. One of them is the system that controls the velocity of the front wheel, i.e. the speed of the bicycle. Another circuit controls the steering angle, i.e. the direction of movement. These two circuits function in accordance with the program of bicycle motion that is assigned at the top control level. The third circuit controls the gyroscopic stabilizer, i.e. it provides stabilization of the vertical position of the bicycle. The current part is dedicated to investigation of the third circuit, that is the most complicated and interesting one. Other control systems are not discussed in frame of this study.

Figure 14.2 illustrates a traditional bicycle; its tilt angle that is denoted as ψ; segment K_1K_2 that connects points K_1 and K_2 where the wheels contact the supporting (horizontal) plane XOY; and a straight line that represents intersection of the front wheel plane and the supporting plane.

The bicycle shown in Figure 14.1, has its front wheel at the same time steering and driving, while the rear wheel is passive. A conventional bicycle shown in Figure 14.2, has a steering front wheel and a driving rear wheel, but the kinematic scheme is the same. When the tilt angle ψ is small, steering angle δ, that is counted counter-clockwise, is close to the angle between segment K_1K_2 and the intersection line of the front wheel plane and the supporting horizontal plane XOY (see Figure 14.3). Unlike a conventional bicycle, the headset axis of the device shown in Figure 14.1 passes through the wheel center, and it is vertical when the frame of the vehicle is vertical.

Trajectory of the contact point of the front wheel K_2 on the supporting plane XOY can be defined using natural coordinates: $\sigma = \sigma(s)$. Here $s \geq 0$ is current distance travelled, and σ is the angle between a tangent line to the motion trajectory of point K_2 and some direction that is fixed with respect to plane XOY, for example, positive direction of axis OX. To unambiguously define the trajectory, Cartesian coordinates of some point of this trajectory must be chosen, for example, of the point that corresponds to value $s = 0$. So the trajectory of vehicle movement and the law of motion

Fig. 14.2. Angle of bicycle tilt ψ, steering angle δ, angle of bicycle heading α.

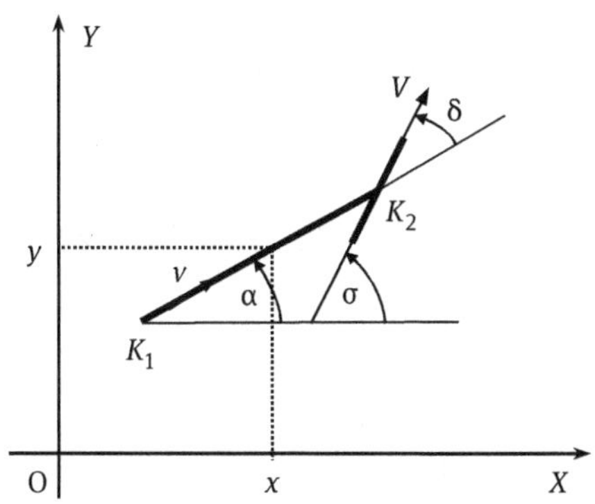

Fig. 14.3. Kinematic diagram of the bicycle with one controlled wheel (top view).

along this trajectory can be defined by equations $\sigma = \sigma(s)$, $\dot{s} = V$. The law of motion of the front wheel along the desired trajectory can be given as a functional dependence of velocity V of contact point K_2 on time t, on travelled distance s, or in some other way. If the rotation plane of the rear wheel cannot turn with respect to the bicycle frame, the motion of the front wheel completely determines the motion of the device as a whole.

Let σ denote angle between the line that is tangent to the trajectory of point K_2 and axis OX. It can be seen in Figure 14.3 that this angle is equal to the sum of steering angle δ and angle α between lines K_1K_2 and OX: $\sigma = \alpha + \delta$.

Since the plane of rotation of the rear wheel coincides with the plane of bicycle frame, velocity v of contact point K_1 is directed along segment K_1K_2 (see Figure 14.3). The instantaneous center of rotation of segment K_1K_2 is located in plane XOY, at the point of intersection of perpendiculars drawn to the wheel planes through their centers. As it is easy to verify, it is located at distance $l/\text{tg}\,\delta$ away from point K_1 and at $l/\sin\delta$ from point K_2, where $l = K_1K_2$. Distance l between points K_1 and K_2 depends on angles ψ and δ. However, when these angles are small, this dependence can be neglected, and it can be assumed that $l = \text{const}$. Velocity of point K_1 is directed along segment K_1K_2, and it is equal to

$$v = \dot{\alpha}l/\text{tg}\,\delta\,. \tag{14.1}$$

The projection of velocity of point K_2 on line K_1K_2 equals $V\cos\delta$. The value of (14.1) is equal to $V\cos\delta$, because segment K_1K_2, being defined in a solid body, does not change its size or shape. Therefore, the following kinematic relation can be written:

$$l\dot{\alpha} = V\sin\delta\,. \tag{14.2}$$

The same relation (14.2) can be derived by taking a component of velocity of point K_2 that is perpendicular to segment K_1K_2. This component, on one hand, is equal to $V\sin\delta$, and on the other hand – to $l\dot{\alpha}$.

The relations above can be presented as a system of differential equations

$$\dot{s} = V, \quad l\dot{\alpha} = V\sin[\sigma(s) - \alpha], \quad \delta = \sigma(s) - \alpha\,. \tag{14.3}$$

Given functions $\sigma(s)$ and $V(s)$ (or $V(t)$), functions $s(t)$, $\alpha(t)$ and $\delta(t)$ can be found by solving system of equations (14.3). So, given a motion trajectory $\sigma(s)$ and a law of movement along this trajectory $V(s)$ (or $V(t)$), one can find the steering angle as a function of time $\delta(t)$, as well as a function of travelled distance $\delta(s)$. Alternatively, given a velocity function $V(s)$ (or $V(t)$) and the steering angle as a function of time $\delta(t)$ or travelled distance $\delta(s)$, then solving differential equations

$$\dot{s} = V, \quad l\dot{\alpha} = V\sin\delta, \quad \sigma = \delta + \alpha\,, \tag{14.4}$$

yields the motion trajectory of the bicycle as a function $\sigma(s)$, or $\sigma(t)$.

14.2 Bicycle with two controlled wheels

Figure 14.4 shows a two-wheel bicycle. Both its wheels are controllable.

The bicycle illustrated in Figure 14.4 has different design from the model shown in Figure 14.1. Each of the wheels of the device discussed in this section is at the same time steering and driving. Thus it has more capabilities than a bicycle with only one steering wheel that is a driving wheel at the same time. A vehicle with two controlled wheels can move in a more complex way. For example, it can move along some path maintaining orientation of its frame, it can move sidewards. It can turn on the spot. Both wheels of the bicycle may remain on a given path when the device is moving along this path, etc.

Fig. 14.4. Bicycle with two controllable wheels: 1 – gyroscopic stabilizer, 2 – video cameras.

Number 1 in Figure 14.4 shows the gyroscopic stabilizer. The stabilizer will be discussed in detail further. Number 2 in the same figure denotes the video cameras. These cameras can be used to make the vehicle move autonomously along some path that is drawn on the supporting plane. The path can be straight or curvilinear.

The bicycle has five control circuits. Two of them are used to control wheel spinning, two more control steering in both of the headsets. The work of these four circuits must be aligned, because the angular velocities of the wheels and the respective steering angles must follow a kinematic constraint described below. Otherwise, the wheels will slip. The fifth circuit provides stabilization of the vertical position of the bicycle.

Figure 14.5 shows the kinematic diagram of the bicycle that has both wheels steering and driving at the same time.

Velocities of points K_1 and K_2 are designated as V_1 and V_2, respectively. Angle δ_1 is the angle of rotation of the rear wheel plane with respect to the bicycle frame (represented by segment K_1K_2), or the rear steering angle. As usually, the angle is considered positive if it represents the counter-clockwise rotation. The front steering angle, i.e. the angle of rotation of the front wheel plane, is denoted as δ_2. Projection of

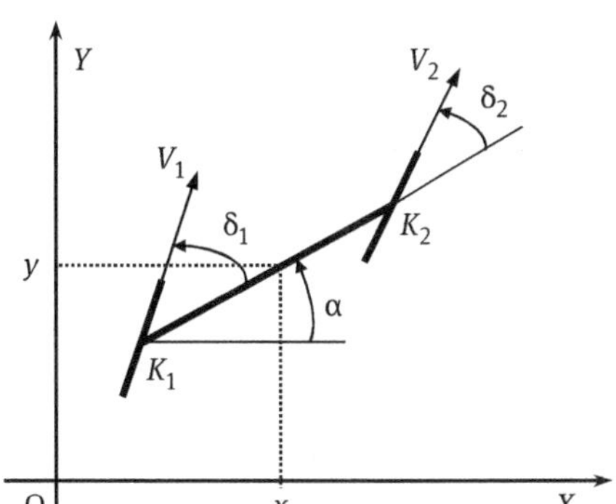

Fig. 14.5. Kinematic diagram of a bicycle with two controlled wheels (top view).

velocity of point K_1 on line K_1K_2 is equal to $V_1 \cos \delta_1$. Projection of velocity of point K_2 on line K_1K_2 is $V_2 \cos \delta_2$. These velocity components must be equal to each other, because segment K_1K_2 does not change its shape or dimension. Thus the following kinematic relation takes place:

$$V_1 \cos \delta_1 = V_2 \cos \delta_2 \,. \tag{14.5}$$

If the radiuses of the wheels are equal, then $V_1 = \omega_1 r$, $V_2 = \omega_2 r$, where ω_1 and ω_2 are angular velocities of the rear wheel and the front wheel, respectively. Then relation (14.5) can be written as

$$\omega_1 \cos \delta_1 = \omega_2 \cos \delta_2 \,. \tag{14.6}$$

The program velocities of the wheels and the corresponding steering angles must comply with constraint (14.5) or (14.6).

14.3 Gyroscopic stabilizer

Gyroscopic stabilization of each of the two described bicycles is done in accordance with a known stabilizer design that bears the name of Scherl and Schilovski. This design was introduced in year 1909 for stabilization of a monorail train car [77, 83–86, 106]. The gyroscopic stabilizer automatically supports the unstable vertical position of the bicycle, and the bicycle can move autonomously following the motion program along some trajectory.

The gyroscopic stabilizer is installed at the bottom part of the bicycle in between the wheels (see Figures 14.1, 14.4). It is schematically shown in Figure 14.6.

The gyrostabilizer includes a set of two identical gyroscopes. The rotor of each gyroscope is contained in a housing that can tilt with respect to the frame (the chassis) of the bicycle about the axis that is perpendicular to the symmetry plane of the vehicle.

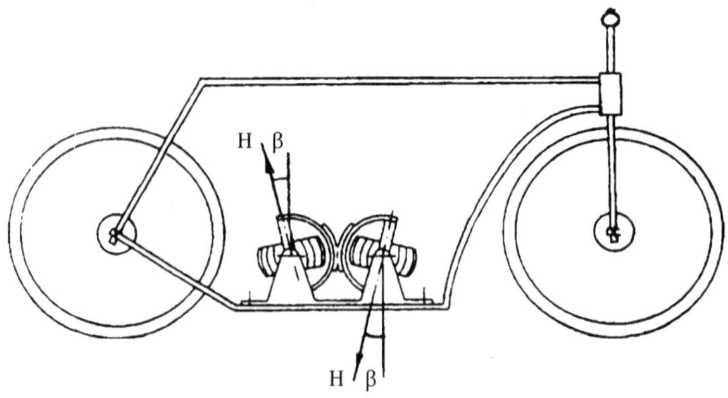

Fig. 14.6. Gyroscopic stabilizer diagram.

The centers of the wheels and the headset axes also belong to this symmetry plane. The gyroscope rotor axes lie in this plane as well. The two gyroscope housings are connected by a gear train so that when one of the gyroscope housings tilts some angle β, the other housing tilts the same angle in the opposite direction (see Figure 14.6).

The gyroscope rotors spin in the opposite directions with the same angular velocity. So their angular momentum vectors $\boldsymbol{H}$ are directed oppositely (Figure 14.6). The gyroscope precession about the housing axes produces gyroscopic torque that counteracts the gravity moment that tends to trip the bicycle. Torque $\boldsymbol{L}$ is applied to the axis of one of the housings. This torque is produced by an electric motor, the rotor is connected to the housing axis via a gearbox, and the armature is firmly attached to the bicycle frame. The voltage supplied to this motor is formed as a function of bicycle tilt angle ψ, precession angle β, precession angular velocity $\dot{\beta}$, steering angle δ and front wheel velocity V. The design of this function (the control law) is discussed in chapter 15.

14.4 Equations of tilt oscillations of the bicycle

In this section, differential equations of the bicycle with a gyroscopic stabilizer will be derived that describe tilt oscillations of the vehicle. These equations are effectively the equations of an inverted pendulum with a gyrostabilizer. The bicycle is affected by the gravity moment that strives to trip the device, and also by inertia forces that take place when the vehicle is moving along a curvilinear trajectory. The equations of motion linearized about state

$$\psi = \beta = \dot{\psi} = \dot{\beta} = 0\,, \tag{14.7}$$

that describe oscillations of the bicycle in terms of tilt angle ψ and oscillations of the gyroscopes in terms of precession angle β can be written as

$$D\ddot{\psi} + 2H\dot{\beta} - Eg\psi = E(\ddot{y}\cos\alpha - \ddot{x}\sin\alpha)\,, \tag{14.8}$$

$$B\ddot{\beta} - 2H\dot{\psi} = L\,. \tag{14.9}$$

In these equations, B and D stand for inertia moments of the gyroscope rotors together with their housings and of the bicycle as a whole, with respect to their corresponding axes; H is the angular momentum of one gyroscope, its value is considered constant; $E = mb$, where m is the mass of the whole vehicle, and b is the distance between the center of mass of the bicycle and segment K_1K_2; x and y denote the coordinates of the midpoint of segment K_1K_2, where, by assumption, the center of mass is projected (see Figures 14.3, 14.5), L denotes the torque produced by the electric motor and applied to the housing of one of the gyroscopes. Term $Eg\psi$ in equation (14.8) describes (in linear approximation) the gravity moment that strives to trip the bicycle, term $2H\dot{\beta}$ in the same equation describes the gyroscopic moment that prevents the bicycle from falling. Expression $E(\ddot{y}\cos\alpha - \ddot{x}\sin\alpha)$ describes the moment with respect to line K_1K_2 of the centrifugal forces, the sum of which is considered to be applied in the mass center of the bicycle. Equations (14.8), (14.9) without term $E(\ddot{y}\cos\alpha - \ddot{x}\sin\alpha)$ and the derivation of these equations can be found in many studies dedicated to the theory of gyroscopes [77, 83–86, 106]. With $H = 0$ equation (14.8) describes (in linear approximation) the motion of an inverted pendulum with a single link.

In the equation of moments with respect to precession axis (equation (14.9)), the friction torque in gyroscope precession axes is not considered – neither dry, nor viscous friction. Counter-EMF in the motor and inductivity of the rotor coil will also be neglected. With these assumptions, the control parameter in the system will be torque L produced by the motor, and not the voltage supplied to it.

The approximate values of some parameters related to the bicycle with a single steering wheel are: $m = 20\,\text{kg}$, $b = 0.2\,\text{m}$, $H = 10\,\text{kg}\cdot\text{m}^2/\text{s}$, $l = 0.75\,\text{m}$. The wheel radius is 0.15 m. The parameters of the bicycle with two steering wheels are also close to these values.

Derivation of motion equations (14.3), (14.8), (14.9) include motion decomposition stage, that is, the vehicle motion is "split" into two motions. Kinematic relations (14.3) describe motion of segment K_1K_2 the trajectory motion of the bicycle, while dynamic equations (14.8), (14.9) describe the bicycle oscillations about axis K_1K_2.

Figure 14.3 helps determine the following kinematic relations:

$$V\cos(\alpha+\delta) = \dot{x} - \frac{1}{2}l\dot{\alpha}\sin\alpha, \qquad V\sin(\alpha+\delta) = \dot{y} + \frac{1}{2}l\dot{\alpha}\cos\alpha\,, \tag{14.10}$$

where $\dot{x}$ and $\dot{y}$ are the components of velocity vector of the midpoint of segment K_1K_2, that is, by assumption, the projection of the mass center of the bicycle when it stands vertically. Applying relations (14.10), taking into consideration equality $l\dot{\alpha} = V\sin\delta$ (see equations (14.4)) and omitting the intermediate formulation, one can see that expression $\ddot{y}\cos\alpha - \ddot{x}\sin\alpha$ that enters the right-hand side of equation (14.8) can be rewritten as

$$\ddot{y}\cos\alpha - \ddot{x}\sin\alpha = \frac{1}{2}\left[\frac{V^2}{l}\sin 2\delta + \frac{d}{dt}(V\sin\delta)\right]. \tag{14.11}$$

Expression $\ddot{y}\cos\alpha - \ddot{x}\sin\alpha$ is effectively the projection of acceleration vector of the midpoint of segment K_1K_2 onto a perpendicular to this segment. Therefore, it is natural that after the transposition it appears to depend only on the front wheel velocity V and on steering angle δ.

Equation (14.8) after substitution of (14.11) becomes as follows:

$$D\ddot{\psi} + 2H\dot{\beta} - Eg\psi = \frac{1}{2}E\left[\frac{V^2}{l}\sin 2\delta + \frac{d}{dt}(V\sin\delta)\right]. \tag{14.12}$$

The right-hand side of equation (14.12) depends on the bicycle trajectory and speed. Substituting functions $\delta(t)$ and $V(t)$ in it yields, whenever control torque L is known, a closed system of differential equations (14.12), (14.9). Solving this system of differential equations (with a given law of controlling torque L) yields functions $\psi(t)$ and $\beta(t)$. That means, oscillations of the bicycle in terms of its tilt and oscillations of the gyroscopes in terms of the precession angle can be found for a particular motion trajectory of the device. Doing this can also help determine whether the desired control law is practical or not; or, in other words, whether the bicycle can maintain its vertical or inclined position, or it trips down.

Motion equations (14.12), (14.9) describe a system that has two degrees of freedom and only one control parameter. Thus the investigated system is underactuated.

15 Designing a control law to stabilize the bicycle tilt

The current chapter suggests a law to control torque L that is applied to precession axes of the gyroscopes in order to maintain the vertical position of the bicycle. The control law that is designed as a feedback involves information on the bicycle tilt angle ψ. This angle can be measured by means of a displacement gyroscope that has three degrees of freedom. This angle can be evaluated precisely enough by means of two accelerometers installed on the vehicle. Below, the principle of their arrangement and the corresponding computation algorithm are provided.

15.1 Bicycle tilt measurement by means of accelerometers

To measure the tilt angle using the accelerometers, they must be attached to the frame of the bicycle in a certain manner.

Figure 15.1 illustrates the bicycle schematically as an inverted pendulum, and it shows the arrangement of the accelerometers.

The figure illustrates a bicycle that is tilted angle ψ away from the vertical. By letter K, line K_1K_2 is denoted. This line is "visible" from the front as a single point. Let point K be actually located in the middle of segment K_1K_2. Axes KX_1 and KZ_1 are fixed in the frame (the chassis) of the bicycle. Axis KX_1 is directed perpendicularly to the chassis of the device. Axis KZ_1 is perpendicular to axis KX_1, and it is located in the vertical plane. Two bi-axial accelerometers are installed on the chassis. They are located on axis KZ_1, at distances l_1 and l_2 away from point K. One of the sensitive axes of each of the accelerometers is parallel to axis KX_1, the other one is parallel to axis KZ_1. When the tilt oscillations of the bicycle occur, a point located on axis KZ_1, at distance l_i ($i = 1, 2$) away from point K, moves with an acceleration which projects onto axes KX_1 and KZ_1 as $-l_i\ddot{\psi}$ and $l_i\dot{\psi}^2$, respectively. Let a_{ix} and a_{iz} denote the accelerometer readings that correspond to their respective sensitive axes that are

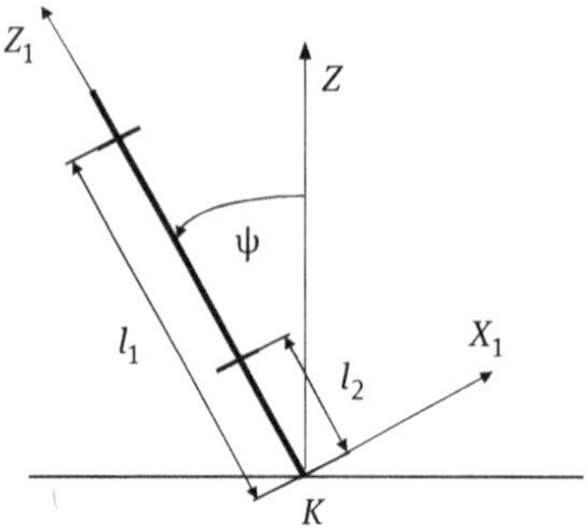

Fig. 15.1. Bicycle arrangement as an inverted pendulum (front view).

parallel to axes KX_1 and KZ_1. Then

$$a_{ix} = -l_i\ddot{\psi} - g\sin\psi, \quad a_{iz} = l_i\dot{\psi}^2 - g\cos\psi \quad (i = 1, 2)\,. \tag{15.1}$$

Consider the first of relations (15.1) with $i = 1, 2$. If readings a_{1x}, a_{2x} of the accelerometers are known, then the tilt angle of the device ψ can be calculated according to expression

$$\sin\psi = \frac{l_2 a_{1x} - l_1 a_{2x}}{g(l_1 - l_2)}\,. \tag{15.2}$$

Expression (15.2) is derived from relations (15.1) by eliminating angular acceleration $\ddot{\psi}$ in them. When angle ψ is small, function $\sin\psi$ in expression (15.2) can be replaced with its argument ψ.

Applying the first expression in (15.1) can also help finding angular acceleration $\ddot{\psi}$

$$\ddot{\psi} = \frac{a_{1x} - a_{2x}}{l_2 - l_1}\,. \tag{15.3}$$

Integration of acceleration (15.3) yields tilt angular velocity $\dot{\psi}$, in case it is necessary for the control law. This angular velocity can also be expressed using the second relation in (15.1)

$$\dot{\psi}^2 = \frac{a_{1z} - a_{2z}}{l_2 - l_1}\,. \tag{15.4}$$

Inspecting expressions (15.2)–(15.4), one can see that if distance $l_1 - l_2$ is “small”, then calculation of the corresponding values will produce “large” errors. So in order to reach appropriate calculation precision, the accelerometers must be installed at a considerable distance away from each other.

In the expressions above it is assumed that the accelerometers are capable of determining the values of acceleration of the respective points precisely, without any errors. However, an accelerometer is a complex device that has its own transients, it can produce errors, so its readings cannot be taken as perfect. Relations (15.2)–(15.4) should be considered as approximate. Yet using, for example, expression (15.2) to measure the tilt angle of the bicycle can proved to be fruitful in practical experiments.

15.2 Bicycle movement along a straight line

First, the stabilizing control law is designed for the case when the bicycle moves along a straight line. If $\delta \equiv 0$, then the bicycle moves straight or it is stationary (when $V \equiv 0$). Equation (14.12) with $\delta \equiv 0$ becomes homogeneous. In addition to equation (14.9), it makes a system:

$$D\ddot{\psi} + 2H\dot{\beta} - Eg\psi = 0\,, \tag{15.5}$$

$$B\ddot{\beta} - 2H\dot{\psi} = L\,. \tag{15.6}$$

Equations (15.5), (15.6) can be written as a system of four equations of the first order, in matrix form

$$\dot{x} = Ax + dL\,, \tag{15.7}$$

where

$$x = \begin{Vmatrix} x_1 \\ x_2 \\ x_3 \\ x_4 \end{Vmatrix} = \begin{Vmatrix} \psi \\ \dot{\psi} \\ \beta \\ \dot{\beta} \end{Vmatrix}, \quad A = \begin{Vmatrix} 0 & 1 & 0 & 0 \\ EgD^{-1} & 0 & 0 & -2HD^{-1} \\ 0 & 0 & 0 & 1 \\ 0 & 2HB^{-1} & 0 & 0 \end{Vmatrix}, \quad d = \begin{Vmatrix} 0 \\ 0 \\ 0 \\ B^{-1} \end{Vmatrix}. \tag{15.8}$$

With $L = 0$, system (15.5), (15.6) or (15.7), (15.8) has a trivial solution (14.7), that can be written as

$$x = 0\,. \tag{15.9}$$

This solution corresponds to the desired vertical position of the bicycle and vertical positions of the gyroscope axes (see Figure 14.6).

The characteristic equation of the open (with $L = 0$) system (15.7), (15.8) is as follows:

$$BD\lambda^4 + (4H^2 - EgB)\lambda^2 = 0\,. \tag{15.10}$$

Equation (15.10) has a double zero root. Besides, when $H = 0$, it has one positive and one negative root, i.e. when $H = 0$, system (15.7), (15.8) is unstable. The system remains unstable while $4H^2 - EgB < 0$. And if $4H^2 - EgB > 0$, then all the roots of equation (15.10) are located on the imaginary axis. The goal is, however, to design such a control law $L = L(\psi, \dot{\psi}, \beta, \dot{\beta})$ that makes solution (15.9) *asymptotically* stable.

Writing out matrix determinant

$$\|d, Ad, A^2d, A^3d\|\,,$$

we can observe that it is nonzero if and only if $H \neq 0$ (we assume that $B \neq 0$). These two inequalities are certainly fulfilled for the bicycle with a gyroscopic stabilizer. Thus system (15.7), (15.8) is completely controllable in Kalman's sense [89–92]. As it is known [89–92], equilibrium (15.9) can be stabilized, that is, it can be rendered asymptotically stable by means of a linear feedback that involves all four phase variables. However, in studies [77, 106], it is shown that equilibrium (15.9) of system (15.7), (15.8) can be stabilized by a feedback that involves only three phase variables

$$L = k_\psi \psi + k_\beta \beta - k\dot{\beta}\,. \tag{15.11}$$

To be more exact, studies [77, 106] show that if angular momentum H is sufficiently large, coefficients k_ψ, k_β, k may be chosen in such a way that Routh–Hurwitz criterion [40, 41, 97] of asymptotic stability of solution (15.9) to system (15.7), (15.8) with control law (15.11) will be fulfilled. It is also shown that for this asymptotic stability, the following conditions are necessary

$$k_\psi > 0, \quad k_\beta > 0, \quad k > 0\,. \tag{15.12}$$

When $k_\beta > 0$, torque $k_\beta \beta$ is directed so that it strives to overturn the gyroscopes and, thus it makes them statically unstable, like a Lagrange gyroscope with center of mass that is located high [83–86, 106]. This situation is similar to a classic problem of stabilization of an unstable equilibrium of a conservative mechanical system by means of gyroscopic forces [40, 77, 106]. In system (15.5), (15.6) the gyroscopic forces are $2H\dot{\psi}$ and $2H\dot{\beta}$. If control law (15.11) is considered given in advance, then the matter of discussion is whether equilibrium (14.7) can be stabilized by these gyroscopic forces. According to Kelvin's theorem [40, 106], an unstable system that is affected by potential forces (and possibly by dissipative forces) cannot be stabilized by gyroscopic forces if its degree of instability is odd. System (15.5), (15.6), (15.11) is in a similar situation, in spite of the fact that a non-conservative force is present, $k_\psi \psi$ [77, 106]. Indeed, if the gyroscopic forces are absent, i.e. $H = 0$, and $k_\beta < 0$, system (15.5), (15.6), (15.11), as it is easy to see, has *one* eigenvalue in the right semiplane. That means, its degree on instability is *odd*. Routh–Hurwitz criterion [40, 41, 97] shows that it is impossible to stabilize the system by gyroscopic forces [40, 106]. If $k_\beta > 0$, the degree of instability of the system is *even*, and it can be stabilized by gyroscopic forces.

15.3 Motion along a circle

Consider a motion of the bicycle at a constant speed with a constant steering angle

$$V(t) = \text{const} \neq 0, \quad \delta(t) = \text{const} \neq 0\,. \tag{15.13}$$

When conditions (15.13) are satisfied, the bicycle can move along a circle. It follows from relations (14.4) that the angular velocity of the bicycle with one controlled wheel that moves along this circle is $\dot{\alpha} = V \sin\delta / l$. The radius of the circle that the front wheel moves along is $V/\dot{\alpha} = l/\sin\delta$, and the radius of the circle that the rear wheel follows is $v/\dot{\alpha} = l/\text{tg}\,\delta$. Radius $l/\text{tg}\,\delta$ is of course less than radius $l/\sin\delta$.

When conditions (15.13) are met, the right-hand side on equation (14.12) has only the first term; equation (14.9) remains unchanged, and system (14.12), (14.9) becomes as follows:

$$D\ddot{\psi} + 2H\dot{\beta} - Eg\psi = \frac{EV^2}{2l}\sin 2\delta\,, \tag{15.14}$$

$$B\ddot{\beta} - 2H\dot{\psi} = L\,. \tag{15.15}$$

Looped by control law (15.11), system (15.14), (15.15) has a steady-state solution

$$\psi = \psi_s = -\frac{V^2}{2gl}\sin 2\delta, \quad \beta = \beta_s = -\frac{k_\psi \psi_s}{k_\beta}, \quad L = 0\,. \tag{15.16}$$

The first equality in (15.16) describes the tilt of the bicycle towards the center of the circle that it moves along. At this tilt, the tripping gravity moment is compensated by the moment of centrifugal force. From the second expression in (15.16) it follows that when the bicycle is moving along a circle, the steady-state value of the precession angle is non-zero. The gyroscopes are known to work best when the precession angle is close to zero. This statement follows from the nonlinear model. Instead of terms $2H\dot{\psi}$ and $2H\dot{\beta}$ that describe the components of the gyroscopic moment in linear equations (15.14), (15.15), there are terms $2H\dot{\psi}\cos\beta$ and $2H\dot{\beta}\cos\beta$ in the nonlinear equations. These terms reach their maximum when $\beta = 0$, and they diminish to zero when $|\beta| \to \pi/2$. From the second equation in (15.16) it follows that the steady-state value of $|\beta_S|$ can be reduced by raising the feedback coefficient k_β. However, this coefficient is limited from above by the stability criteria.

The static error of the precession angle β can be reduced to zero by introducing a target angle ψ_S into expression (15.11)

$$L = k_\psi(\psi - \psi_S) + k_\beta\beta - k\dot{\beta}\,. \tag{15.17}$$

The steady-state solution of system (15.14), (15.15), (15.17) with conditions (15.13) is as follows:

$$\psi = \psi_S = -\frac{V^2}{2gl}\sin 2\delta, \quad \beta = \beta_S = 0, \quad L = 0\,. \tag{15.18}$$

The steady-state values (15.18) of the tilt angle and the control torque apparently remain the same as for feedback (15.11). As for the precession angle in the steady mode, it is zero when control law is as in (15.17).

If coefficients k_ψ, k_β, k are chosen in such a way that steady-state mode (14.7) become an asymptotically stable solution to system (15.5), (15.6), (15.11), then the same values of these coefficients make (15.18) an asymptotically stable solution to system (15.14), (15.15), (15.17). The fact is that the equations of system (15.14), (15.15) written in deviations about steady-state solution (15.18), coincide with equations (15.5), (15.6), (15.11).

Suppose that there is an error in measuring the bicycle tilt angle. This error $\Delta\psi$ will be considered constant ($\Delta\psi$ = const) during the entire motion process of the vehicle. Such error occurs in experiments, for example, when the vertical was not determined precisely before the start. When such error is present, feedback (15.17), takes signal $\psi + \Delta\psi$ instead of ψ. The actual control law, instead of (15.17), becomes as follows:

$$L = k_\psi(\psi + \Delta\psi - \psi_S) + k_\beta\beta - k\dot{\beta}\,. \tag{15.19}$$

With law (15.19), the steady-state value of the tilt angle ψ, as well as torque L, are still determined by expressions (15.18), and the steady-state value of the precession angle is equal to

$$\beta = \beta_S = -\frac{k_\psi\Delta\psi}{k_\beta}\,. \tag{15.20}$$

So when an error is present in tilt angle measurements, the angle of precession in steady-state mode is nonzero.

To eliminate deviation (15.20) of the precession angle, an integral of the precession angle is introduced into control law (15.19)

$$L = k_{\psi}(\psi + \Delta\psi - \psi_s) + k_{\beta}\beta - k\dot{\beta} + k_{\rho}\rho, \quad \dot{\rho} = \beta \,. \tag{15.21}$$

If the system is controlled by law (15.21), then in the steady-state motion

$$\beta = \beta_s = 0, \quad \rho = \rho_s = -\frac{k_{\psi}\Delta\psi}{k_{\beta}}, \tag{15.22}$$

but the tilt angle and the torque remain the same as in expressions (15.18). So the steady-state value of the precession angle is zero when the control law is as in (15.21), even though an unknown error $\Delta\psi$ is still present.

Applying Routh–Hurwitz criterion [40, 41, 97], it can be verified that the necessary conditions of asymptotic stability of the steady-state solution to system (15.14), (15.15) looped by feedback law (15.21) are inequalities

$$k_{\psi} > 0, \quad k_{\beta} > 0, \quad k > 0, \quad k_{\rho} > 0 \,.$$

Unlike a typical PID-controller, feedback (15.21) has the coefficient sign at term with derivative $\dot{\beta}$ different from the coefficient sign at position variable β, as well as the sign at its integral. It is known from the control theory that introducing an integral term to the control law eliminates the static error, but the control system tends to become unstable. Therefore, coefficient k_{ρ} cannot be too large. Routh–Hurwirtz criterion [40, 41, 97] helps establish the upper limitation to the value of this coefficient.

In the end, the following law can be used to control the system:

$$L = k_{\psi}\left(\psi - \frac{V^2}{2gl}\sin 2\delta\right) + k_{\beta}\beta - k\dot{\beta} + k_{\rho}\int_0^t \beta(\tau)d\tau \,. \tag{15.23}$$

Control law (15.23) was designed having in mind stabilization of the steady-state motion of the bicycle. If $\delta = 0$, then the bicycle motion becomes straight. In this case, the "target" angle ψ_s in the control law becomes zero. In order to use control law (15.23), angles ψ and β, as well as precession angular velocity $\dot{\beta}$, bicycle linear speed V and steering angle of the front wheel δ, must be measured during the motion.

15.4 Numerical and practical experiments

The bicycle is fully autonomous during its motion. All control systems are implemented in onboard circuits that involve programmable controllers and power amplifiers. All control laws, including (15.23) are implemented as programs that are input to these controllers. The power is supplied from onboard accumulator batteries.

Practical experiments were carried out on a track used for mobile robot competitions that were held in France in years 1997–2000 and in Portugal in 1997, in frame of

Fig. 15.2. The track used for experiments.

the International Festival of Science and Technologies. The competition track is illustrated in Figure 15.2.

The track consists of segments of straight lines and circle curves. The radius of the curves is 1 m. Figure 15.2 shows this track as white line on black squares and black line on white squares.

The same track as shown in Figure 15.2 was used for numerical experiments. Function $\sigma = \sigma(s)$ (see expressions (14.3), (14.4)), that corresponds to this trajectory is piecewise-linear. When a line segment is straight, the angle corresponding to this segment σ = const, and on the arc segments this angle changes linearly as path s grows. If the arc is half-circle, then angle σ changes by the value of π, and if it is a quarter-circle, then angle σ changes by $\pi/2$. Whenever a straight segment connects to a curvilinear segment, at the contact point derivative $d\sigma/ds$ has a discontinuity, i.e. function $\sigma = \sigma(s)$ on this trajectory is not smooth.

Coefficients k_ψ, k_β, k, k_ρ in control law (15.23) are chosen so that to ensure asymptotic stability of the steady-state motion along both the straight line and the circle. Mathematical modeling shows that if speed V is "small", then the vehicle is able to switch to a steady mode when entering a straight segment as well as an arc segment. The tilt and precession angles deviate from the desired values insignificantly. If the bicycle moves along the track at a "high" speed V, it virtually does not switch to a steady-state motion during cornering (when it enters an arc segment) – it just does not have enough time. The control system functions in transient mode. Nevertheless, as it is shown by the modeling, control law (15.23) is capable of stabilizing the vertical position of the vehicle. If the speed is up to 1 m/s, it is capable of moving along the track not falling down. Along with equations (14.8), (14.9), a nonlinear model was built that describes oscillations of the bicycle tilt and gyroscope precession angles. Numerical integration of the oscillation equations of a bicycle moving at speed of nearly

1 m/s along a track shown above produce results close to those obtained from the linear model.

Thus numerical and practical experiments carried out for the track shown above prove that the gyroscopic stabilization system and the proposed control law (15.23) are capable of stabilizing the vertical position of a two-wheel bicycle when it moves at speeds up to 1 m/s.

Videos have been shot to illustrate the experiments carried out with both bicycles. The video clips are available online at web site [164].

Part V: **Avoiding undesired vibrations**

The current part discusses mechanical systems that include compliant elements. A compliant element may be, for example, a compliant platform with a controlled object (plant) installed on it, or/and an elastic gear that connects a motor with this moving plant. The control parameter (force or torque) is limited in its magnitude. Only the first (lowest) resonance frequency is taken into account. Thus the system under consideration has two degrees of freedom and one control input. The linear mathematical model of this system has a double zero eigenvalue and two complex eigenvalues. So the system without control is unstable. Control laws to drive the plant from the given initial state to the given final state in finite time are designed. For these laws, intervals are determined; on some intervals the control function is linear with respect to time, on other intervals it is constant. The length of time intervals where the control function is non-constant is equal to the period of natural vibrations of the system. If there is no damping, this helps *completely* avoid vibration on time intervals where the control is constant, including intervals where the control signal is identically equal to zero. Another way to avoid vibrations is to set the total time of the transient equal to a multiple of the doubled period of natural vibrations.

16 Bang-bang control and fluent control

Many mechanisms of tools (for example, metalworking machinery, robotic manipulators, walking robots, and exoskeletons) during their operation are subject to undesirable vibrations that reduce the overall performance of these devices. Such vibrations typically arise due to compliant elements used in the design of these devices. Compliant elements of the environment may also come in contact with such plants. For example, the base of a plant, or a gear connecting the electric motor to the moving body can be compliant. Elements of a sensor that measures motor velocity can also be compliant.

The problem of control design for a system containing compliant elements implies contradictory requirements. The response of the system has to be as fast as possible. But a system with compliant parts usually has many different resonance frequencies, and its mathematical model is complex. In order to avoid this complexity, it is possible to take into account only main frequencies. If the order of the corresponding mathematical model is not too high and its parameters are well known, it is possible to design the time-optimal control law using optimal control theory. However, the parameters of the model are often known only approximately. In this case, the design of the optimal control based on the mathematical model makes no sense because such a control would not be optimal for the real plant. Besides, the time-optimal problem can be difficult to solve from theoretical point of view, especially for systems with many degrees of freedom. And what is more important, time-optimal control is discontinuous (the so-called bang-bang control) for many systems. Under such a control, large elastic vibrations may occur that correspond to ignored frequencies. This is not acceptable for the real plant. Therefore, time-optimal control is not really welcome. For a plant mounted on a compliant platform, a quasi-optimal control (with respect to time) is considered. To implement such a control law, only the natural frequency of the system is to be known. Note that the natural frequency can often be determined experimentally. The mathematical model of a motor connected with a moving body by an elastic gear has the same structure with the mathematical model of the plant installed on a compliant platform. Therefore, the approach proposed here can be also used for systems that contain compliant gears.

There are many studies (see, for example [81, 117, 140, 170]) devoted to the problem of suppressing undesirable vibrations. But here the control law is designed so that *to avoid* generation of undesirable vibrations, to *prevent* them from being excited [66, 70].

16.1 Mathematical model

Figure 16.1 shows the approximated experimental Bode gain-frequency plots for two different mechanisms – the pivot mechanism of the SCARA-robot with harmonic drive 1 and the support drive mechanism of a metalworking machine with a ball screw 2 (see [80, 144]). The input signal is the current that flows in the motor. Its frequency, measured in Hertzs (Hz) is presented on a logarithmic scale on the horizontal axis, and the ratio of the input magnitude to output velocity in decibels (dB) is presented on the vertical axis also on a logarithmic scale.

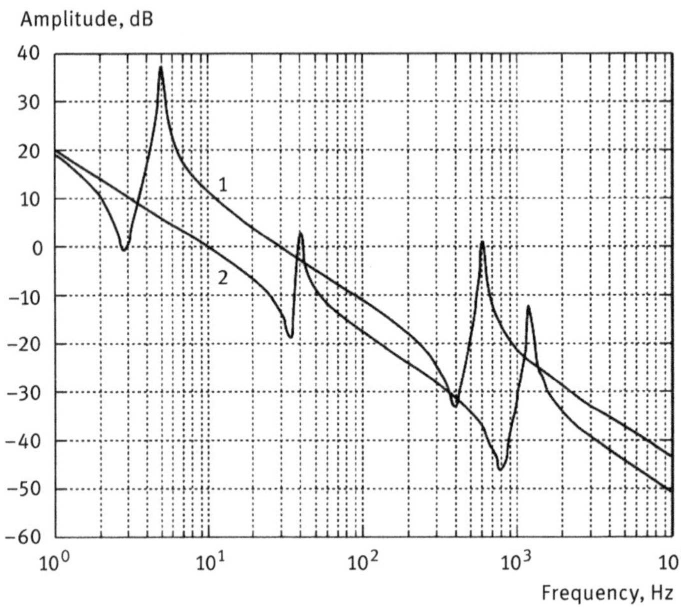

Fig. 16.1. Bode gain-frequency diagram for two mechanisms.

It can be seen in Figure 16.1 that the first resonance frequency for the robot pivot mechanism is about 5 Hz, and the second one is about 600 Hz. For the machine tool, the first resonance frequency is near 40 Hz, and the second one is greater than 1000 Hz. In both mechanisms, the second resonance frequency is much higher than the first one. The analysis of these and other mechanisms with a compliant gear or a compliant platform shows that the main source of undesirable vibrations is the first resonance. Furthermore, high-frequency vibrations are typically suppressed by filters. For this reason, the two-mass system with two degrees of freedom is considered as the mathematical model of a controlled system with compliant elements. It is assumed that the natural frequency of the model is equal to the first resonance frequency of the real system. The motion equations of such system can be written in the following form:

$$M\ddot{\xi} + e\dot{\xi} + k\xi = -F, \quad m(\ddot{\xi} + \ddot{x}) = F\,. \tag{16.1}$$

Equations (16.1) describe, for example, the motion of a plant installed on a compliant platform (see Figure 16.2).

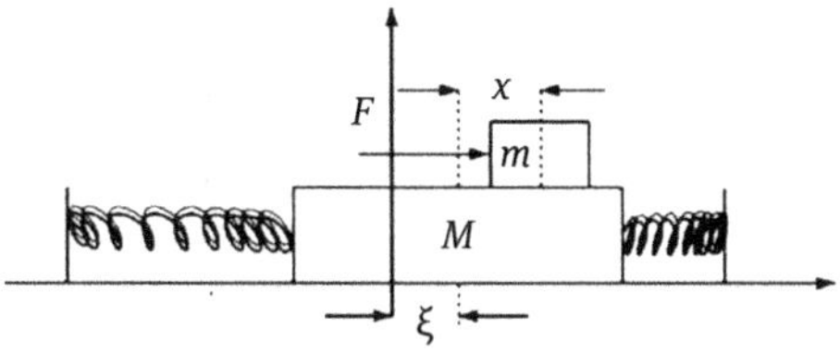

Fig. 16.2. Controlled body of mass m on a compliant platform of mass M.

In equations (16.1) and in Figure 16.2, M is the mass of the platform, and m is the mass of the plant. In real life, in order to suppress external disturbances, the platform is sometimes installed on a base that provides passive vibration insulation on both sides of the platform. In Figure 16.2, two identical springs represent compliance of such a structure. The spring on the left of the platform is shown stretched, and the spring on its right is compressed. The plant is installed on top of the platform. The deviation of the platform from the neutral position is denoted as ξ, and the displacement of the plant with respect to the platform is denoted as x. In the neutral position $\xi = 0$, and the forces developed by both left and right springs that act on the platform are balanced. Sum $\xi + x$ describes the position of the plant in a fixed frame of the reference. In equations (16.1), e is the damping coefficient of the springs or the coefficient of the viscous friction of the platform against the fixed surface, or the total damping coefficient, and k is the reduced stiffness coefficient of both springs together. A drive is installed on the platform (this drive may be linear or rotary with ball-and-screw assembly). This drive develops force F applied to the plant. At the same time, counteracting force $-F$ is applied to the platform, which is directed oppositely. The friction between the platform and the plant is neglected.

Without control ($F = 0$), system (16.1) has a double zero eigenvalue and a pair of complex conjugate eigenvalues. So, if no control forces are applied, the considered system is unstable. Such a system has been studied, for example, in [39, 46, 142]. Equations that have structure like in (16.1) are applicable not only to a plant on a compliant platform, but they can as well describe a system of an electric motor and a plant connected by a compliant gear (see for example [142]).

If the damping coefficient e is sufficiently small, then the system has complex conjugate eigenvalues. The corresponding natural frequency ω of the system is described by the following expression:

$$\omega = \sqrt{\frac{k}{M} - \frac{e^2}{4M^2}}\,. \tag{16.2}$$

With $e = 0$, it becomes $\omega = \omega_0 = \sqrt{k/M}$. Let the period of natural vibrations be denoted as θ, then

$$\theta = 2\pi/\omega \,. \tag{16.3}$$

Assume that the absolute value of force F developed by the motor is limited to a given value F_0:

$$|F| \le F_0 \quad (F_0 = \text{const}) \,. \tag{16.4}$$

Limitation (16.4) imposed on force F typically comes from the fact that the electric current that flows through the motor is limited, and the torque developed by an electric motor is proportional to the current.

16.2 Formulation of the problem

Let all the initial values be equal to zero,

$$\xi(0) = 0, \quad \dot{\xi}(0) = 0, \quad x(0) = 0, \quad \dot{x}(0) = 0 \,. \tag{16.5}$$

Under initial conditions (16.5), displacement ξ of the platform and the velocity of this displacement vanish. The objective of the control is to drive system (16.1) to desired state

$$\xi = 0, \quad \dot{\xi} = 0, \quad x = x_d, \quad \dot{x} = 0 \,. \tag{16.6}$$

Here x_d is the desired position of the plant. In state (16.6), as well as in initial state (16.5), the displacement of the platform and its velocity vanish.

If the system under consideration does not contain compliant elements, that is, if the platform is fixed ($\xi \equiv 0$), then system (16.1) transposes into a simple equation of the second order

$$m\ddot{x} = F \,. \tag{16.7}$$

The design of time-optimal control for equation (16.7) is a classical problem in the theory of optimal processes. It is known [27, 130], that if $x_d > 0$, then system (16.7) under constraint (16.4) is can be driven to state $x = x_d$, $\dot{x} = 0$ in least possible time by means of bang-bang control (see Figure 16.3).

$$F = \begin{cases} F_0 & \text{when} \quad 0 \le x < x_d/2, \\ -F_0 & \text{when} \quad x_d/2 \le x < x_d \,, \\ 0 & \text{when} \quad x = x_d \,. \end{cases} \tag{16.8}$$

Minimal time T needed to drive plant (16.7) to desired state $x = x_d$, $\dot{x} = 0$ by means of control (16.8) is (see [27, 130])

$$T = 2\left(\frac{x_d m}{F_0}\right)^{1/2} \tag{16.9}$$

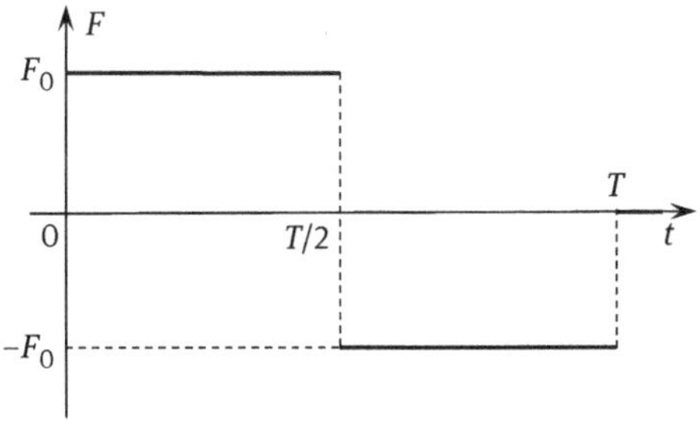

Fig. 16.3. Time-optimal control (16.8) for system (16.7).

The "profile" of acceleration $\ddot{x}$ (16.7) under control (16.8) is as follows:

$$\ddot{x} = \begin{cases} \dfrac{F_0}{m} & \text{when} \quad 0 \le x < x_d/2, \\ -\dfrac{F_0}{m} & \text{when} \quad x_d/2 \le x < x_d\,, \\ 0 & \text{when} \quad x = x_d\,. \end{cases} \tag{16.10}$$

16.3 Bang-bang control versus fluent control

Discontinuous control (16.8) is optimal for model (16.7). Current section estimates its effect on original model (16.1).

Consider the first part of control (16.8)

$$F = \begin{cases} 0 & \text{for} \quad t < 0\,, \\ F_0 & \text{for} \quad t \ge 0\,. \end{cases} \tag{16.11}$$

System (16.1) under control (16.11) has a steady-state solution:

$$\xi = -\frac{F_0}{k}, \quad \ddot{x} = \frac{F_0}{m}\,. \tag{16.12}$$

It can be seen from (16.12) that the direction of displacement ξ is opposite to acceleration $\ddot{x}$. If $\ddot{x} > 0$ ($\ddot{x} < 0$), then $\xi < 0$ ($\xi > 0$). This fact has a clear physical sense, and it also happens during control of an elastic manipulator (see [9, 44]).

If

$$\xi(0) = 0, \quad \dot{\xi}(0) = 0\,, \tag{16.13}$$

then the solution to system (16.1), (16.11) (for $e = 0$) is as follows:

$$\xi(t) = -\frac{F_0}{k}(1 - \cos\omega_0 t), \quad \ddot{x}(t) = \frac{F_0}{m}\left(1 + \frac{m}{M}\cos\omega_0 t\right)\,. \tag{16.14}$$

Expressions (16.14) describe vibrations of system (16.1) about steady-state motion (16.12). These vibrations do not fade if no damping is present (i.e. if $e = 0$). The magnitude of these vibrations depends on the system parameters. Compliant elements often have low natural damping, the fact that justifies the investigation of case $e = 0$.

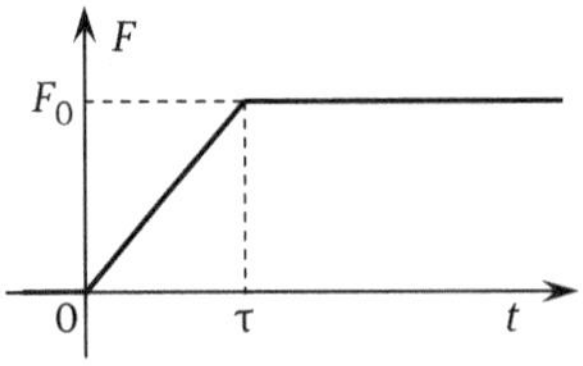

Fig. 16.4. Fluent control (16.15).

Instead of control law (16.11), which is discontinuous at $t = 0$, consider a continuous control law derived from function (16.11). An interval is inserted where the control function grows linearly with respect to time from zero to maximal value F_0. Then the control function maintains that value (see Figure 16.4):

$$F = \begin{cases} 0 & \text{when} \quad t < 0\,, \\ \dfrac{F_0}{\tau} t & \text{when} \quad 0 \le t \le \tau\,, \\ F_0 & \text{when} \quad t \ge \tau\,. \end{cases} \tag{16.15}$$

Here $\tau = \text{const} > 0$. Continuous control law (16.15), in contrast with (16.11), can be called *fluent control* (see [2]).

On interval $0 \le t < \tau$ system (16.1), (16.15) ($e = 0$) has a particular solution

$$\xi(t) = -\frac{F_0}{k\tau} t, \quad \ddot{x}(t) = \frac{F_0}{m\tau} t\,, \tag{16.16}$$

which, however, does not meet initial conditions (16.13). Solution (16.16) describes linear with respect to time variation of displacement ξ and acceleration $\ddot{x}$. Under initial conditions (16.13) system (16.1), (16.15) has the following solution

$$\xi(t) = \begin{cases} -\dfrac{F_0}{k\tau}\left(t - \dfrac{\sin \omega_0 t}{\omega_0}\right) & \text{when} \quad 0 \le t \le \tau\,, \\ -\dfrac{F_0}{k}\left[1 - \dfrac{2 \sin \dfrac{\omega_0 \tau}{2}}{\omega_0 \tau} \cos \omega_0 \left(t - \dfrac{\tau}{2}\right)\right] & \text{when} \quad t \ge \tau\,; \end{cases} \tag{16.17}$$

$$\ddot{x}(t) = \begin{cases} \dfrac{F_0}{m\tau} t + \dfrac{F_0 \omega_0}{k\tau} \sin \omega_0 t & \text{when} \quad 0 \le t \le \tau\,, \\ \dfrac{F_0}{m} + \dfrac{2 F_0 \omega_0 \sin \dfrac{\omega_0 \tau}{2}}{k\tau} \cos \omega_0 \left(t - \dfrac{\tau}{2}\right) & \text{when} \quad t \ge \tau\,. \end{cases} \tag{16.18}$$

Expressions (16.17) and (16.18) are derived by combining the solutions to equations (16.1), (16.15) obtained for $0 \le t \le \tau$ and for $t \ge \tau$. The intermediate derivation is omitted due to its complexity.

A dimensionless variable χ is introduced as follows:

$$\chi = \frac{\tau}{\theta} = \frac{\tau \omega}{2\pi} \quad \left(\tau = \chi \theta = \chi \frac{2\pi}{\omega}\right). \tag{16.19}$$

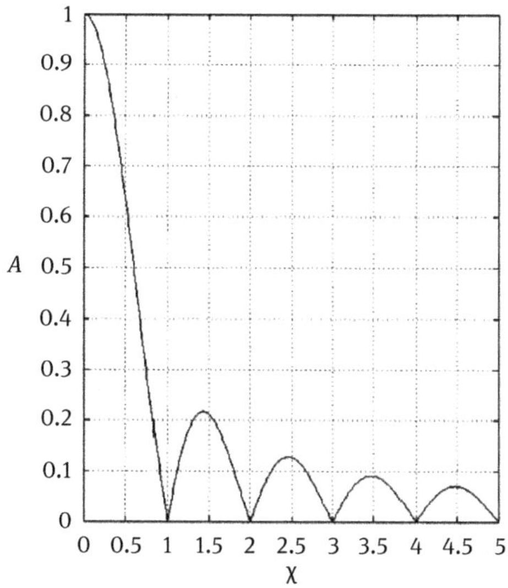

Fig. 16.5. Plot of function $A = A(\chi)$ (16.20).

Consider the following function:

$$A = A(\chi) = \left|\frac{2\sin(\omega_0\tau/2)}{\omega_0\tau}\right| = \left|\frac{\sin\pi\chi}{\pi\chi}\right| . \tag{16.20}$$

Function (16.20) characterizes the relative magnitude (see the second line in expression (16.17)) of the platform vibrations about the steady-state displacement (see the first expression in (16.12)). It is known that $\lim_{\chi\to 0} A(\chi) = 1$. So, if $\chi \to 0$ ($\tau \to 0$), then the expressions in the second lines of formulas (16.17) and (16.18) turn into expressions (16.14). If $\chi \to \infty$ ($\tau \to \infty$), then $A(\chi) \to 0$. The plot of function $A = A(\chi)$ is shown in Figure 16.5.

Vibrations of the platform (see (16.17)) and vibrations of the plant acceleration (see (16.18)) about their stationary values (16.12) do not damp on interval where control (16.15) is constant, that is, for $t \geq \tau$. This is similar to control law (16.11). However, it follows from expressions (16.17) and (16.18) that the magnitude of these vibrations tends to zero if $\tau \to \infty$ ($\chi \to \infty$). Therefore, this magnitude can be made arbitrarily small under control (16.15) by choosing a sufficiently large time interval τ (value χ), during which force F increases linearly with respect to time (see also Figure 16.5). This assertion is in agreement with the results obtained in [2]. That study investigates the motion of a system that contains perfectly rigid bodies connected by elastic elements with distributed characteristics. However, it should be kept in mind that the transients take longer time to settle when time τ increases.

Thus, by changing parameter τ in (16.15), one can affect the magnitude of vibrations that occur in the system. The vibrations can be controlled regardless of the system parameters. This statement also refers to system (16.1) with damping ($e \neq 0$); the corresponding expressions are not presented here due to their complexity. Taking particular sets of parameters (with $e \neq 0$), it can be shown (for example, numerically) that the magnitude of the vibrations decreases as time τ grows.

Considering expressions (16.17) and (16.18), it is possible to draw another important conclusion for case $e = 0$. Assume that $\tau = \theta l$ $(l = 1, 2, \dots)$. Then expressions (16.17) and (16.18) take form (16.12) for $t \geq \tau$. In order words, in case $\tau = \theta l$, there are *no vibrations* in the system on the time interval when the control force is constant. Thus, the following theorem can be stated.

Theorem 1. *If the control force changes linearly during time interval that is equal to the period of natural oscillations of the uncontrolled system, or to a multiple of this period, then no vibrations occur in the system on time interval where the control force is constant.*

So, if $e = 0$ and $\tau = \theta = 2\pi/\omega_0$, then under control (16.15) with "jerk" (see [80]) on interval $0 \leq t \leq \tau$, the vibrations of variables ξ and $\ddot{x}$ do not appear at time $t \geq \tau$.

The closer is quantity τ (the time of the jerk) to the period of natural vibrations θ (see (16.3)) or to value $\tau = \theta l$ $(l = 2, 3, \dots)$, the less is the magnitude of vibrations on interval where control force F is constant.

The above Theorem 1 is true for a system without damping, that is, if $e = 0$. If damping is present, then Theorem 1 is not true. However, since solutions of differential equations depend on parameters in a continuous way, the magnitude of vibrations on the interval where the control force is constant is "small" if damping coefficient e is "close" to zero.

17 Trapezoidal control for a system with compliant elements

Bang-bang control (16.8) with a single switching is time-optimal for second-order equation (16.7). This equation describes the motion of a perfectly rigid body. If this control is used for compliant system (16.1), the vibrations in system do not damp by time instant (16.9). The time-optimal control for system (16.1) has, in general, three or more points where switching occurs, because system (16.1) with $F = 0$ has a double zero eigenvalue and two complex eigenvalues. The minimum time interval needed to drive system (16.1) to the desired state is longer than time (16.9). The time-optimal control can be designed if the parameters of the system are known. However, if these parameters are unavailable, or they cannot be determined accurately enough, it is reasonable to consider a control law derived from control (16.8) that has additional intervals where the control function depends linearly on time.

17.1 Trapezoidal fluent control

Consider the following continuous control of trapezoid shape:

$$F(t) = \begin{cases} \dfrac{F_0}{\tau} t & \text{when} \quad 0 \le t \le \tau \,, \\ F_0 & \text{when} \quad \tau \le t \le \dfrac{T}{2} - \tau \,, \\ \dfrac{F_0}{\tau}\left(\dfrac{T}{2} - t\right) & \text{when} \quad \dfrac{T}{2} - \tau \le t \le \dfrac{T}{2} \,, \end{cases} \tag{17.1}$$

$$F(t) = \begin{cases} -F(T-t) & \text{when} \quad \dfrac{T}{2} \le t \le T \,, \\ 0 & \text{when} \quad t \ge T \,. \end{cases} \tag{17.2}$$

Formula (17.2) indicates that control (17.1), (17.2) is symmetric with respect to point $(t = T/2, F = 0)$. The diagram of this symmetric control (17.1), (17.2) is shown in Figure 17.1.

The trapezoidal profile of the control force used to decrease the magnitude of elastic vibrations is studied in [2, 10]. The restriction imposed on the rate of change (increase and decrease) of the control function is a common practice in modern drive controllers (see [78, 116]). It was shown in the preceding section that a control function F that changes fluently on interval $(0, \tau)$ decreases the magnitude of the system vibrations on interval $\tau \le t \le T/2 - \tau$. A fluent change of function F on interval $(T/2 - \tau, T/2 + \tau)$ decreases the magnitude of system vibrations on $T/2+\tau \le t \le T-\tau$. Finally, a fluent change of function F on interval $T - \tau \le t \le T$ decreases the magnitude of system vibrations for all $t > T$. By increasing the value of τ, the magnitude of

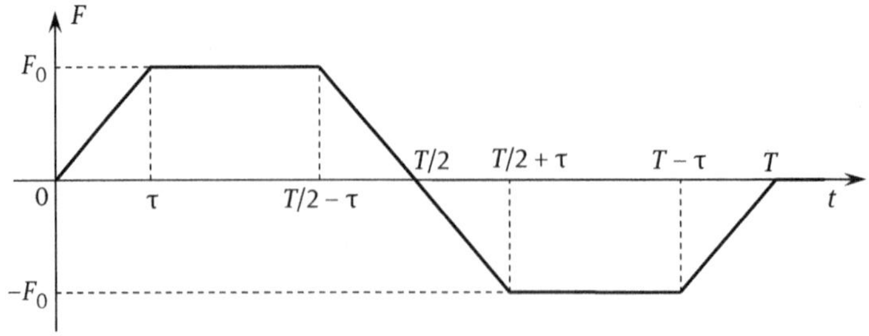

Fig. 17.1. Trapezoidal control law (17.1), (17.2).

system vibrations can be made arbitrary small not only after time $t = \tau$, but also after final time $t = T$.

Notice that control law (17.1), (17.2) is optimal for equation (16.7) in terms of time consumption if constraint (16.4) is imposed on the magnitude of the control force, along with a constraint imposed on its derivative:

$$|\dot{F}(t)| \leq F_0/\tau\,. \tag{17.3}$$

For given values F_0 and τ, control law (17.1), (17.2) can be realized only if

$$T \geq 4\tau\,. \tag{17.4}$$

If $T = 4\tau$, then control law (17.1), (17.2) attains its terminal values $F = F_0$ and $F = -F_0$ only at points $t = \tau$ and $t = 3\tau$, respectively (see Figure 17.2).

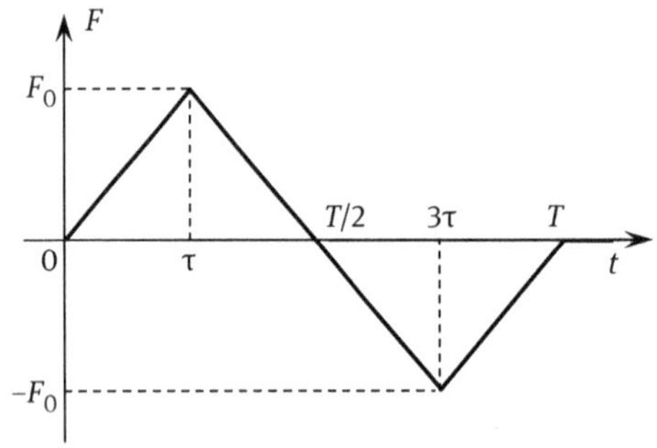

Fig. 17.2. For $T = 4\tau$ trapezoidal control (17.1), (17.2) turns into a sawblade shape.

The time intervals during which the control signal has a slope, i.e. it grows or falls, are regulated by parameter τ. The author suggests that the value of τ is set equal to the period of natural oscillations of the system, the value that is presented in (16.3). If there is no damping, such choice helps *completely* avoid elastic vibrations in the system after the control process is finished. If the damping coefficient is “sufficiently small”, then the vibrations are reduced significantly, “almost” avoided.

Under control (17.1), (17.2) with $\tau = \theta$, the solution to equations (16.1) with initial conditions (16.13) is like

$$\xi(t) = \begin{cases} -\dfrac{F_0}{k\tau}\left(t - \dfrac{\sin\omega_0 t}{\omega_0}\right) & \text{when} \quad 0 \le t \le \tau\,, \\ -\dfrac{F_0}{k} & \text{when} \quad \tau \le t \le T/2 - \tau\,, \\ \dfrac{F_0}{k\tau}\left[t - \dfrac{T}{2} - \dfrac{\sin\omega_0(t - T/2)}{\omega_0}\right] & \text{when} \quad T/2 - \tau \le t \le T/2\,; \end{cases} \tag{17.5}$$

$$\ddot{x}(t) = \begin{cases} \dfrac{F_0}{m\tau}t + \dfrac{F_0\omega_0}{k\tau}\sin\omega_0 t & \text{when} \quad 0 \le t \le \tau\,, \\ \dfrac{F_0}{m} & \text{when} \quad \tau \le t \le T/2 - \tau\,, \\ \dfrac{F_0}{m\tau}\left(\dfrac{T}{2} - t\right) - \dfrac{F_0\omega_0}{k\tau}\sin\omega_0\left(t - \dfrac{T}{2}\right) & \text{when} \quad T/2 - \tau \le t \le T/2. \end{cases} \tag{17.6}$$

on interval $0 \le t \le T/2$. Expressions (17.5), (17.6) imply that, for $0 \le t \le T/2$,

$$\xi(t) = \xi(T/2 - t), \quad \ddot{x}(t) = \ddot{x}(T/2 - t)\,;$$

that is, the plots of functions $\xi(t)$ and $\ddot{x}(t)$ are symmetric about line $t = T/4$ on time interval $0 \le t \le T/2$. Furthermore, $\xi(T/2) = \xi(0) = 0$, $\ddot{x}(T/2) = \ddot{x}(0) = 0$, and $\dot{\xi}(T/2) = -\dot{\xi}(0) = 0$. When $T/2 \le t \le T$, motion (17.5), (17.6) is repeated, except that the sign of the deformation and acceleration is opposite. In other words, the solution to equations (16.1) with initial conditions (16.13) under control (17.1), (17.2), where $\tau = \theta$, is symmetric, and control function (17.1), (17.2) is symmetric as well:

$$\xi(t) = -\xi(T - t), \quad \ddot{x}(t) = -\ddot{x}(T - t)\,. \tag{17.7}$$

Derivative $\dot{x}(t)$ satisfies equation $\dot{x}(t) = \dot{x}(T - t)$. The plot of function $x(t)$ is symmetric with respect to point $(t = T/2, x(T/2))$:

$$x(t) = -x(T - t) + 2x\left(\frac{T}{2}\right)\,. \tag{17.8}$$

It follows from equations (17.7) that, under control (17.1), (17.2) with $\tau = \theta$,

$$\xi(T) = -\xi(0) = 0, \quad \dot{\xi}(T) = \dot{\xi}(0) = 0, \quad \text{and also} \quad \dot{x}(T) = 0\,.$$

It follows from (17.8), that, if $x(T) = x_d$, then

$$x\left(\frac{T}{2}\right) = \frac{x_d}{2}\,. \tag{17.9}$$

Theorem 2 concludes the consideration above:

Theorem 2. *Trapezoidal control law (17.1), (17.2) with $\tau = \theta$ drives system (16.1) (with $e = 0$) from initial state of rest (16.5) to the final state of rest (16.6) in a finite time interval; on intervals where the control function is constant, including $t \ge T$, there are no vibrations in the system.*

So, if $e = 0$ and $\tau = \theta = 2\pi/\omega_0$, then under control (17.1), (17.2), vibrations of variables ξ and $\ddot{x}$ do not occur on intervals $\tau \le t \le T/2 - \tau$, $T/2 + \tau \le t \le T - \tau$, and, what is most important, at $t > T$.

17.2 Trapezoidal control with shorter transients

A different fluent control law can be used when $e = 0$ and $\tau = \theta = 2\pi/\omega_0$. This law is closer to the time-optimal control law than (17.1), (17.2):

$$F(t) = \begin{cases} \dfrac{F_0}{\tau} t & \text{when } 0 \le t \le \tau\,, \\ F_0 & \text{when } \tau \le t \le \dfrac{T}{2} - \dfrac{\tau}{2}\,, \\ \dfrac{2F_0}{\tau}\left(\dfrac{T}{2} - t\right) & \text{when } \dfrac{T}{2} - \dfrac{\tau}{2} \le t \le \dfrac{T}{2}\,, \end{cases} \tag{17.10}$$

$$F(t) = \begin{cases} -F(T - t) & \text{when } \dfrac{T}{2} \le t \le T\,, \\ 0 & \text{when } t \ge T\,. \end{cases} \tag{17.11}$$

The profile of this symmetric control law (17.10), (17.11) is shown in Figure 17.3.

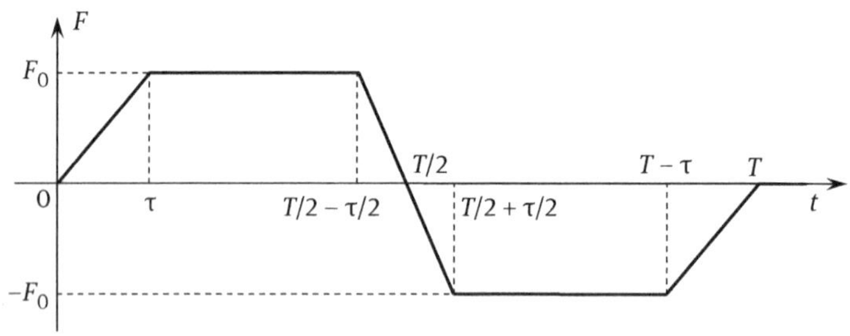

Fig. 17.3. Profile of control (17.10), (17.11).

Control law (17.10), (17.11) is different from control (17.1), (17.2) on time interval $(T/2 - \tau, T/2 + \tau)$ only. Unlike control (17.1), (17.2), function (17.10), (17.11) takes constant F_0 on interval $(T/2 - \tau, T/2 - \tau/2)$, and it takes constant value $-F_0$ on interval $(T/2 + \tau/2, T/2 + \tau)$. On interval $(T/2 - \tau/2, T/2 + \tau/2)$ this function decreases linearly in time from value F_0 to value $-F_0$, and the duration of this jerk is τ instead of 2τ for the jerk in function (17.1), (17.2) (compare Figures 17.3 and 17.1). So, control function (17.10), (17.11) switches from its maximal value F_0 to minimal value $-F_0$ faster than function (17.1), (17.2). It will be proved further that law (17.10), (17.11) does not produce vibrations, as well as control (17.1), (17.2). To do that, it is sufficient to prove that control law (17.10), (17.11) does not produce vibrations up to time instant $t = T/2 + \tau/2$.

When law (17.10), (17.11) is applied,

$$\xi\left(\frac{T}{2}-\frac{\tau}{2}\right)=-\frac{F_0}{k},\quad \dot{\xi}\left(\frac{T}{2}-\frac{\tau}{2}\right)=0\,. \tag{17.12}$$

Now consider this control law on interval

$$\frac{T}{2}-\frac{\tau}{2}\le t\le\frac{T}{2}+\frac{\tau}{2}, \tag{17.13}$$

Function (17.10), (17.11) on interval (17.13) is linear in time

$$F(t)=\frac{2F_0}{\tau}\left(\frac{T}{2}-t\right)=\frac{\omega_0 F_0}{\pi}\left(\frac{T}{2}-t\right). \tag{17.14}$$

Under control (17.14), solution to the first equation in (16.1) (with $e = 0$) that corresponds to initial (at time instant $t = T/2 - \tau/2$) conditions (17.12) is as follows:

$$\xi(t)=\frac{F_0}{k\pi}\sin\left[\omega_0\left(t-\frac{T}{2}\right)\right]+\frac{\omega_0 F_0}{k\pi}\left(t-\frac{T}{2}\right),\quad \dot{\xi}(t)=\frac{\omega_0 F_0}{k\pi}\cos\left[\omega_0\left(t-\frac{T}{2}\right)\right]+\frac{\omega_0 F_0}{k\pi}. \tag{17.15}$$

It follows from expressions (17.15) that

$$\xi\left(\frac{T}{2}+\frac{\tau}{2}\right)=\frac{F_0}{k},\quad \dot{\xi}\left(\frac{T}{2}+\frac{\tau}{2}\right)=0\,.$$

Thus, under control law (17.10), (17.11), on interval $T/2+\tau/2\le t\le T-\tau$, deformation

$$\xi(t)\equiv\frac{F_0}{k},$$

and no vibrations occur. So, Theorem 2 proves true not only for control law (17.1), (17.2) with $\tau=\theta=2\pi/\omega_0$, but also for control law (17.10), (17.11).

For given values F_0 and τ, control law (17.10), (17.11) can be realized only if

$$T\ge 3\tau\,.$$

If $T = 3\tau$, then this control takes its terminal values $F = F_0$ and $F = -F_0$ only at points $t = \tau$ and $t = 2\tau$, respectively. In this case, its plot is shown in Figure 17.4 (compare to Figure 17.3).

If total time T is given, then system (16.1) (with $e = 0$) under control law (17.10), (17.11), reaches larger displacement x_d than under control (17.1), (17.2). It is obvious that if total time T increases, then corresponding displacement $x(T)$ increases strictly monotonically when any of control laws (17.1), (17.2) or (17.10), (17.11) is used. Consequently, if desired displacement x_d is given, then system (16.1) (with $e = 0$) controlled by law (17.10), (17.11), reaches this displacement x_d faster than under control (17.1), (17.2).

Now consider a particular case when final time T is a multiple of the doubled period (16.3) of natural vibrations, that is $T = 2\theta n$ $(n = 1, 2, \ldots)$. Besides, let control

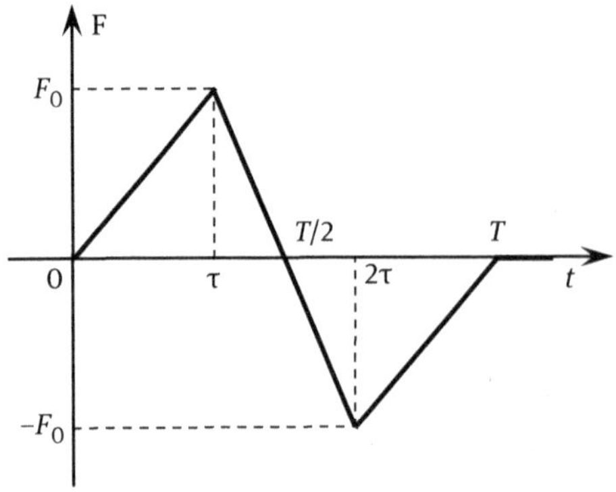

Fig. 17.4. For $T = 3\tau$ trapezoidal control (17.10), (17.11) turns into a sawblade shape.

function $F(t)$ satisfy the symmetry condition with respect to point ($t = T/2 = \theta n$, $F = 0$) on interval $0 \leq t \leq T$,

$$F(t) = -F(T - t)\,, \tag{17.16}$$

and let this function $F(t)$ become zero when $t \geq T$ (see (17.2) or (17.11)). Consider the solution to system (16.1) with some control function $F(t)$ on interval $0 \leq t \leq T/2$ with initial conditions $\xi(0) = \dot{\xi}(0) = \dot{x}(0) = 0$. At instant $t = T/2 = \theta n$ this solution comes to some state

$$\xi(T/2), \quad \dot{\xi}(T/2), \quad \dot{x}(T/2)\,. \tag{17.17}$$

Consider the solution to system (16.1) on interval $T/2 \leq t \leq T$ with initial (at instant $t = T/2$) conditions (17.17) under control $F(t)$ satisfying symmetry condition (17.16). If $e = 0$, then at instant $t = T$ this solution comes back to the state $\xi = \dot{\xi} = \dot{x} = 0$. For $t \geq T$, control $F(t) \equiv 0$, and consequently, $\xi(t) \equiv \dot{\xi}(t) \equiv \dot{x}(t) \equiv 0$. This consideration can be concluded by the following Theorem 3:

Theorem 3. *Assume that $e = 0$ in the system (16.1), final time $T = 2\theta n$ $(n = 1, 2, \ldots)$, control function $F(t)$ satisfies symmetry condition (17.16), and $F(t) = 0$ for $t \geq T$. Then this control function drives system (16.1) from initial state of rest (16.5) to final state of rest (16.6) in a finite time interval.*

Symmetry condition (17.16) is satisfied by trapezoidal control (17.1), (17.2) (or (17.10), (17.11)) for any value of parameter τ; it is also satisfied by bang-bang control (16.8). Then, for control (17.1), (17.2) (or (17.10), (17.11)) and control (16.8) the statement of Theorem 3 is also true. On time intervals where control signal (17.1), (17.2) (or (17.10), (17.11)) is constant, there are no vibrations (for $0 \leq t \leq T$) only if τ is a multiple of the period (16.3) of natural oscillations ($\tau = \theta n$, $n = 1, 2, \ldots$). As for bang-bang control, on time intervals where control function (16.8) is constant, vibrations do occur.

17.3 Relationship between time T and displacement x_d

The goal of this section is to find the relationship between final time T and desired displacement of the object x_d when system (16.1) is controlled by law (17.1), (17.2) with $\tau = \theta$. It is assumed that $e = 0$. Time T is not restricted here to be a multiple of the double period of natural oscillations. Relation (17.6), integrated twice from $t = 0$ to $t = T/2$, taking into account the second equation in (16.1) and equalities

$$\xi(0) = \dot{\xi}(0) = \xi(T/2) = \dot{\xi}(T/2) = 0\,,$$

then, using relation (17.9), yields the following expression:

$$x\left(\frac{T}{2}\right) = \frac{F_0 T}{4m}\left(\frac{T}{2} - \tau\right) = \frac{x_d}{2} \quad (\tau = \theta)\,. \tag{17.18}$$

Solving (17.18) as an equation for T yields

$$T = \tau + \sqrt{\frac{4x_d m}{F_0} + \tau^2} \tag{17.19}$$

Considering that $\tau = \theta = 2\pi/\omega$ and $\omega = \omega_0 = \sqrt{k/M}$, the following expression can be obtained

$$T = 2\sqrt{\frac{x_d m}{F_0} + \pi^2\frac{M}{k}} + 2\pi\sqrt{\frac{M}{k}}\,. \tag{17.20}$$

If $\omega_0 \to \infty \quad (k/M \to \infty)$, then $\tau \to 0$ and value (17.19) (or (17.20)) converges to value (16.9). Thus, if frequency ω_0 of natural vibrations is high, control law (17.1), (17.2) is close to the time-optimal control (it becomes quasi-optimal).

Relation (17.18) and, consequently, (17.19) and (17.20), can alternatively be obtained by integrating the second equation in (16.1) twice from $t = 0$ to $t = T/2$, assuming that the control function is as in (17.1) and the initial conditions are $\xi(0) + x(0) = 0$, $\dot{\xi}(0) + \dot{x}(0) = 0$.

Expressions (17.9), (17.18), and (17.19) are true under condition (17.4). In case when $\tau = \theta = 2\pi/\omega_0 = 2\pi\sqrt{M/k}$, inequality (17.4) becomes as follows:

$$T \ge 8\pi\sqrt{\frac{M}{k}} \quad (T \ge 4\tau)\,. \tag{17.21}$$

Substituting expression (17.20) into inequality (17.21) yields:

$$x_d \ge 8\pi^2\frac{MF_0}{mk}\,. \tag{17.22}$$

In order to be able to move the plant a given distance x_d away by means of control law (17.1), (17.2) with $\tau = \theta = 2\pi/\omega_0 = 2\pi\sqrt{M/k}$, the parameters of system (16.1) and value F_0 must satisfy inequality (17.22).

Assume that the given distance x_d, which the plant must be moved, is such that inequality (17.22) turns into equality

$$x_d = 8\pi^2 \frac{MF_0}{mk} . \quad (17.23)$$

Then control function (17.1), (17.2) with $\tau = \theta$ has no intervals where the control signal is constant. In this case, the control function takes the shape of a sawblade (see Figure 17.2), and it attains the terminal values $F = \pm F_0$ only at points $t = \tau$ and $t = 3\tau$.

Now, assume that the given distance x_d is less than value (17.23):

$$x_d < 8\pi^2 \frac{MF_0}{mk} . \quad (17.24)$$

Then, control function (17.1), (17.2) that has linear slopes on four time intervals of length $\tau = \theta$ and at the same time attains terminal values $F = \pm F_0$ cannot be realized. Some requirements must be sacrificed, and so, under condition (17.24), control law (17.1), (17.2) can be realized only if constant F_0, that is taken as the terminal value (see inequality (16.4)), is replaced with a lower value. In other words, when condition (17.24) takes place, control law (17.1), (17.2) is possible to use, but the control resource will only be used partly.

For control law (17.10), (17.11) with $\tau = \theta$, relationship between final time T and desired displacement of the object x_d similar to (17.19), (17.20) can also be found. The reader is encouraged to do it manually.

17.4 Numerical analysis of the open-loop system

Let the parameters of system (16.1), (16.4) take the following numerical values:

$$M = 144.7\,\text{kg}, \quad m = 75\,\text{kg}, \quad k = 142857\,\text{N}\cdot\text{m}^{-1}, \quad F_0 = 187.5\,\text{N} . \quad (17.25)$$

With parameter values (17.25) and $e = 0$, frequency (see expression (16.2)) $\omega = \omega_0 = \sqrt{k/M} = 31.42\,\text{s}^{-1}$, and the corresponding period of natural vibrations is $\theta = 2\pi/\omega_0 = 0.2\,\text{s}$.

Figure 17.5 (the top plot) shows the dynamics of acceleration F/m (dashed line) and $\ddot{x}$ (solid line). The middle plot shows deformation ξ, and the bottom plot shows displacement x. All plots were obtained for control function as in (17.1), (17.2) with $e = 0$, $\tau = \theta = 0.2\,\text{s}$. Final time T is assumed to be 1.1 s. The acceleration is presented in m/s^2, the deformation – in millimeters (mm), the displacement in meters (m).

The plots in Figure 17.5 can be obtained analytically by means of relations (17.5), (17.6) and symmetry conditions (17.7). Alternatively, they can be found by integrating equations (16.1), (17.1), (17.2) numerically.

The plot of function $F(t)/m$ naturally has a trapezoid shape because control function $F(t)$ (17.1), (17.2) has such a trapezoid shape (see Figure 17.1). Fraction $F(t)/m$ represents the absolute acceleration of the mechanical system under control. It is seen

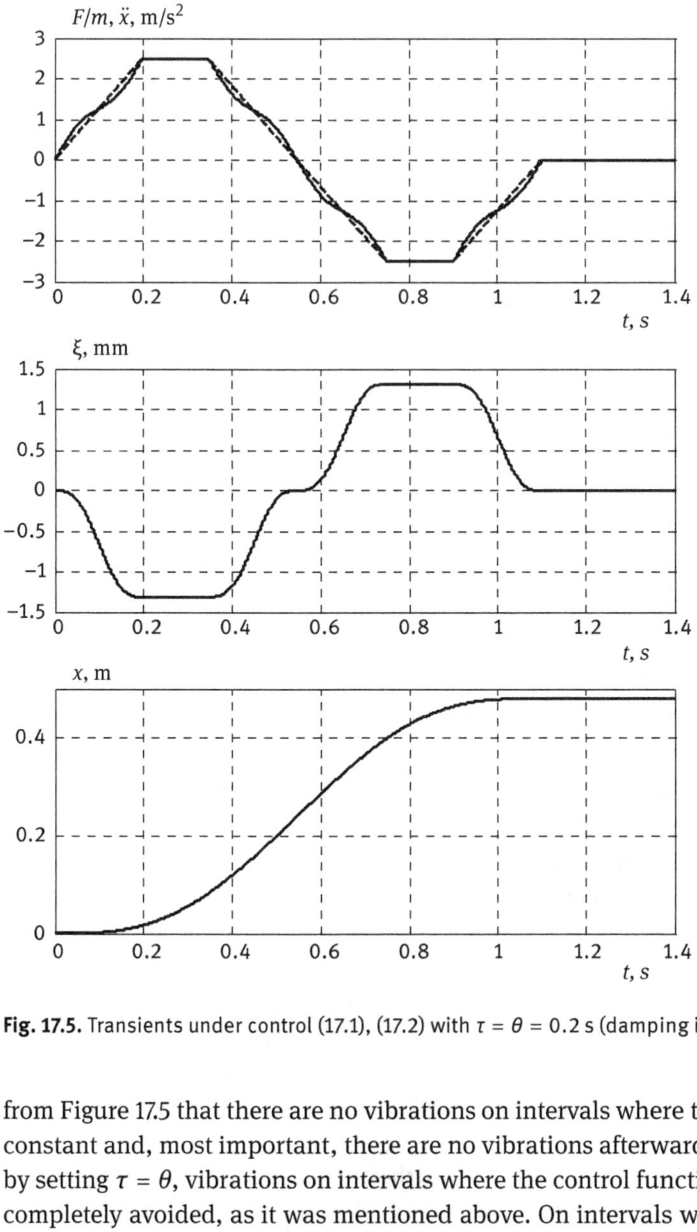

Fig. 17.5. Transients under control (17.1), (17.2) with $\tau = \theta = 0.2$ s (damping is absent).

from Figure 17.5 that there are no vibrations on intervals where the control function is constant and, most important, there are no vibrations afterwards at $t \geq T$. Therefore, by setting $\tau = \theta$, vibrations on intervals where the control function is constant can be completely avoided, as it was mentioned above. On intervals where control function (17.1), (17.2) changes linearly in time, acceleration $\ddot{x}$ also changes almost linearly – its deviations from acceleration $F(t)/m$ are small. The acceleration and the deformation are counter-phase.

Figure 17.6 (top plot) illustrates dynamics of acceleration F/m (dashed line) and $\ddot{x}$ (solid line) in system (16.1) with damping ($e = 4500\,\mathrm{N \cdot s \cdot m^{-1}}$) controlled by law (17.1), (17.2). The bottom plot of Figure 17.6 shows dynamics of deformation ξ. At $e =$

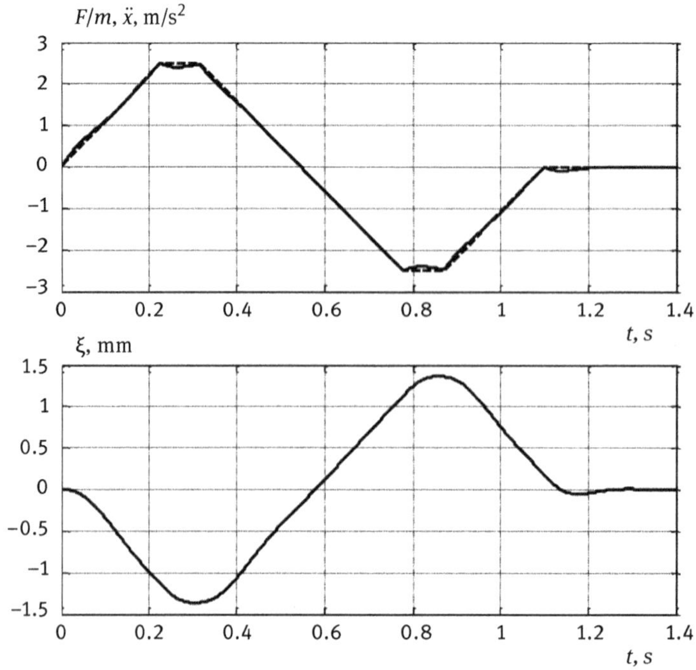

Fig. 17.6. Transients under control (17.1), (17.2) with $\tau = \theta = 0.23$ s (damping is present).

$4500\,\mathrm{N\cdot s\cdot m^{-1}}$, the frequency of natural vibrations is $\omega = 27.30\,\mathrm{s}^{-1}$ (see (16.2)) and $\tau = \theta = 2\pi/\omega = 0.23$ s.

Figure 17.6 shows that the difference between acceleration $\ddot{x}$ and F/m is small. In this case, when the control function is constant, the deformation is not constant. At $t \geq T$, the vibrations of $\ddot{x}$ are close to zero, and the deformations are small. In addition, these vibrations fade away with time because of natural damping in the system.

Numerical experiments show the following phenomenon. Suppose that the time of jerks $\tau = \theta$, and the period of natural vibrations θ is calculated for a particular value of the damping coefficient e (according to expressions (16.2) and (16.3)). In this case, the pattern of transient process is *insignificantly* influenced by the value of damping coefficient.

If the damping coefficient vanishes ($e = 0$), then deformation ξ and acceleration $\ddot{x}$ under control law (17.1), (17.2) with $\tau = \theta = 2\pi/\omega_0$ are constant ($\xi = \mp F_0/k$, $\ddot{x} = \pm F_0/m$) on the intervals where the control function is constant; furthermore, for $t \geq T$ they become zero. For $t \geq T$, the position of the plant is $x(t) \equiv x_d$. Thus control law (17.1), (17.2) with $\tau = \theta = 2\pi/\omega_0$ drives system (16.1) (with $e = 0$) to desired terminal state (16.6) in a finite time interval. Note that the higher frequency ω_0 is, the less is time $\tau = \theta = 2\pi/\omega_0$ and consequently, final time T. If $\omega_0 \to \infty$ ($\theta \to 0$), then trapezoidal control (17.1), (17.2) converges to bang-bang control (16.8). Therefore, if frequency ω_0

is high, the time during which the control signal changes (increases or decreases) can be short. For example, if $\omega_0 = 40$ Hz (see Figure 16.1), then $\tau = \theta = 2\pi/\omega_0 \approx 0.157$ s. When frequency ω_0 is high, time interval T is close to (16.9), and law (17.1), (17.2) is close to the time-optimal control law.

It should be mentioned that typically it is difficult to identify all the parameters of the system, and it is easier to determine its first resonance frequency experimentally.

17.5 Feedback control

Suppose that frequency ω of system (16.1) is known, and the desired displacement of the controlled object x_d is given. Then, upon calculation of $\tau = \theta$, expression (17.19) (or (17.20)) can be used to find time T and to design control function (17.1), (17.2). Let this open-loop function be designated as $F_d(t)$. It will be used as a feedforward signal in the closed loop. After integrating the second equation in (16.1) (the mass of the plant m is supposed to be known) alone under control $F_d(t)$, functions $\dot{z}(t) = \dot{\xi}(t) + \dot{x}(t)$ and $z(t) = \xi(t) + x(t)$, are found. These functions are denoted as $\dot{z}_d(t)$ and $z_d(t)$. Function $z_d(t)$ is the desired absolute displacement of the plant, and function $\dot{z}_d(t)$ is its desired absolute velocity. On intervals where control function $F_d(t)$ is constant, the following identities apply: $\dot{\xi}(t) \equiv 0$, $\dot{z}_d(t) \equiv \dot{x}(t)$. If deformation $\xi(t)$ and its derivative $\dot{\xi}(t)$ under control (17.1), (17.2) are small on entire time interval $[0, T]$, then $z_d(t) \approx x(t)$ and $\dot{z}_d(t) \approx \dot{x}(t)$ on this time interval (remember that $\dot{z}_d(t) \equiv \dot{x}(t)$ wherever control $F_d(t) \equiv$ const and $z_d(t) \equiv x(t) \equiv x_d$ wherever control signal $F_d(t) \equiv 0$). Now consider the following feedback control:

$$F = \alpha_1[z_d(t) - x(t)] + \alpha_2[\dot{z}_d(t) - \dot{x}(t)] + F_d(t)\,. \tag{17.26}$$

Here, α_1 and α_2 are constant feedback gains, that must be chosen so that to make solution (16.6) to system (16.1) asymptotically stable (e.g., Routh–Hurwitz criteria [40, 41, 97] can be used to determine their values). Term $F_d(t)$ is typically called the feedforward. Note that functions (17.5), (17.6) are a solution to system (16.1) with open-loop control $F_d(t)$, but they are not a solution to system (16.1) with closed-loop control (17.26).

Deformation $\xi(t)$ of the platform and displacement $x(t)$ of the plant were obtained as time functions by integrating system (16.1) numerically with feedback control (17.26) ($\alpha_1 = 1$ N/m, $\alpha_2 = 1$ N · s/m). For such numerical integration, the parameters were taken as in (17.25), and the damping $e = 0$.

At first glance, the transient dynamics in terms of variables ξ and x does not differ from the corresponding transient in the system with open-loop control function $F_d(t)$ shown in Figure 17.5. Both closed-loop control law (17.26) and open-loop control law (17.1), (17.2) drive the system to state $x_d = 0.441$ m in the same time $T \approx 1.1$ s, and afterwards the transients settle. Term $F_d(t)$, and signals $z_d(t)$, $\dot{z}_d(t)$ in feedback control (17.26) reduce the errors in the system and make it more robust.

To summarize the current part, the following can be said. In order to reduce the magnitude of natural vibrations in system that includes compliant elements, control laws were suggested that use the jerk time as a parameter. Setting jerk time equal to the period of the first (dominant) resonance, eliminate vibrations completely when no damping is present in the system. If damping is present, the vibrations can be significantly reduced. The results are applicable for a system with a compliant platform, as well as for a system with a compliant gear. The first resonance frequency that is used in calculations can be easily determined in experiments. Attempts to shorten the transient and to make it settle faster by reducing the jerk time can lead to the opposite effect – the transients will take longer time to settle, and therefore, the system positioning will take longer time. The feedback control can be designed using feedforward term (17.1), (17.2) or (17.10), (17.11).

Bibliography

[1] Aguilar L, Boiko L, Fridman L, Freidovich L. Generating self-excited oscillations in an inertia wheel pendulum via two-relays controllers. St. Louis, Missouri, Proc. of American Control Conference, CD, 2009, 3039–3044.

[2] Akulenko LD. Quasi Stationary Finite Motion Control of Hybrid Oscillatory Systems. Applied Mathematics and Mechanics 1991, 55(2), 183–192.

[3] Akulenko LD. Problems and methods of optimal control. Mathematics and its Applications Series, 286, Springer Science+Business Media, 1994.

[4] Alexandrov VV, Zhermolenko VN. On absolute stability of second-order systems. Moscow University Mechanics Bulletin 1972, 27(5), 91–97.

[5] Andre J, Seibert P. Uber stuckweise lineare differential gleichungen, die bei regelungsproblemen auftreten; I und II, Archiv d. Math., 1956, 7, 148–156, 157–164.

[6] Andre J, Seibert P. Motion after final point and its stability analysis for discontinuous control systems in general case. Proc. of 1st IFAC Congress in Moscow, Moscow, Publishing House of USSR Academy of Science, 1961, 1, 691–698. In Russian.

[7] Andrievsky BR. Global stabilization of the unstable reaction-wheel pendulum. Automation and Remote Control 2011, 72(9), 1981–1993.

[8] Andronov AA, Vitt AA, Khaikin SE. Theory of oscillators. Oxford, UK, Pergamon Press, 1966. Translated from Russian.

[9] Aoustin Y and Formal'sky A. On the Synthesis of a Nominal Trajectory for Control Law of a One-Link Flexible Arm. International Journal of Robotics Research 1997, 16(1), 36–46.

[10] Aoustin Y, Formalskii A. Simple anti-swing feedback control for a gantry crane. Robotica 2003, 21, 655–666.

[11] Aoustin Y, Formal'sky A, Martynenko Y. Stabilization of unstable equilibrium postures of a two-link pendulum using a flywheel. Journal of Computer and Systems Sciences International 2006, 45(2), 204–211.

[12] Aoustin Y, Formalskii A. An original ball-and-beam system: stabilization strategy under saturating control with large basin of attraction. Proc. of European Control Conference, Kos Greece, July 2–5, 2007, 4833–4838.

[13] Aoustin Y, Formalskii A. Ball on a beam: stabilization under saturated input control with large basin of attraction. Multibody System Dynamics 2009, 21(1), 71–89.

[14] Aoustin Y, Formalskii A. Beam-and-ball system under limited control: stabilization with large basin of attraction. St. Louis, Missouri, Proc. of American Control Conference, CD, 2009, 555–560.

[15] Aoustin Y, Formalskii A, Martynenko Yu. Pendubot: combining of energy and intuitive approaches to swing up, stabilization in erected pose. Multibody System Dynamics 2010, 5, 1–16.

[16] Appel P. Traite de mécanique rationnelle: Tome 2, Dynamique des systémes mécanique analytique. Paris, Gauthier–Villars, 1953.

[17] Astrom KJ, Furuta K. Swinging up a pendulum by energy control. Automatica 2000, 36(2), 287–295.

[18] Beletskii VV. Optimal transfer of an earth satellite to a gravitationally stable position. Cosmic Research 1971, 9(3), 337–344.

[19] Beletskii VV. Bipedal walking: Model problems of dynamics and control. Moscow, Publishing House "Nauka", 1984. In Russian.

[20] Belotelov VN, Martynenko YuG. Control of spatial motion of an inverted pendulum mounted on a wheel pair. Mechanics of Solids 2006, 41(6), 6–21.

[21] Beznos AV, Formalskii AM, Gurfinkel EV, Jicharev DN, Lensky AV, Savitsky KV, Tchesalin LS. Control of autonomous motion of two-wheel bicycle with gyroscopic stabilization. Leuven, Belgium, Proc. of IEEE Intern. Conference on Robotics and Automation 1998, 2670–2675.

[22] Beznos AV, Grishin AA, Lenskii AV, Okhotsimsky DE, Formalskii AM. A pendulum controlled using flywheel. Doklady Mathematics 2003, 392(6), 743–749.

[23] Beznos AV, Grishin AA, Lensky AV, Okhotsimsky DE, Formalskii AM. A flywheel use-based control for a pendulum with a fixed suspension point. Journal of Computer and Systems Sciences International 2004, 43(1), 22–33.

[24] Beznos AV, Grishin AA, Lensky AV, Okhotsimsky DE, Formalskii AM. Control of the pendulum by means of a flywheel. Practicum on the theoretical and applied mechanics. Moscow, Publishing House of Lomonosov Moscow State University, 2009, 170–195. In Russian.

[25] Block D, Astrom K, Spong M. The reaction wheel pendulum. Princeton NJ, Synthesis Lectures on Control and Mechatronics, Morgan & Claypool Publishers, 2007.

[26] Bolotnik NN, Zeidis IM, Zimmermann K, Yatsun SF. Dynamics of controlled motion of vibration-driven systems. Journal of Computer and Systems Sciences International 2006, 45(5), 831–840.

[27] Boltyanskii VG. Mathematical methods of optimal control. NY, USA, Holt, Rinehart and Winston, 1971. Translated from Russian.

[28] Bortoff S, Spong MW. Pseudolinearization of the acrobot using spline functions. USA, Westin LA Paloma, Tucson, Arizona, Proc. of IEEE Conference on Decision and Control, 1992, 593–598.

[29] Brockett RW, Hongyi Li. A light weight rotary double pendulum: maximizing the domain of attraction. USA, Maui, Hawaii, Proc. of IEEE Conference on Decision and Control 2003, 3299–3304.

[30] Brown HB, Xu Y. A single wheel gyroscopically stabilized robot. USA, Minneapolis, Minnesota, Proc. of IEEE Intern. Conference on Robotics and Automation, 1996, 4, 3658–3663.

[31] Buchin VA. Solution of the problem of a substantial increase in the critical load of a compressed elastic rod using boundary conditions leading to a multipoint boundary-value problem. Soviet Physics Doklady 1982, 27(4), 355–357.

[32] Buchin VA, German VO, Lyubimov GA, Morozov VM. A method to preserve stability of an extended body being under compression in its longitudinal direction and a device implementing this method. Author's certificate USSR No. 658252, issued by USSR State Committee for Inventions and Discoveries, Bulletin No. 15, Published April, 25, 1979. In Russian.

[33] Butenin NV, Neimark YuI, Fufaev NA. Introduction to the theory of nonlinear oscillations. Moscow, Publishing House "Nauka", 1987. In Russian.

[34] Butkovskii AG. Phase portraits of controlled dynamical systems. Moscow, Publishing House "Nauka", 1985. In Russian.

[35] Cambrini L, Chevallereau C, Moog CH, Stojic R. Stable trajectory tracking for biped robots. Sydney, Australia, Proc. of IEEE Conference on Decision and Control, 2000, 5, 4815–4820.

[36] Chechurin SL. Parametric vibrations and stability of periodic motion. Leningrad, Publishing House of Leningrad University, 1983. In Russian.

[37] Chernous'ko FL, Akulenko LD, Sokolov BN. Control of vibrations. Moscow, Publishing House "Nauka", 1980. In Russian.

[38] Chernous'ko FL, Ananievski IM, Reshmin SA. Control of nonlinear dynamical systems: methods and applications. Berlin, Heidelberg, Springer-Verlag, 2008. Translated from Russian.

[39] Chernousko FL, Dobrynina IS. Constrained Control in a Mechanical System with Two Degrees of Freedom. IUTAM Symposium on Optimization of Mechanical Systems / Eds D. Bestle and W. Schiehlen. Stuttgart: Kluwer Academic Publisher, 1995, 57–64.

[40] Chetaev NG. The stability of motion. Oxford, UK, Pergamon Press, 1961. Translated from Russian.

[41] Chetaev NG. Theoretical mechanics (eds. Rumyantsev VV, Yakimova KE). Moscow, Publishing House "Nauka", 1987. In Russian.

[42] Chilikin MG, Kluchev VI, Sandler AS. Theory of automated electric drive. Moscow, Publishing House "Power", 1979. In Russian.

[43] Courant R, Hilbert D. Methoden der mathematishen physik: Band 1. Berlin, Springer-Verlag, 1931.

[44] De Luca A, Siciliano B. Trajectory Control of a Non-Linear One-Link Flexible Arm. International Journal on Control 1989, 50(5), 1699–1715.

[45] Demidovich BP. Lectures on the mathematical theory of stability. Moscow, Publishing House "Nauka", 1967. In Russian.

[46] Dobrynina IS, Chernousko FL. Constrained Control for Linear System of Fourth Order. Journal of Computer and Systems Sciences International 1994, 4, 108–115.

[47] El'sholts LE, Norkin SB. Introduction to the theory of differential equations with deviating argument. Moscow, Publishing House "Nauka", 1971. In Russian.

[48] Fantoni I, Losano R, Spong MW. Energy based control of the pendubot. IEEE Transaction on Automatic Control 2000, 45(4), 725–729.

[49] Fantoni I, Losano R. Non-linear control for underactuated mechanical systems. London, Springer-Verlag, 2002. 295 p.

[50] Filimonov YuM. Optimal control of a mathematical pendulum. Differential Equations 1965, 1(8), 783–789.

[51] Filippov AF. Differential equations with discontinuous right-hand sides. In: American Mathematical Society Translations, Series 2, Ann Arbor, 1964, 42, 199–231.

[52] Filippov AF. Differential equations with discontinuous right-hand sides. Kluver Academic Publishers, Dodrecht, The Netherlands, 1988. Translated from Russian.

[53] Formalskii AM. Controllability and stability of systems with limited resources. Moscow, Publishing House "Nauka", 1974. In Russian.

[54] Formalskii AM. Control of a pendulum with minimum expenditure of mechanical energy. Mechanics of Solids 1977, 12(2), 19–27.

[55] Formalskii AM. Moving anthropomorphic mechanisms. Moscow, Publishing House "Nauka", 1982. In Russian.

[56] Formalskii AM. On the corner points of the boundaries of regions of attainability. Journal of Applied Mathematics and Mechanics 1983, 47(4), 466–472.

[57] Formalskii AM. Stabilization of an inverted pendulum with a fixed or movable suspension point. Doklady Mathematics 2006, 73(1), 152–156.

[58] Formalskii AM. An inverted pendulum on a fixed and a moving base. Journal of Applied Mathematics and Mechanics 2006, 70(1), 56–64.

[59] Formalskii AM. On stabilization of an inverted double pendulum with one control torque. Journal of Computer and Systems Sciences International 2006, 45(3), 337–344.

[60] Formalskii AM. Global stabilization of a double inverted pendulum with control at the hinge between the links. Mechanics of Solids 2008, 43(5), 687–697.

[61] Formalskii A, Aoustin Y. Stabilization of a ball (wheel) on a beam with large basin of attraction. Advanced in Mechanics, Dynamics and Control. Proc. of the 14th Intern. Workshop on Dynamics and Control, Moscow–Zvenigorod, Russia, May 28 – June 2, 2007, Moscow, Publishing House "Nauka", 2008, 90–99.

[62] Formalskii AM. On the synthesis of optimal control for second-order systems. Doklady Mathematics 2010, 81(1), 164–167.

[63] Formalskii AM. On the design of optimal feedback control for systems of second order. Applied Mathematics 2010, 1(4), 301–306.
[64] Formalskii AM. Stabilization of unstable mechanical systems. Journal of Optimization Theory and Applications 2010, 144(2), 227–253.
[65] Formalskii AM. Ballistic walking design via impulsive control. ASCE, Journal of Aerospace Engineering 2010, 23(2), 129–138.
[66] Formalskii A, Gannel L. Control to avoid vibrations in systems with compliant elements. Journal of Vibration and Control, 2014, DOI: 10.1177/1077546313517587.
[67] Freidovich L, Robertsson A, Shiraev A, Johansson R. Periodic motions of the pendubot via virtual holonomic constraints: theory and experiments. Automatica 2008, 44(3), 785–791.
[68] Furuta K. Control of pendulum: from super mechano-system to human adaptive mechatronics. USA, Maui, Hawaii, Proc. of IEEE Conference on Decision and Control, 2003, 1498–1507.
[69] Gabrielian MS, Krasovskii NN. On the problem of the stabilization of a mechanical system. Journal of Applied Mathematics and Mechanics 1964, 25(8), 979–990.
[70] Gannel LV, Formal'skii AM. Control for minimizing vibration in systems with compliant elements. Journal of Computer and Systems Sciences International 2013, 52(1), 117–128.
[71] Gantmacher FR. Applications of the theory of matrices. AMS Chelsea Publishing. Reprinted by American Mathematical Society, 2000. Translated from Russian.
[72] Gnoenskii LS, Kamenskii GA, El'sholts LE. Mathematical foundations of controlled systems theory. Moscow, Publishing House "Nauka", 1969. In Russian.
[73] Golubev YuF. Foundations of theoretical mechanics. Moscow, Publishing House "Nauka", 2000. In Russian.
[74] Golubev YuF. A robot-balancer on a cylinder. Journal of Applied Mathematics and Mechanics 2003, 67(4), 539–552.
[75] Golubev YuF, Hairulin RZ. Optimal control of double link pendulum rocking. Moscow, Keldysh Institute of Applied Mathematics, Russian Academy of Science, 1999, Preprint No. 27. In Russian.
[76] Gorinevsky MD, Formalsky AM, Schneider AYu. Force control of robotics systems. Boca Raton, NY, USA, CRC Press, 1997. Translated from Russian.
[77] Grammel R. Der kreisel: Seine theorie und anwendungen. Berlin, Springer, 1950.
[78] Gravot F, Hirano Y and Yoshizava S. Generation of "optimal" speed profile for motion planning. Proc. of 2007 IEEE/RSI International Conference on Intelligent Robots and Systems. San-Diego. CA, USA, 2007.
[79] Grishin AA, Lenskii AV, Okhotsimsky DE, Panin DA, Formalskii AM. A control synthesis for an unstable object. An inverted pendulum. Journal of Computer and Systems Sciences International 2002, 41(5), 685–694.
[80] Gudzenko AB, Gannel LV, Smotrov EA. The synthesis of modal control in high bandwidth transistor's electric drives. Electricity 1987, 1, 40–46.
[81] Hakan Yavuz, Selçuk Mistikoğlu, Sadettin Kapucu. Hybrid input shaping to suppress residual vibration of flexible systems. Journal of Vibration and Control 2012, 18(1), 132–140.
[82] Hauser J, Sastry S, Kokotovic P. Nonlinear control via approximate input-output linearization. IEEE Transactions on Automatic Control 1992, 37(3), 392–398.
[83] Ishlinskii AYu. Mechanics of gyroscopic systems. Jerusalem, Israel Program for Scientific Translation, 1965. Translated from Russian.
[84] Ishlinskii A. Orientation, gyroscopes et navigation par inertie. Cinematique des mobiles gyrostabilises. Tome I. Moscou, "Mir", 1984. Translated from Russian.
[85] Ishlinskii A. Orientation, gyroscopes et navigation par inertie. Systemes gyroscopiques de navigation par inertie. Tome II. Moscou, "Mir", 1984. Translated from Russian.

[86] Ishlinskii AYu, Borzov VI, Stepanenko NP. Lectures on the theory of gyroscopes. Moscow, Publishing House of Lomonosov Moscow State University, 1983. In Russian.
[87] Kalenova VI, Morozov VM, Sheveleva EN. Stability and stabilization of motion of a monocycle. Mechanics of Solids 2001, 36(4), 40–47.
[88] Kalenova VI, Morozov VM. Linear nonstationary systems and their applications to problems of mechanics. Moscow, "Fizmatlit", 2010. In Russian.
[89] Kalman RE. Contribution to the theory of optimal control. Bulletin of Society Mathematics Mexicana 1960, 5, 102–119.
[90] Kalman RE. On the general theory of control systems. Proc. of 1st IFAC Congress in Moscow, Butterworth, London, 1960, 1, 481–492.
[91] Kalman RE. Lectures on controllability and observability. Bologna, Centro Intern. Matematico Estivo (C.I.M.E.), 1969, 1–149.
[92] Kalman RE, Falb PL, Arbib MA. Topics in mathematical system theory. New York, San Francisco, St. Louis, Toronto, London, Sydney, MC Graw–Hill Book Company, 1969.
[93] Kapitsa PL. Dynamic stability of pendulum with vibrating suspension point. Journal of Experimental and Theoretical Physics 1951, 21(5), 588–597. In Russian.
[94] Kayumov OR. Globally controlled mechanical systems. Moscow, Publishing House "Fizmatlit", 2007. In Russian.
[95] Khaled Gamal Eltohamy, Chen-Yuan Kuo. Nonlinear generalized equations of motion for multi-link inverted pendulum systems. Intern. Journal of Systems Science 1999, 30(5), 505–513.
[96] Khalil HK. Nonlinear systems. USA, New Jersey, Prentice–Hall, 2002.
[97] Korn GA, Korn TM. Mathematical handbook for scientists and engineers. New York, Toronto, London, McGraw–Hill Book Company, Inc. 1961.
[98] Krasovskii NN. Theory of motion control. Linear Systems. Moscow, Publishing House "Nauka", 1968. In Russian.
[99] Lam S, Davison EJ. The real stabilizability radius of the multi-link inverted pendulum. USA, Minneapolis, Minnesota, Proc. of American Control Conference, 2006, 1814–1819.
[100] Lavrovskii EK, Formalskii AM. Optimal control of the pumping and damping of a swing. Journal of Applied Mathematics and Mechanics 1993, 57(2), 311–320.
[101] Lavrovskii EK, Formalskii AM. The optimal control synthesis of the swinging and damping of a double pendulum. Journal of Applied Mathematics and Mechanics 2001, 65(2), 219–227.
[102] Lenskii AV, Formalskii AM. Two-wheel robot-bicycle with a gyroscopic stabilizer. Journal of Computer and Systems Sciences International 2003, 42(3), 482–489.
[103] Lenskii AV, Formalskii AM. Gyroscopic stabilization of a two-wheeled robot- bicycle. Doklady Mathematics 2004, 70(3), 993–997.
[104] Lee EB, Markus L. Foundations of optimal control theory. NY, USA, Wiley, 1967.
[105] Letov AM. Stability of nonlinear controlled systems. Moscow, "Fizmatlit", 1962. In Russian.
[106] Magnus K. Kreisel: Theorie und anwendungen. Berlin, Heidelberg, Springer- Verlag, 1971.
[107] Magnus K. Schwingungen: Eine einfuhrung in die theoretische behandlung von schwingungsproblemen. Stuttgart, J. Teubner, 1976.
[108] Martynenko YuG. Analytical dynamics of electromechanical systems. Moscow, Publishing House of Moscow Power Institute, 1984. In Russian.
[109] Martynenko YuG, Lenskii AV, Kobrin AI. The decomposition of control problem for single wheel mobile robot with an unperturbed gyroscopically stabilized platform. Doklady Mathematics 2002, 386(6), 767–770.
[110] Martynenko YuG, Formalskii AM. A control of the longitudinal motion of a single- wheel robot on an uneven surface. Journal of Computer and Systems Sciences International 2005, 44(4), 662–670.

[111] Martynenko YuG, Formalskii AM. Stabilization methods of unstable objects. Gyroscopy and Navigation 2005, 2(49), 7–18. In Russian.
[112] Martynenko YuG, Formalskii AM. Control problems for unstable systems. Successes in Mechanics 2005, 3(2), 73–135. In Russian.
[113] Martynenko YuG, Formalskii AM. The theory of the control of a monocycle. Journal of Applied Mathematics and Mechanics 2005, 69(4), 516–528.
[114] Martynenko YuG, Formalskii AM. Pendulum on a movable base. Doklady Mathematics 2011, 84(1), 594–599.
[115] Martynenko YuG, Formalskii AM. Controlled pendulum on a movable base. Mechanics of Solids 2013, 48(1), 6–18.
[116] Meckl PH. Optimized S-Curve Motion Profiles for Minimum Residual Vibration. Proc. of American Control Conference. Philadelphia. Pennsylvania. USA, 1998.
[117] Ming-Chang Pai. Robust input shaping control for multi-mode flexible structures. International Journal of Control, Automation and Systems 2011, 9(1), 23–31.
[118] Mori S, Nishihara H, Furuta K. Control of unstable mechanical systems. Control of Pendulum. International Journal on Control 1976, 23(5), 673–692.
[119] Nakawaki D, Joo S, Miyazaki F. Dynamic modeling approach to gymnastic coaching. Leuven, Belgium, Proc. of IEEE Intern. Conference on Robotics and Automation, May 1998, 1069–1076.
[120] Neimark YuI. Dynamical systems and control processes. Moscow, Publishing House "Nauka", 1978. In Russian.
[121] Nemytskii VV, Stepanov VV. Qualitative theory of differential equations. NY, USA, Princeton University Press, 1960. Translated from Russian.
[122] Okhotsimskii DE, Grishin AA, Lenskii AV, Formalskii AM. On stabilization of unstable systems. Collection of methodological articles on Theoretical mechanics, Issue 24, Moscow, Publishing House of Lomonosov Moscow State University, 2003, p. 39–52. In Russian.
[123] Okhotsimskii DE, Gurfinkel' EV, Lavrovskii EK, Lenskii AV, Tatarskii SL, Formalskii AM. Stabilizing of upright bicycle position by means of flywheel. Papers of scientific school-conference "Mobile robots and mechatronics systems". Moscow, Publishing House of Lomonosov Moscow State University, 1999, p. 14–30. In Russian.
[124] Olfati-Saber R. Control of underactuated mechanical systems with two degrees of freedom and symmetry. USA, Chicago, Proc. of American Control Conference, 2000, 4092–4096.
[125] Olfati-Saber R. Nonlinear control of underactuated mechanical systems with application to robotics and aerospace vehicles. Ph.D. thesis, Department of EECS, Massachusetts Institute of Technology, February 2001.
[126] Olfati-Saber R. Normal forms for underactuated mechanical systems with symmetry. IEEE Transactions on Automatic Control 2002, 47(2), 305–308.
[127] Orlov V, Aguilar LT, Acho L, Ortiz A. Swing up and balancing control of pendubot via model orbit stabilization: algorithm synthesis and experimental verification. USA, San Diego, California, Proc. of IEEE Conference on Decision and Control, 2006, 6138–6144.
[128] Panovko YaG, Gubanova I. Stability and oscillations of elastic systems: Modern concepts, paradoxes and mistakes. Moscow, Publishing House "Nauka", 1987. In Russian.
[129] Pinney E. Ordinary difference-differential equations. Berkeley and Los Angeles, University of California Press, 1958.
[130] Pontryagin LS, Boltyanskii VG, Gamkrelidze RV, Mishchenko EF. The mathematical theory of optimal processes. NY, USA, John Wiley and Sons, 1962. Translated from Russian.
[131] Reshmin SA. Decomposition method in the problem of controlling an inverted double pendulum with the use of one control moment. Journal of Computer and Systems Sciences International 2005, 44(6), 861–877.

[132] Reshmin SA, Chernous'ko FL. Synthesis of a control in a non-linear dynamical system based on decomposition. Journal of Applied Mathematics and Mechanics 1998, 62(1), 115–122.
[133] Reshmin SA, Chernous'ko FL. Time-optimal control of an inverted pendulum in the feedback form. Journal of Computer and Systems Sciences International 2006, 45(3), 383–394.
[134] Reshmin SA, Chernous'ko FL. A time-optimal control synthesis for a nonlinear pendulum. Journal of Computer and Systems Sciences International 2007, 46(1), 9–18.
[135] Roitenberg YaN. Gyroscopes. Moscow, Publishing House "Nauka", 1975. In Russian.
[136] Roitenberg IN. Theorie du controle automatique. Moscou, Publishing House "Mir", 1974. Translated from Russian.
[137] Schaefer IF, Cannon RH. On the control of unstable mechanical systems. IFAC, 3d Congress, London, 1966, 601.
[138] Schmidt Chr. An autonomous self-rising pendulum. Germany, Karlsrue, Proc. of European Control Conference, Invited Paper F1022–3, Karlsruhe, 1999.
[139] Sepulchre R, Jankovic M, Kokotovic P. Constructive nonlinear control. Berlin, Springer, 1997.
[140] Shahruz SM. Active Vibration Suppression in Multi-degree-of-freedom Systems by Disturbance Observers. Journal of Vibration and Control 2009, 15(8), 1207–1228.
[141] Shiriaev AS, Egeland O, Ludvigsen H, Fradkov AL. Vss-version of energy based control for swinging up a pendulum. Systems and Control Letters 2001, 44(1), 41–56.
[142] Slotine JJE, Li W. Applied nonlinear control. USA, New Jersey, Prentice–Hall, 1991.
[143] Smolnikov BA. Problems in mechanics and optimization of robots. Moscow, Publishing House "Nauka", 1991. In Russian.
[144] Smotrov EA, Gannel LV, Shehvitz EI et al. Experimental determination of the dynamic parameters of an industrial robot with electric drive. In: Nakhapetyan EG Monitoring and diagnosis of automatic equipment. Moscow: Publishing House "Nauka", 1990, 91–94. In Russian.
[145] Spong MW. The swing up control problem for the acrobot. IEEE Control System Magasin 1995, 14(1), 49–55.
[146] Spong MW, Block DJ. The pendubot: a mechatronic system for control research and education. USA, New Orleans, Louisiana, Proc. of IEEE Conference on Decision and Control, 1995, 555–557.
[147] Spong MW, Corke P, Lozano R. Nonlinear control of the inertia wheel pendulum. Automatica 2001, 37, 1845–1851.
[148] Stepan'yants GA, Tararoshchenko NS. Structure of control rules ensuring asymptotic stability of control systems with an unstable object. Soviet Physics Doklady, 1970/71, 15, 717–718.
[149] Stephenson A. On a new type of dynamical stability. Memoirs and Proc. of Manchester Literary and Philosophical Society 1908, 52(8), Pt 2, 1–10.
[150] Stojic R, Chevallereau C. On walking with point foot-ground contacts. In: Virk GS et al., eds. Proc. of 2nd Intern. Conference Climbing and Walking Robots (CLAWAR 99), Portsmouth: Professional Engineering Publ., 1999, 463–471.
[151] Strizhak TG. Methods of investigation of dynamical systems such as "pendulum". Alma-ata, Publishing House "Nauka", 1981. In Russian.
[152] Sultanov IA. Control processes described by equations with incompletely determined functional parameters. Automation and Remote Control 1980, 41(10), 1356–1364.
[153] Sussmann HJ, Sontag ED, Yang YD. A general result on the stabilization of linear systems using bounded controls. IEEE Transactions on Automatic Control 1994, 39(12), 2411–2425.
[154] Teel AR. Using saturation to stabilize a class of single-input partially linear composite systems. IFAC NOLCOS'92 Symposium, May 1992, 369–374.
[155] Teel AR, Praly L. Tools for semiglobal stabilization by partial state and output feedback. SIAM Journal Control Optimization 1995, 33, 1443–1488.

[156] Tingshu Hu, Zongli Lin, Li Qiu. Stabilization of exponentially unstable linear systems with saturating actuators. IEEE Transactions on Automatic Control 2001, 46(6), 973–979.
[157] Tingshu Hu, Zongli Lin. Control systems with actuator saturation: analysis and design. Boston, Birkhauser, 2001.
[158] Tsien HS. Engineering: Cybernetics. NY, USA, McGraw–Hill Book Company, 1954.
[159] Utkin VI. Sliding modes in control and optimization. Berlin, Heidelberg, New York, Springer-Verlag, 1992. Translated from Russian.
[160] Voronkov VS, Pozdeev OD. Dynamics of the stabilization system of the magnetic suspension of a gradiometer sensor. Mechanics of Solids 1995, 30(1), 23–31.
[161] Voronov AV, Lavrovskii EK. Determination of mass-inertia characteristics of a human leg. Human Physiology 1998, 24(2), 210–220.
[162] www.jackiechabanais.com
[163] www.dself.dsl.pipe
[164] www.msurobot.com
[165] www.segway.com
[166] Xin X, She JH, Yamasaki T, Liu Y. Swing-up control based on composite links for n-link under-actuated robot with passive first joint. Automatica 2009, 45, 1186–1194.
[167] Yangsheng Xu, Kwok-Wai AuS. Stabilization and path following of a single wheel robot. IEEE/ASME Transactions on Mechatronics 2004, 9(2), 407–419.
[168] Zatsiorskii VM, Aruin AS, Seluyanov VN. Biomechanics of human musculoskeletal system. Moscow, "Physical Culture and Sport", 1981. In Russian.
[169] Zhang M, Tarn TJ. Hybrid control for pendubot. IEEE/ASME Transaction on Mechatronics 2002, 7(1), 79–86.
[170] Zhang R, Tong C. Torsional Vibration Control of the Main Drive System of a Rolling Mill Based on an Extended State Observer and Linear Quadratic Control. Journal of Vibration and Control 2006, 12(3), 313–327.
[171] Zimmermann K, Zeidis I, Behn C. Mechanics of terrestrial locomotion: with a focus on non-pedal motion systems. Berlin, Heidelberg, Springer-Verlag, 2009.

Index

Also of Interest

Volume 32
Victor V. Przyjalkowski, Thomas Coates, 2017
Laurent Polynomials in Mirror Symmetry
ISBN 978-3-11-031109-9, e-ISBN (PDF) 978-3-11-031110-5,
e-ISBN (EPUB) 978-3-11-038299-0, Set-ISBN 978-3-11-031111-2

Volume 31
Christian Bizouard, 2016
Geophysical Modelling of the Polar Motion
ISBN 978-3-11-029804-8, e-ISBN (PDF) 978-3-11-029809-3,
e-ISBN (EPUB) 978-3-11-038913-5-9, Set-ISBN 978-3-11-029810-9

Volume 30
Alexander Getling, 2015
Solar Hydrodynamics
ISBN 978-3-11-026666-5, e-ISBN (PDF) 978-3-11-026746-4,
e-ISBN (EPUB) 978-3-11-039085-8, Set-ISBN 978-3-11-026747-1

Volume 29
Michael V. Vesnik, 2015
The Method of the Generalised Eikonal
ISBN 978-3-11-031112-9, e-ISBN (PDF) 978-3-11-031129-7,
e-ISBN (EPUB) 978-3-11-038301-0, Set-ISBN 978-3-11-031130-3

Volume 28
Carlo Rovelli (Ed.), 2015
General Relativity: The most beautiful of theories
ISBN 978-3-11-034042-6, e-ISBN (PDF) 978-3-11-034330-4,
e-ISBN (EPUB) 978-3-11-038364-5, Set-ISBN 978-3-11-034331-1

Volume 27
Ivan A. Lukovsky, 2015
Nonlinear Dynamics
ISBN 978-3-11-031655-1, e-ISBN (PDF) 978-3-11-031657-5,
e-ISBN (EPUB) 978-3-11-038973-9, Set-ISBN 978-3-11-031658-2

CPSIA information can be obtained
at www.ICGtesting.com
Printed in the USA
LVHW062033051021
699616LV00003B/108